Web–Based Engineering Education:
Critical Design and Effective Tools

Donna L. Russell
University of Missouri, USA

A.K. Haghi
University of Guilan, Iran

ENGINEERING SCIENCE REFERENCE

Hershey · New York

Director of Editorial Content: Kristin Klinger
Director of Book Publications: Julia Mosemann
Acquisitions Editor: Lindsay Johnston
Development Editor: Joel Gamon
Publishing Assistant: Tom Foley
Typesetter: Deanna Jo Zombro
Production Editor: Jamie Snavely
Cover Design: Lisa Tosheff
Printed at: Yurchak Printing Inc.

Published in the United States of America by
        Engineering Science Reference (an imprint of IGI Global)
        701 E. Chocolate Avenue
        Hershey PA 17033
        Tel: 717-533-8845
        Fax: 717-533-8661
        E-mail: cust@igi-global.com
        Web site: http://www.igi-global.com

Library of Congress Cataloging-in-Publication Data

Web-based engineering education : critical design and effective tools / Donna L. Russell and A.K. Haghi, editors.
      p. cm.
   Includes bibliographical references and index.
   Summary: "This book evaluates the usefulness of advanced learning systems in delivering instructions in a virtual academic
environment for different engineering sectors. The learning process discussed in each chapter will include a walk-through of
a case problem-solving exercise in the virtual company environment"--Provided by publisher.
   ISBN 978-1-61520-659-9 (hardcover) -- ISBN 978-1-61520-660-5 (ebook)
1. Engineering--Study and teaching (Higher) 2. Web-based instruction. 3. Educational Web sites. 4. Distance education. 5.
Engineers--In-service training. I.Russell, Donna II. Haghi, A. K.
   T65.5.C65W43 2010
   620.0078'5--dc22
                                          2009053465
British Cataloguing in Publication Data
A Cataloguing in Publication record for this book is available from the British Library.

Editorial Advisory Board

Table of Contents

Detailed Table of Contents

Chapter 1
> *B. Noroozi, University of Guilan, Iran & University of Cincinnati, USA*
> *M. Valizadeh, University of Guilan, Iran*
> *G. A. Sorial, University of Cincinnati, USA*

Traditional education for engineers has shifted towards new methods of teaching and learning through the proliferation of Information and Communication Technologies (ICT). The continuous advances in technology enable the realization of a more distributed structure of knowledge transfer. This becomes critically important for developing countries that lack the resources and infrastructure for implementing engineering education practices. The two main themes of technology in designing e-Learning for engineering education in developing countries focus either on aspects of technological support for traditional methods and localized processes, or on the investigation of how such technologies may assist distance learning. Commonly such efforts are threefold, relating to content delivery, assessment and provision of feedback. This chapter is based on the authors ten years' experience in e-Learning, and reviews the most important key issues and success factors regarding the design of e-Learning for engineering education in developing countries.

Chapter 2
> *Attila Somfai, "Széchenyi István" University, Hungary*

This chapter describes the use of a teaching web portal at a university in Hungary. The Internet has created potential new and effective ways of cooperation between lecturers and students of the university and other institutions of higher education. The teaching web portal of the Faculty of Architecture at Széchenyi István University (www.arc.sze.hu/indexen.html) realizes the diversity and complexity of architecture with efficient grouping of information and being attentive to high professional standards. Computer Aided Architectural Modelling (www.arc.sze.hu/cad) is one of the new types of online lecture notes, where many narrated screen capture videos show the proper usage of CAD software instead of texts and figures. This interactive type of learning assists students to become more independent learners. This type of teaching modality provides the opportunity for students who need more time to acquire subject matter by viewing video examples again. Additionally, the success of our departments' common web initiations can be measured through internet statistics and feedback of the students and external professionals.

In this book chapter, the authors summarize their retrospections as an engineering educator for more than 20 years. Consideration is given to a number of educational developments to which the authors has contributed during their career in academia and the contribution made to engineering and technological education. Increasing emphasis is being placed on establishing teaching and learning centers at the institutional level with the stated objective of improving the quality of teaching and education. The results of this study provide information for the revision of engineering curricula, the pedagogical training of engineering faculty and the preparation of engineering students for the academic challenges of higher education in the field. The book chapter provides an in-depth review of a range of critical factors liable to have a significant effect and impact on the sustainability of engineering as a discipline. Issues such as learning and teaching methodologies and the effect of E-development and the importance of communications are discussed.

Virtual Reality and Virtual Learning Environments have become increasingly ambiguous terms in recent years because of changes in essential elements facilitating a consistent environment for learners. Three-dimensional (3D) environments have the potential to position the learner within a meaningful context to a much greater extent than traditional interactive multimedia environments. The term 3D environment has been chosen to focus on a particular type of virtual environment that makes use of a 3D model. 3D models are very useful to make acquainted students with features of different shapes and objects, and can be particularly useful in teaching younger students different procedures and mechanisms for carrying out specific tasks. This chapter explains that the 3D Virtual Reality is mature enough to be used for enhancing communication of ideas and concepts and stimulate the interest of students when compared to 2D education.

Chapter 5

Siddhartha Ghosh, G. Narayanamma Institute of Technology and Science, India

Automated Essay Grading (AEG) or Scoring (AES) systems are more a myth they are reality. Today the human written (not hand written) essays are corrected not only by examiners and teachers but also by machines. The TOEFL exam is one of the best examples of this application. The students' essays are evaluated both by human & web based automated essay grading system. Then the average is taken. Many researchers consider essays as the most useful tool to assess learning outcomes, implying the ability to recall, organize and integrate ideas, the ability to supply merely than identify interpretation and application of data. Automated Writing Evaluation Systems, also known as Automated Essay Assessors, might provide precisely the platform we need to explicate many of the features those characterize good and bad writing and many of the linguistic, cognitive and other skills those underline the human capability for both reading and writing. This chapter focuses on the existing automated essay grading systems, the basic technologies behind them and proposes a new framework to show that how best these AEG systems can be used for Engineering Education.

Chapter 6

Rochelle Jones, Lockheed Martin Corporation, USA
Chandre Butler, ,University of Central Florida, USA
Pamela McCauley-Bush, University of Central Florida, USA

Mobile learning is becoming an extension of distance learning, providing a channel for students to learn, communicate, and access educational material outside of the traditional classroom environment. Because students are becoming more digitally mobile, understanding how mobile devices can be integrated into existing learning environments is advantageous however, the lack of social cues between professors and students may be an issue. Understanding metrics of usability that address the concern of student connectedness as well as defining and measuring human engagement in mobile learning students is needed to promote the use of mobile devices in educational environments. The purpose of this chapter is to introduce contemporary topics of applied mobile learning in distance education and the viability of mobile learning (m-learning) as an effective instructional approach.

Chapter 7

Doug Holton, Utah State University, USA
Amit Verma, Texas A&M – Kingsville, USA

Over the past decade, our research group has uncovered more evidence about the difficulties undergraduate students have understanding electrical circuit behavior. This led to the development of an AC/DC Concept Inventory instrument to assess student understanding of these concepts, and various software tools have been developed to address the identified difficulties students have when learning about electrical circuits. In this chapter two software tools in particular are discussed, a web-based dynamic assessment environment (Inductor) and an animated circuit simulation (Nodicity). Students showed gains over time when using Inductor, and students using the simulation showed significant improvements on half of the questions in the AC/DC Concept Inventory. The chapter concludes by discussing current and future work focused on creating a more complete, well-rounded circuits learning environment suitable for supplementing traditional circuits instruction. This work includes the use of a contrasting cases strategy that presents pairs of simulated circuit problems, as well as the design of an online learning community in which teachers and students can share their work.

Chapter 8

Kauser Jahan, Rowan University, USA
Jess W. Everett, Rowan University, USA
Gina Tang, Rowan University, USA
Stephanie Farrell, Rowan University, USA
Hong Zhang, Rowan University, USA
Angela Wenger, New Jersey Academic for Aquatic Sciences, USA
Majid Noori, Cumberland County College, USA

Engineering educators have typically used non-living systems or products to demonstrate engineering principles. Each traditional engineering discipline has its own products or processes that they use to demonstrate concepts and principles relevant to the discipline. In recent years engineering education has undergone major changes with a drive to incorporate sustainability and green engineering concepts into the curriculum. As such an innovative initiative has been undertaken to use a living system such as an aquarium to teach basic engineering principles. Activities and course content were developed for a freshman engineering class at Rowan University and the Cumberland County College and K-12 outreach for the New Jersey Academy for Aquatic Sciences. All developed materials are available on a dynamic website for rapid dissemination and adoption.

Chapter 9

B.M. Trigo, Polytechnic School of the University of São Paulo, Brazil
G.S. Olguin, Polytechnic School of the University of São Paulo, Brazil
P.H.L.S. Matai, Polytechnic School of the University of São Paulo, Brazil

Since the sprouting of the personal computers, around 1980, computer science has been gaining importance, becoming a supporting instrument for other daily activities of ours. With the introduction of computing in a certain area or activity, barriers and difficulties can be overcome. Consequently, new paradigms, possibilities and challenges are created. In education it is not different, computing is more and more present, assisting the learning process in a variety of ways, creating new challenges that compel us to re-think the way education is performed, considering new delivery methods instead of only the traditional chalk-and-talk method. "While we may feel comfortable with traditional approaches, the new technologies provide us with the tools to challenge these positions, and open up the teaching/ learning questions for some rethinking" (Roy & Lee, 1999). With the sprouting of the Internet, it was created what we today understand as the web-based teaching, a method that brings innumerable benefits and challenges for the educators. This chapter describes the development of Applets in a General Technological Chemistry (QTG) course covering the topics of the discipline and joining them in a main "host" Applet, creating a virtual chemistry laboratory, which would be available for the students in the discipline's website.

Chapter 10

Developing engineering study programs of high quality, able to satisfy customized needs, with flexible paths of study, with easy and rapid access to the most appropriate educational facilities and lecturers is a critical and challenging issue for the future of engineering education. The latest developments in communication and information technologies facilitate the creation of reliable solutions in this respect. Provision of web-based courses in engineering education represents one of these solutions. However, the absence of physical interactions with the training facilities and the specificity of remote collaboration with lecturers rise up additional challenges in designing a high-quality web-based engineering course. In order to define superior solutions to the complex set of requirements expressed by several stakeholders (e.g. students, lecturers, educational institutions and companies), a comprehensive planning of quality and an innovative approach of potential conflicting problems are required during the design process of web-based engineering courses. In this context, the present chapter introduces a generic roadmap for optimizing the design process of web-based engineering courses when a multitude of requirements and constrains are brought into equation. Advanced tools of quality planning and innovation are considered to handle the complexity of this process. The application of this methodology demonstrates that no unique, best-of-the-world solution exists in developing a web-based engineering course; therefore customized approaches should be considered for each course category to maximize the impact of the web-based educational process.

Chapter 11

Many researchers have shown that blended learning can more effectively enhance motivation, communication skills, and learning achievement compared to single-form teaching methods. However, the crucial issue that needs to be addressed in blended learning is the question of how to integrate the selected blended format, technology, and teaching strategy into a coherent learning model, and while maintaining interaction between the teacher and students both inside and outside the classroom. Most useful functions of e-learning tools have not been meaningfully integrated into teaching and learning strategies, or into the course management systems that have been used in most campuses. This chapter introduces a highly interactive strategy, the WIRE model, for blended learning that incorporates web-based and face-to-face learning environments into a course by linking the Warm-up before class, Interactive teaching in class, and Review and Exercise after class, creating a sustained learning experience through meaningful use of technology in engineering curricula.

Chapter 12

Giancarlo Anzelotti, University of Parma, Italy
Masuomeh Valizadeh, University of Guilan, Iran

The critical tenet of engineering education reform is the integral role of virtual environment capabilities that is provided by fabulous advances in information technology. Current technological progresses combined with changes in engineering content and instructional method require engineering instructors to be able to design intensive and concentrated lessons for exploration and discovery of the engineering concepts through appropriate computer applications. In actual practice, however, most computer applications provided for engineering education consist of software designed for a specific educational purpose. Furthermore, economical constraints often stand in the way of incorporating such special purpose software into an instructional setting. This chapter discusses an alternative to the traditional approach that shifts the instructional focus from specific computer applications to more sophisticated uses of general purpose software. In particular, educational uses of purpose-oriented small software that can be implemented in multi purpose software are exampled as an introduction to this approach.

Chapter 13

Mehregan Mahdavi, University of Guilan, Iran
Mohammad H. Khoobkar, Islamic Azad University of Lahijan, Iran

Learning Management Systems (LMS) enable effective design and delivery of learning materials. They are Web-based software applications used to plan, implement, and assess a specific learning process. LMSs allow learners to connect to and interact with the educational material through the Internet. They enable tools for authors (instructors) to design learning materials that include text, html, audio, video, etc. They also enable learner activity management in the learning process. Moreover, they provide tools for effective and efficient assessment of the learners. This chapter explores learning management systems and their key components that enable instructors organize and monitor learning activities of the learners. It also introduces the authoring features provided by such systems for preparing learning material. Moreover, it presents assessment methods and tools that enable evaluation of the learners in the learning process. Furthermore, existing challenges and issues in this field are explored.

Chapter 14

Masoumeh Valizadeh, University of Guilan, Iran
Giancarlo Anzelotti, University of Parma, Italy
Salesi Sedigheh, Bekaert, Kortrijk, Belgium

Due to its singular capabilities, web-based learning has entered and is widely used in every field of science and technology. It is not only warmly welcomed at schools and universities, but also in factories and houses. However utilization of web-based learning technique requires tender and comprehensive attentions in designing, applying and assessing configurations, directed by when, where and for which purpose it is being employed. Among different branches of human knowledge and sciences, engineering as well as medicine is more involved in practical and daily-life aspects, where the virtual utilities and educational software can be utilized to consummate the practical features of engineering education. Furthermore the virtual environment of e-learning courses can provide cheaper, safer, more comprehensive and more inclusive approaches to engineering educational material. What usually the students need to learn in laboratories and workshops and it is costive and demanding for the universities and the schools. The aim of this chapter is to count the requirements of engineering education and to accord the facilities and inadequacies of e-learning as training technique in engineering instruction.

Chapter 15

The successful diagnostic activity has an important role in the changes of the repair costs and the efficient elimination of the damages. The aim of the general building diagnostics is to determine the various visible or instrumentally observable alterations, to qualify the constructions from the suitability and personal safety (accidence) points of view. Our diagnostic system is primarily based on a visual examination on the spot; its method is suitable for the examination of almost all-important structures and structure changes of the buildings. During the operation of the diagnostic system a large number of data – valuable for the professional practice – was collected and will be collected also in the future, the analysis of which data set is specially suitable for revaluing construction and the practical application of the experiences later during the building maintenance and reconstruction work. For using the system a so-called "morphological box" has been created, that contains the hierarchic system of constructions, which is connected with the construction components' thesaurus appointed by the correct structure codes of these constructions' place in the hierarchy. The thesaurus was not only necessary because of the easy surveillance of the system, but to exclude the usage of structure-name synonyms in the interest of unified handling. The analysis of which data set is specially suitable for revaluing earlier built constructions and which data can help to create knowledge based new constructions for the future is the topic of this chapter.

Preface

INTRODUCTION

In recent years the interest in applying information technology (IT) for teaching and learning in a wide range of subject areas at all levels has grown rapidly. This development has been accelerated by the significant reduction in cost of the Internet infrastructure and the easy accessibility of the World-Wide-Web. Rapid advances in computer technology and the Internet have created new opportunities for delivering instruction and have revolutionized the learning environment. It is anticipated that these technologies will dramatically change the way instruction will be imparted throughout the educational system. For example, in a digital environment an engineering student would be able to participate in the interactive problem solving process. A virtual industrial sector could be designed for engineering students to consult the personnel in any department, to listen to the comments given by the employees through the Voice Lists, and to browse available online company documents in the form of hypertext and video presentations. Engineering students would be able to work in virtual engineering laboratory systems on the Internet enabling them to adjust to employment and conditions of the world of work and preparing them for productive employment after training. As these emerging technologies impact the design and implementation of web-based engineering education programs at the university level, it is also important to consider the multifaceted interrelated aspects of the educational system that will be impacted, for example how k-12 educators can respond to the potential of these technologies. In this preface I will give a review of an engineering education program in a Midwestern USA city that identifies some of the issues facing recruitment and retention of students in university engineering programs followed by overviews of the fifteen chapters in the book.

ARROWS: Achieving Recruitment, Retention & Outreach with STEP

This description will review the professional development aspect of the ARROWS (Achieving Recruitment, Retention and Outreach With STEP) program. ARROWS is a program that provides high school students, especially under-represented groups such as women and minorities, career mentoring opportunities. This project is funded by a US National Science Foundation grant and designed to develop new understandings about educational opportunities in the science fields for urban high school students. ARROWS provides urban students in a Midwestern USA city with an overview of the varied careers computer scientists and engineers can pursue and connects their high school curriculum with engineering applications. This grant is a collaborative program between the School of Education and the School of Computing and Engineering at the University of Missouri-Kansas City (UMKC). The goal of this program is to develop student understanding of the varied careers of engineers by providing this information in an engaging and relevant manner that increases student interest in pursuing engineering and computer science degrees.

This grant tries to address the need to increase the number of minority students interested in STEM areas in the university. Student attitude and perception about the field of engineering does impact retention in the university (Besterfield-Sacre, Atman & Shuman, 1997). Many urban high school students do not conceptualize engineering as career and therefore do not consider it when they examine career options (Yates, Vos, & Tsai, 1999). Only 25% of high school students surveyed could name five engineering disciplines (Hirsch, Gibbons, Kimmel, Rockland & Bloom, 2003). However, 65% of high school students who completed an introductory engineering course offered by the Infinity Project said they wanted to be engineers (Delissio, 2006). The ARROW project exposes students to the career possibilities from a computer science or engineering undergraduate degree.

A lack of role models is a serious psychological barrier for minority students and is a significant factor in minority student recruitment and retention (May & Chubin, 2003). 41% of females attending Discover Engineering, a summer program designed to expose females to careers in engineering, stated that direct exposure to female role models had a impact on their choice to study engineering (Anderson & Gilbride, 2006). Frequent feedback and interaction with mentors has been shown to increase student self-confidence and expectations of success in engineering (Colbeck, Cabrera, & Terenzini, 2001). Self-esteem and self-confidence, for both males and females, is improved when students become involved in student leadership activities (Astin & Kent, 1983). The ARROWS program's focus on under-represented groups such as women and minorities provides a positive contact and career mentoring opportunities with role models including minority university students currently enrolled in engineering and computer science programs.

The ARROWS project involves students in engineering and computer science projects that demonstrate the relevance and application of their high school science and mathematics curriculum along with exposing the students to the types of team work and design environments they would be likely to work in should they choose a career in computer science or engineering. Introduction to Engineering, a laboratory module experience at the University of Florida increased retention of women and minorities in their undergraduate program (Hoit & Ohland, 1998). A University of Massachusetts Dartmouth review of literature revealed consistent supporting evidence that collaborative learning techniques result in better understanding of material and higher levels of response in an integrated engineering program (Fortier, Fowler, Laoulach, Pendergrass, Sims-Knight & Upchurch, 2002). The ARROWS program recreates the academic and social community by engaging students in collaborative modules in the summer Engineering Essentials.

Engineering Essentials

The ARROWS program provides multiple forums for recruitment and outreach with urban students, however, this description will focus on the summer program. The summer program, called Engineering Essentials, consists of seven days where high school students and high school teachers come to the university. The students engage in three activities 1) discussing careers with engineers, 2) working on the problem-solving modules and 3) tours of the university and engineering firms in the city. The high school teachers that attend the labs with the students in the morning go to a professional development program focusing on the design of problem-based learning units in the afternoon. The School of Computing and Engineering (SCE) faculty developed integrated laboratory modules for use in this pre-college program summer program. After the laboratory module sessions, students attend sessions focusing on the intellectual, social supports and campus life students have available at SCE.   These modules are designed around real-world science problem solving activities focused on four modules including: 1) Biometric Personal Identification (Electrical Engineering), 2) Understanding and preventing ACL among athletes (Mechanical) 3) Engineering), Building Structures to withstand Earthquakes (Civil and Mechanical) and 4) Web Page Design

Biometric Personal Identification

Biometrics is the science of determining or verifying the identity of an individual based on her or his unique physiological or behavioral traits. Fingerprints, face, hand geometry, iris, voice, signature, gait, keystroke dynamics, and palm and eye vein patterns are examples of such traits utilized for Biometric identification. Applications range from law enforcement, border control, and financial transactions to computer login and building access. An ideal biometric trait must have the following characteristics, universality, permanence, acceptability, uniqueness, practicality and must be spoof-proof. In this unit the participants are introduced to the concept of Biometric identification. First, the participants are asked to brainstorm different recipes for automated personal identification through observation and monitored discussion. Then, the actual devices built upon such concepts and methods are presented. The Bio-Identification lab, where four computers are set up running the following Biometric includes 1) a hand geometry system; 2). Two fingerprint systems (one using optical and one using CMOS technology, 3). an Iris-scan system, 4). a face-recognition system. The participants are divided into groups and will be rotated across the aforementioned test platforms. They study and investigate each technology according to the six required characteristic of a Biometric system. This module will conclude with a moderated discussion on the pros and cons of Biometric systems in practice.

Multimedia and Web Page Design (Computer Science Module)

In this module they investigate how to create digital multimedia and integrate this media into a web page to understand how multimedia is created and how web pages are designed. During the laboratory component the students divide themselves into groups and use digital equipment (digital still cameras, MP3 recorder, digital HDTV camcorder, lapel microphones and a green screen) to record multimedia. Once a group has generated media, they then design web pages using that media.

Understanding and Building Structures to withstand Earthquakes (Civil Engineering Module)

The energy released at the epicenter of an earthquake is transmitted to the surface of the earth in the form of waves. This wave motion results in a ground motion that is oscillatory in nature. The response of a building or a bridge to an earthquake depends on a number of factors. They include: a) the nature of the ground motion, b) the stiffness of the structure, c) the mass of the structure, and d) the height and number of stories (if it is a building). The response is in the form of predominantly horizontal accelerations. The accelerations when multiplied by the mass results in forces in the buildings and bridges. Buildings are designed to ensure that they are able to withstand these horizontal forces. The didactic portion of the module is to explain these concepts in a simple manner. The laboratory component involves two types of small-scale building frames that have been built. These include a one story and a two-story building. Accelerometers are attached to the base and each of the stories and measurements can be made using a data acquisition system. The goal is give a certain ground motion and measure the input and response accelerations. Different masses will be placed while performing the experiment in order to measure the influence of the mass factor.

Understanding and Preventing ACL Injury Among Athletes (Bio-Mechanics Engineering Module)

Female athletes have an eight-times-greater risk of tearing their anterior cruciate ligament (ACL), a fibrous band that connects the shinbone (tibia) to the thighbone (femur), than males. Basketball, soccer and other sports that require cutting moves or jumping put female athletes at risk for ACL injury. The biomechanics lab will use the tools and techniques of mechanical engineering to explore possible causes for the discrepancy in female to male ACL injury risk. Measurements will include the activation level of the quadriceps and hamstring muscles using electromyography (EMG). Muscles are "activated" by an electrical signal that is sent from the central nervous system to the muscle. This electrical signal in the muscle can be measured with a sensor placed on the skin close to the activated muscle. In addition, knee flexion angle and ground reaction forces will be simultaneously measured along with muscle activation while the test subjects jump in the air. Drawing from knowledge of the anatomy and function of knee joint tissues and from the lab measurements collected on both male and female volunteers, participants in the lab will evaluate possible reasons for the eight-fold difference in ACL injury between male and female athletes.

Teacher Professional Development

The ARROWS Engineering Essentials program includes high school teachers from participating school districts in a professional development program that focuses on the design of a problem-based learning curriculum. The professional development program and the supporting online forums were designed and implemented by Donna Russell at the UMKC School of Education.

Problem-Based Learning

Problem-based learning units provide students with the opportunity to engage in simulations of real-world computer science and engineering problems (Savery & Duffy, 1996). Problem-based learning develops higher-order thinking abilities in students by asking them to respond to authentic issues from the field (Jonassen, 2000). Additionally, studies have shown that in underrepresented and underserved student populations inquiry-oriented strategies as part of a problem-based learning project enhanced scientific ways of thinking, talking and writing for language learners and helped them acquire language and reasoning skills (Rosebery, Warren, & Conant, 1992). In a problem-based learning unit students are asked to develop new ideas and share knowledge by completing research on the problem space. These types of inquiry-based projects can be used to illustrate concepts and connections in science (Anderson, Reder, & Simon, 1997). During a problem-based unit students are asked to respond to simulations of a real-world issue thus situating the learning in an authentic context (Lave & Wenger, 1991). This provides the students with a meaningful, authentic, and collaborative learning environment that develops the advanced cognitive processes, communicative skills and functional knowledge needed to be successful in science and engineering careers.

The ARROWS program provides teachers with professional development that they can use in their teaching practice. During the summer Engineering Essentials program teachers design a problem-based unit. Since teachers come from both Kansas and Missouri the units develop science, technology and engineering standards from both the Kansas science standards and Missouri Show-Me standards for science as well as the local districts' science and engineering objectives. These standards are part of these state's compliance with the Federally-mandated No Child Left Behind Act. All the units created are shared on the ARROWS professional development website.

The high school teachers attend the modules with students in the morning and participate in professional development activities in the afternoon. The teachers receive a brief overview of the principle characteristics of problem-based learning. The teachers use a Problem-Based Learning (PBL) Design Template created by Dr. Russell to create their PBL units. The teachers create potential science and technology units for their classrooms using the template. The teachers use a wiki site designed by Dr. Russell to download and upload documents and communicate using a discussion board. The teachers post their new units to the wiki. Dr. Russell designed a syllabus for a 3-credit Continuing Education course from the School of Education that the teachers can enroll in as part of their participation in the ARROWS program.

The teachers complete two pre and post surveys to identify changes in their ideas and beliefs about problem-based learning and science instruction methods in their classroom. The teachers respond to their interactions in the ARROWS lab activities in the morning by completing Reflection Notes each afternoon using guiding questions. Additionally the teachers used the ARROWS Professional Development Blog site to respond to surveys after each day's workshops. The main ARROWS Professional Development website is located at http://education.umkc.edu/arrows. The main teacher PD site is shown below as Figure 1.

Additionally the blog site was used by the ARROWS teachers to dialog about important issues in the design and implementation of advanced science programs in urban settings. The blog site is shown below in Figure 2.

The ARROWS PD websites includes a wikispace. In this site ARROWS teachers can develop collaborative curriculum documents and access documents created in past workshops. The PBL Design Template is accessible for use in the design of future curriculum. The curriculum loaded into this site in the summer of 2009 included a collaborative curriculum where students in two classes would design an interactive gaming environment using the core concepts of physics, and a curriculum that engages students in designing robots. The main page of the wiki is shown below as Figure 3.

The purpose of the discussion forums is to identify major concepts and also issues that aid or detract from the ability of these high school teachers from implementing PBL STEM-based units based on the modules in their classrooms. Examples of the teachers' dialogs and comments from the wiki and blogs are shown below.

Why I Liked ARROWS

1.   The use of technology – Using technology or information systems in any capacity is vital to the success of students and teachers today. In ARROWS, we created web pages and we practiced using photo editing software, voice analysis programs, and green screen technology. Teachers used a weblog to write reflections. In addition, we all use computers to complete online assessments and surveys.

Figure 1. PD Website

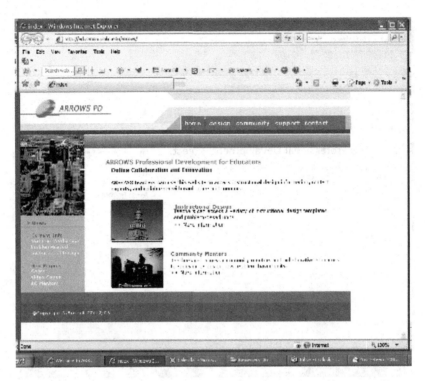

Figure 2. PD Blog Site

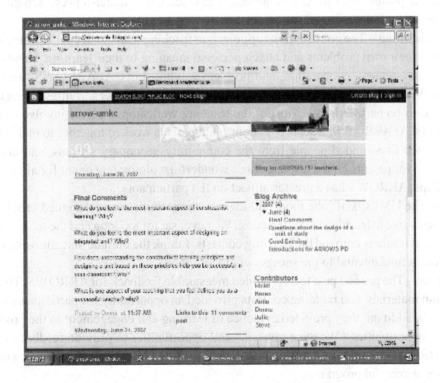

Figure 3. PD Wiki space

2. The emphasis on diversity – The ARROWS workshop supported a diverse group of students, teachers, and professors. I think it is important to reach out to under-represented groups, and the attitude of the staff is the same.

3. Hands-on learning – The constructivist approach to teaching leads to deeper learning because students solve their own problems and attach personal meaning to their learning. Hands-on learning promotes engagement, cooperation, and problem solving skills. At ARROWS, we built towers, created web pages, wrote curriculum, and participated in several other activities during the modules.

4. The emphasis on partnerships – Most of the teaching workshops I attend involve teachers teaching teachers. At ARROWS, teachers, students, and parents worked together to make new connections between like-minded people from the community, secondary schools, and university. The presentation for parents on the last day was a wonderful, culminating event for all. Because of the partnerships, ARROWS has a greater impact on the participants.

5. Climate - The UMKC staff, the food, and the way the activities were designed provided an atmosphere conducive to learning and interacting. We never sat for too long and the trivia game was fun. Students and teachers enjoyed the building contests. I think the climate and atmosphere created by the leaders is fundamental to the success of any program.

6. Assessment – The pre and post tests provided measurable feedback for ARROWS. The web pages, curriculum materials, and building contests provided an opportunity for participants to internalize concepts. In addition, they provided evidence of learning and engagement in the program.

7. Feedback opportunities - We were given several opportunities to provide feedback and input for future ARROWS programs. Clearly, the UMKC staff is flexible and willing to continuously improve an already successful program.

An example of another teacher's comments on the blog are listed below.

So far, I have participated in 3 modules. I have enjoyed them all in terms of gaining new knowledge. In earthquakes, I enjoyed the teamwork aspect of designing a building that could withstand the forces of an earthquake. In biometrics, learning the evolution of the field and what is currently available was interesting, and I liked seeing the actual devices. In web page design, we learned quite a few aspects of design that I had not done before, so that was enjoyable to see a product after just a few hours. In terms of what I can take to my students, I think a unit incorporating building structures would be something to develop teamwork and problem solving skills. For biometrics, many of my students voice an interest in CSI and crime, so I think they would be curious about biometrics and learning some of the technologies would be an interesting hands-on lab to kick off discussion of genetics in biology class. The web page development module is also fascinating, and could serve to get students working on computers in order to develop the communication aspect of all sciences. The bridge module was good in that there was little lecture, a short demonstration, brief criterion, and then we were able to create and build. The biometrics module was interesting and information heavy, but we did not create anything, or produce a product. That probably would be hard based on the nature of the subject. A project where we identified someone using the technology would be engaging. Perhaps two teams, one as criminals trying to enter, the other as security.

Each afternoon the teachers were asked to respond to the engineering and technology modules form the morning activities. The teachers were asked to identify the module, the module topic and are this viable for your classroom in a daily discussion board. The teachers enjoyed the modules and found them to be viable for their classrooms. The teachers also used the reflection discussion board to share resources such as websites for their unit development. Some examples of their responses are below:

Module: Robotics
Topic: How to build a staccitto robot. This was over the top fun. I enjoyed this a lot and would love to try this with my students.
Is this Viable
Yes!!!! Need more of this!!!!

Module: Mechanical Engineering
Topic: Human powered vechicles and biomechanics of the ACL of the knee. Very interesting presentation using the video's of actual HPV projects showcased. We also were introduced to the gait lab to see how pressure plates and video camera's could capture human movement.
Is this Viable
Yes. Applied science at its best.

Module: Web/ Multimedia
Topic: How to implement Multimedia (webpages/ sound/ video) into the classroom. I will definitely use this topic, since I am in a 1-1 laptop school and I need much assistance in this area. I will be using multimedia for projects and presentations. I am interested in learning how to get this access for my students to make projects for my classes.
Is this Viable
Yes!!!! Need more of this!!!!

Module: PBL

Topic: Discussion: What projects are being implemented?

I have changed my topic to roller coasters since this will address more standards and benchmarks for my subject of physical science.I saw that another teacher was working on this topic for physics and after thinking about it overnight I realized this would be a better fit for me.

Module: Biometrics

Topic: Measuring the physiology of human beings.

Very fascinating. I am trying to think of how to implementing this topic since I know it will generate student interest.

Is this Viable

Yes

What didn't work

N/A Everything was appropriate

Module: PBL

Topic: PBL Project and Using Second Life in the classroom I selected the unit sound and energy to work on. So I can work on two benchmarks

standards at once.

Is this Viable

Second Life...I fear that my school will not allow this usage because of possible abuse by students. (Cyberbullying, etc.) But if this is allowed I would like to learn how to use it. It sounds like fun.

Is this Viable

Yes!!!

The teacher's feedback included responses to the surveys in the discussion board. Below are examples of teacher responses to the daily surveys in the ARROWS PD discussion board.

In response to our discussions yesterday, I would like to get feedback on your ideas concerning STEM (science, technology, engineering and math) areas.

Do you currently teach in a STEM area? If yes please briefly describe your area.

I teach Biology, Anatomy & Physiology, and an integrated science class that combines geology, astronomy, physics and technology. I work with the students a great deal outside the school day on a FIRST robotics team, helping them build and design a competitive, programmable mechanical device.

Are you more or less able to design your own course when compared to 5 years ago – or previous years? Explain why yes or no.

I have been a high school teacher at the same small charter school for three years I was given the freedom and responsibility to develop the science curriculum. I essentially framed it within the state standards, implementing it using the constructivist learning principles I had learned during my graduate work.

What is the single most important reason that you do, or do not, design your own curriculum? Explain why.

I design my own curriculum because my school needed someone to do it and they felt I was qualified to do it. We are expected to have it aligned to the state standards (the newly released "course level expectations") and help prepare students for the state test. While I respect the expectorations of the state and I feel that most of their standards are valid/ relevant, it does limit our ability as teachers to create rich, hands-on, integrated, cross-curricular projects.

If you wanted to develop integrated STEM courses, are you aware of any professional development programs available to teach you how to design such a unit?

Yes. FORD PAS, Geoworlds and Project Lead the Way are some examples. These may or may not help the schools meet the AYP they need to keep afloat.

Did the ARROWS modules and PD help you think about potential STEM activities or units in your classroom? If yes, briefly describe your idea for a unit or activity?

For my integrated science class and robotics team, I could use much of the info presented in the modules. I could also implement a PD designed unit in my integrated class provided that I actually teach it this coming year, which is still up in the air. My unit idea is actually for Biology, which I know I will be teaching. It relates to biotechnology application in the "real world."

An example of another teacher's response is below.
Do you currently teach in a STEM area? Yes. If yes please briefly describe your area.

I teach biology and integrate both technology and mathematics.

Are you more or less able to design your own course when compared to 5 yearsago – or previous years?

Less likely.

Explain why yes or no.

I am no longer likely to design my own course due to the standards in the district as well as standardized testing at the state and national levels that require students to learn particular information that five years ago I would not have considered content of a biology class.

What is the single most important reason that you do, or do not, design your own curriculum. Explain why.

From past experience I have seen teachers who do not teach to a standardized test fired or reprimanded due to their students not performing well on state end of instruction tests.

If you wanted to develop integrated STEM courses, are you aware of any professional development programs available to teach you how to design such a unit?

No, I am not.

Did the ARROWS modules and PD help you think about potential STEM activities or units in your classroom? If yes, briefly describe your idea for a unit or activity?

Yes, they did. Because I am no longer teaching environmental science and zoology, I will have honors biology. Having my students realize the importance of biometrics and how it will affect their lives and what exactly this technology is will challenge my students. They will need to understand its importance, who should use this system, and the ethical dilemmas that come from using it.

Problem-Based Learning Units

Each afternoon the teachers used the PBL Design Template to develop a PBL unit. PBL Design Template is included in supplemental materials for this book. Only one teacher had previously written a PBL curriculum. This teacher works in a gifted education program in the Rockwood School District, a suburban district outside of St. Louis. She had heard about the ARROWS program and knew Dr. Russell. She scheduled a long visit to Kansas City to coincide with the Engineering Essentials program and enrolled.

During discussions of problem-based learning characteristics and design, Dr. Russell emphasized the importance of developing a project that involves resources, organizations and activities outside of the classroom including guest speakers, presentations, and field trips. Most teachers are given standards as well as goals and objectives for their content areas. In Missouri these are tested at the end of the year in grade level exams. However Missouri is switching to course level exams that will be given at the end of each course requiring a much more specific 'fit' between the student activities, the curriculum and the testing process.

When asked to develop a unifying theme for the science, engineering and technology units all the teachers were able to develop a concept based on these the state standards for science, technology and math using the PBL Design Template. Some of the units created included a water quality unit, an examination of bacteria and virus, a unit where the students design a new game in Second Life, an amusement park unit to teach physics and an environmental awareness unit. The teacher working on the Second Life game unit teaches in a one-on-one laptop program in Kansas City Kansas School District. She wants her students to be able to work collaboratively to design a game in Second Life. She would set them up in groups to develop the ideas for the educational games, such as teaching history to fourth graders, and they would then design the interactions in Second Life using the open coding system in Second Life. The students would interview the teachers and the students then write the code and create the objects in Second Life. The bacteria and virus unit would develop the science concepts identifying differences between the characteristics and functions of bacteria and viruses focusing on the issue of human diseases and treatment for both.

Another unit that was developed was a water quality unit. The teacher teaches at an inner city high school. The unit was developed to include interactions with a local water quality monitoring program already in place in Kansas City metro area. In the PBL Design Template the teachers are asked to list community organizations that can provide expertise and mentor the students. Below are examples of the curriculum she developed during professional development with ARROWS. She contacted a local agency via email during her professional development time and they responded that they would work with her to implement the unit during the next school year. Her response to the questions in the Problem-Based Unit Design Template are listed below:

Why is this unit meaningful?

Water quality issues are continually being discussed in the community. These issues are easy to conceptualize and the solutions are quickly experienced. Making this problem something that students can easily relate to on many levels.

What is the nature of the problem students will tackle in this unit? Students will tackle a complex problem: water quality in the urban environment. They will learn background on water quality (causes of pollution, how we know the water quality, etc.).

List the benchmarks/objectives/standards addressed in this project, based on my district goals relate to the science and technology benchmark unit, see below:

Standard 5 Science and Technology, Benchmark 1, Indicator 1 (5.1.1)

The student understands technology is the application of scientific knowledge for functional purposes.

Benchmarks
- Technology is driven by the need to meet human needs and solve human problems.
- Engineering is the practical application of science to commerce and industry.
- Medicine is a practical application of science to human health.
- All technological advances contain a potential for both gains and risks for society.

Objectives
- I can …
    o   Differentiate between science and technology.
    o   Relate scientific discoveries to the development of a technology.
    o   Describe the benefits or risks of a particular technology.
    o   Read an article about technology, describe the technology and analyze the benefits and risks of that technology.

Why do you think students will find the problem in this unit meaningful?

Students will find this meaningful because it is problem they see every day walking around and living in their neighborhood. They may know someone that has been negatively affected by pollution in the form health problems like asthma, cancer, or other health issues. In addition, they will learn how this pollution can cycle into their own drinking water, even though it has been "treated", for example, caffeine, oral contraceptive hormones, and antibiotics are now found at trace levels in tap water.

In addition, work on this project has already begun. I will post the email I send to the EPA recently regarding findings that a local creek is quite poor water quality. I am currently trying to follow up.

The teachers, in all cases, were able to design an overview of a PBL unit of study based on the design template given to them by Dr. Russell and using the morning modules as guides for an interactive learning experience.

Professional Development Research

The ARROWS program includes a research model designed to understand changes in the teachers' concepts about teaching and learning in science, technology and engineering areas. The goals of the research program for the teachers was to 1) identify changes in the teachers' attitudes about problem-based learning in STEM areas, 2) identify changes in the teachers' attitudes about science teaching and learning in their classroom, 3) define the issues that impact the development of PBL units in STEM areas in their classrooms. The teachers participating in several online forums including uploading new curriculum in the wiki site, a discussion board with daily questions, a reflection on their daily ARROWS activities and two surveys. All of this information was reviewed in order to identify changes resulting from the professional development activities and to analyze their responses for issues impacting the development of STEM courses based on PBL principles. The professional development data has been collected since 2005 when the ARROWS program was funded.

Teacher Surveys

In order to understand changes of PBL design principles as a result of participation in the professional development program and the design of a problem-based learning unit Dr. Russell designed the ARROWS Teacher Survey of Attitudes and Beliefs based on six areas described in constructivist education research as characteristic of PBL units. The areas assessed in the survey include 1) scaffolding, 2) mediation, 3) goal-directed, 4) meaningful, 5) constructivist and 6) inquiry (Jonassen, 2000). The PBL Survey was scaled form 1-5 to identify the teacher's awareness of these key characteristics of PBL.

The mean answer to all the scaled questions in the pre test was 3.5. The mean answer to the scaled questions in the posttest was 3.96. The sum of the averages for all scaled questions was 3.76. This increase is a reflection of the teachers' increased awareness of the terminology as well as the characteristics of a constructivist-based learning response. The area that scored lowest on both the pre and posttest was the area titled collaboration. In this area the scaled questions totaled only 3.4. The teachers expressed a lack of comfort with the purpose and assessment of the collaborative processes built into a constructivist-learning unit of study. The area that scored highest on both pre and post was mediation. The teachers scored mediation as 4.0. Mediation, which entails the use of scientific tools and the development of the scientific process, was rated highly by the teachers and included as a major influence in their new unit.

The survey included an open response at the end of each section. Some of their comments to the open-ended sections are copied below:

How do you get students interested in learning that is outside their comfort zone. How do you encourage students to explore different levels of problem solving- how not to only explore the first solution but to explore other options.

The idea of PBL is familiar to me, but the jargon is not. I come from a chemistry/biology background, and worked in industry for 3 years before becoming a secondary science educator. One difficulty in my district is that every teacher is required to teach the same 14+ units with the same 2 exams per unit, but with relatively few resources both in terms of curriculum, money, and supplies. These benchmarks and objectives do not necessary match the textbook, and the exams often involve vocabulary and concepts beyond the scope of the current course. This is frustrating because our curriculum is becoming "a mile wide and an inch deep". We compartmentalize knowledge to the point that the big picture is hazy, fuzzy, or not there. In addition, there is not enough time for the number of labs I would prefer to offer. Half of the 28 exams from August to December or January to May (due to our college-like semester schedule) are supposed to be Performance, but in reality this has not yet happened due to a variety of restraints above. I think our district IS making progress, but many teachers now have their hands tied and cannot teach to their strengths as much. It has some benefits, but I disagree with the fact that our students can retake tests as many times as they want. This is not preparing them for reality.

Science in the Classroom Survey

The Science in the Classroom survey was designed to identify the teachers' concepts about student interest and abilities in a science classroom. These concepts included Learning to communicate, Learning to learn, Learning to speak out, Learning about science, Teacher support in learning science, Interest in learning science and Learning about the world. The responses are scaled 1-5 from almost always to almost never. There are six responses in each of the seven categories with a total of 42. The teachers were asked to respond as a student in their own class would respond. Of the seven categories the lowest level of response was in the category of Interest in learning science with an overall mean response of 3.3. Next was the Teacher support in learning with 3.9. Then Learning to learn with 4.0. Learning to speak out with 4.0. Learning about science with 4.1 and finally Learning about the world with the highest level of response at 4.3.

In some ways the results of the student perspective survey contradicts the teacher reflections and discussions. In online and face-to-face discussions, the teachers expressed concern over the logistics and assessment of group work and discussions but in this survey, responding from the perspective of their students in their classrooms, they included talking about science as an important aspect of their classroom. Additionally all the teachers except one had never developed a PBL science unit based on a real world issue or problem but they scored this type of learning by the student, meaningful learning, at the highest level because they felt that students would be making the connections between science in their classrooms and the real-world of science and technology outside of their classroom. The teachers noted the need to make science meaningful and important to students in their PBL design survey but scored student interest in science as the lowest category of student science response. Perhaps this denotes a disconnect between their ideas about the design of a meaningful learning environment by the teacher and what they perceive as the students' ability to pay attention, be motivated and engage collaboratively during these units.

Summary of Teacher Responses

The teachers responded to several daily forums and two surveys during their professional development program. As a result they discussed several issues impacting the implementation of a PBL unit to develop STEM concepts and knowledge in their classrooms.

Professional Development Issues

In summary, the teachers have good ideas about how science can be taught using PBL principles but the lack of training on the design and implementation of these constructivist-based units inhibits their ability to implement them. The teachers described a high degree of frustration with the fragmentation of the curriculum in their schools that is needed to prepare students for standardized testing. The isolation of content for the standardized assessment decreased the potential to implement an integrated unit, like a PBL unit, in their schools. The teachers wanted to be able to coalesce the curriculum standards into meaningful units of study however only one teacher had received prior training in how to design an integrated unit of study based on a unifying theme such as a PBL unit. In their discussions and surveys, the teachers understood the importance in the type and quality of learning that is possible in a PBL unit, but they were unsure how to design the interactions and the subsequent assessments that would develop the learning. One teacher discussed her interest in using groups in her course but was afraid of the lack of discipline that could result.

Assessment

The teachers identified assessment as a major issue. Assessment of problem-based learning is a core issue that must be dealt with if teachers are asked to develop constructivist-based learning environments. They are unsure how to assess the group work, collaborative projects, inquiry processes and other student responses in a PBL unit. Additionally, standardized testing makes it much more difficult for teachers to implement these innovative units as teachers find it difficult to match the assessment of the PBL processes possible with the traditional testing assessment model. Teachers noted that there are penalties for themselves and their students if they do not raise test scores. Implementing innovative new units are difficult in this atmosphere. Additionally, one state, Missouri, has changed assessment from grade level evaluations to course level evaluations. This requires the teacher to teach specific content during each course reducing the potential to integrate content.

Concepts of Learner Characteristics

The two surveys, one from the teacher's perspective on the characteristics of PBL units and the other from the perspective of a student in their classroom identified differences between their concepts of PBL and their classroom. They identified a disconnect between their response to the need for collaborative work but their lack of use of this instructional process. Although identifying student engagement as a primary aspect of PBL, they were unsure how to manage this including questioning whether students would be motivated to engage in these activities. The classroom management tasks of focusing group work, identifying and assessing individual responses and managing behaviors were mentioned as difficulties that ultimately contradicted their responses in the PBL survey.

Lack of Resources

The teachers were concerned with the cost of a PBL or science unit. The modules that they participated in during ARROWS Engineering Essentials program all used expensive equipment or software that they and the district cannot afford. Another resource that is limited is their time. They are not given a lot of professional development time and would have to develop a PBL unit on their own using their own resources.

Scheduling

The high school teachers also discussed their schedules at the school as an impediment to implementation. Because they are all departmentalized it is difficult to find the time to develop an interdisciplinary topic in depth. Several were teaching in different science areas from year to year such as teaching biology one year and earth science the next based on enrollment and other issues. This makes it difficult to design large integrated PBL units.

CONCLUSION

The ARROWS program is a long-term outreach program in an urban community with the goal of motivating more computer science and engineering undergraduate students to achieve the skills needed for careers in computer science and engineering. The professional development aspect was designed to develop new understanding and knowledge about problem-based learning as a model for teaching STEM content knowledge. The teachers were asked to develop a PBL unit and respond to multiple forums for discussion on the potential of these units to teach advanced STEM content knowledge to their students. The analysis of the teachers' responses identified several issues including the lack of professional development, lack of resources, standardized testing, scheduling, and the teachers' concepts of their students' ability to engage in problem-based learning.

There is an urgent need for more minority students in science and engineering programs at the university level. Problem-based learning units support the development of the advanced cognitive abilities needed to be develop the STEM content knowledge needed to be successful in science and engineering programs. It is important for urban high school teachers to be able to implement PBL units in the science and engineering courses to develop these students' capabilities to enter and be successful in university programs. This research identified several important issues that impede the ability of urban high school teachers to develop these units.

Ultimately, it is important to understand that both sides of the educational continuum in the US, the k-12 educational system and the university system, need to focus on responding to this need by developing a coherent forum for dialog and the implementation of collaborative efforts, such as the ARROWS program, that focus on increasing the number of minority students capable of successfully completing a degree in STEM areas.

Overviews of the Chapters

This book was conceptualized to provide a comprehensive review of web-based engineering education focusing on the development of real-world problem-based learning skills. The chapters provide in-depth descriptions of multiple educational settings.

Chapter 1. *Designing of E-Learning for Engineering Education in Developing Countries: Key Issues and Success Factors* by B. Noroozi, M. Valizadeh, & G. A. Sorial. Traditional education for engineers has shifted towards new methods of teaching and learning through the proliferation of Information and Communication Technologies (ICT). The continuous advances in technology enable the realization of a more distributed structure of knowledge transfer. This becomes critically important for developing countries that lack the resources and infrastructure for implementing engineering education practices.

The two main themes of technology in designing e-Learning for engineering education in developing countries focus either on aspects of technological support for traditional methods and localized processes, or on the investigation of how such technologies may assist distance learning. Commonly such efforts are threefold, relating to content delivery, assessment and provision of feedback. This chapter is based on the authors ten years' experience in e-Learning, and reviews the most important key issues and success factors regarding the design of e-Learning for engineering education in developing countries.

Chapter 2. *Architectural Web Portal and Interactive CAD Learning in Hungary* by Attila Somfai. This chapter describes the use of a teaching web portal at a university in Hungary. The Internet has created potential new and effective ways of cooperation between lecturers and students of the university and other institutions of higher education. The teaching web portal of the Faculty of Architecture at Széchenyi István University (www.arc.sze.hu/indexen.html) realizes the diversity and complexity of architecture with efficient grouping of information and being attentive to high professional standards. Computer Aided Architectural Modelling (www.arc.sze.hu/cad) is one of the new types of online lecture notes, where many narrated screen capture videos show the proper usage of CAD software instead of texts and figures. This interactive type of learning assists students to become more independent learners. This type of teaching modality provides the opportunity for students who need more time to acquire subject matter by viewing video examples again. Additionally, the success of our departments' common web initiations can be measured through Internet statistics and feedback of the students and external professionals.

Chapter 3. *Adapting Engineering Education to the New Century* by A.K. Haghi, & B. Noroozi. In this book chapter, the authors summarize their retrospections as an engineering educator for more than 20 years. Consideration is given to a number of educational developments to which the authors has contributed during their career in academia and the contribution made to engineering and technological education. Increasing emphasis is being placed on establishing teaching and learning centers at the institutional level with the stated objective of improving the quality of teaching and education. The results of this study provide information for the revision of engineering curricula, the pedagogical training of engineering faculty and the preparation of engineering students for the academic challenges of higher education in the field. The book chapter provides an in-depth review of a range of critical factors liable to have a significant effect and impact on the sustainability of engineering as a discipline. Issues such as learning and teaching methodologies and the effect of E-development and the importance of communications are discussed.

Chapter 4. *3D Virtual Learning Environment for Engineering Students* by M. Valizadeh, B. Noroozi, & G. A. Sorial. Virtual Reality and Virtual Learning Environments have become increasingly ambiguous terms in recent years because of changes in essential elements facilitating a consistent environment for learners. Three-dimensional (3D) environments have the potential to position the learner within a meaningful context to a much greater extent than traditional interactive multimedia environments. The term 3D environment has been chosen to focus on a particular type of virtual environment that makes use of a 3D model. 3D models are very useful to make acquainted students with features of different shapes and objects, and can be particularly useful in teaching younger students different procedures and mechanisms for carrying out specific tasks. This chapter explains that the 3D Virtual Reality is mature enough to be used for enhancing communication of ideas and concepts and stimulate the interest of students when compared to 2D education.

Chapter 5. *Online Automated Essay Grading System-As a Web Based Learning (WBL) Tool in Engineering Education* by Siddhartha Ghosh. Automated Essay Grading (AEG) or Scoring (AES) systems are more a myth they are reality. Today the human written (not hand written) essays are corrected not only by examiners and teachers but also by machines. The TOEFL exam is one of the best examples of this application. The students' essays are evaluated both by human & web based automated essay grading system. Then the average is taken. Many researchers consider essays as the most useful tool to assess learning outcomes, implying the ability to recall, organize and integrate ideas, the ability to supply merely than identify interpretation and application of data. Automated Writing Evaluation Systems, also known as Automated Essay Assessors, might provide precisely the platform we need to explicate many of the features those characterize good and bad writing and many of the linguistic, cognitive and other skills those underline the human capability for both reading and writing. This chapter focuses on the existing automated essay grading systems, the basic technologies behind them and proposes a new framework to show that how best these AEG systems can be used for Engineering Education.

Chapter 6. *Future Challenges of Mobile Learning in Web-Based Instruction* by Chandre Butler, Rochelle Jones, & Pamela McCauley-Bush. Mobile learning is becoming an extension of distance learning, providing a channel for students to learn, communicate, and access educational material outside of the traditional classroom environment. Because students are becoming more digitally mobile, understanding how mobile devices can be integrated into existing learning environments is advantageous however, the lack of social cues between professors and students may be an issue. Understanding metrics of usability that address the concern of student connectedness as well as defining and measuring human engagement in mobile learning students is needed to promote the use of mobile devices in educational environments. The purpose of this chapter is to introduce contemporary topics of applied mobile learning in distance education and the viability of mobile learning (m-learning) as an effective instructional approach.

Chapter 7. *Designing Animated Simulations and Web-Based Assessments to Improve Electrical Engineering Education* by Doug Holton, & Amit Verma. Over the past decade, our research group has uncovered more evidence about the difficulties undergraduate students have understanding electrical circuit behavior. This led to the development of an AC/DC Concept Inventory instrument to assess student understanding of these concepts, and various software tools have been developed to address the identified difficulties students have when learning about electrical circuits. In this chapter two software tools in particular are discussed, a web-based dynamic assessment environment (Inductor) and an animated circuit simulation (Nodicity). Students showed gains over time when using Inductor, and students using the simulation showed significant improvements on half of the questions in the AC/DC Concept Inventory. The chapter concludes by discussing current and future work focused on creating a more complete, well-rounded circuits learning environment suitable for supplementing traditional circuits instruction. This work includes the use of a contrasting cases strategy that presents pairs of simulated circuit problems, as well as the design of an online learning community in which teachers and students can share their work.

Chapter 8. *Use of Living Systems to Teach Basic Engineering Concepts* by Kauser Jahan, Jess W. Everett, Gina Tang, Stephanie Farrell, Hong Zhang, Angela Wenger & Majid Noori. Engineering educators have typically used non-living systems or products to demonstrate engineering principles. Each traditional engineering discipline has its own products or processes that they use to demonstrate concepts and principles relevant to the discipline. In recent years engineering education has undergone major changes with a drive to incorporate sustainability and green engineering concepts into the curriculum. As such an innovative initiative has been undertaken to use a living system such as an aquarium to teach basic engineering principles. Activities and course content were developed for a freshman engineering class at Rowan University and the Cumberland County College and K-12 outreach for the New Jersey

Academy for Aquatic Sciences. All developed materials are available on a dynamic website for rapid dissemination and adoption.

Chapter 9. *The Use of Applets in an Engineering Chemistry Course: Advantages and New Ideas* by B.M. Trigo, G.S. Olguin, & P.H.L.S. Matai. Since the sprouting of the personal computers, around 1980, computer science has been gaining importance, becoming a supporting instrument for other daily activities of ours. With the introduction of computing in a certain area or activity, barriers and difficulties can be overcome. Consequently, new paradigms, possibilities and challenges are created. In education it is not different, computing is more and more present, assisting the learning process in a variety of ways, creating new challenges that compel us to re-think the way education is performed, considering new delivery methods instead of only the traditional chalk-and-talk method. "While we may feel comfortable with traditional approaches, the new technologies provide us with the tools to challenge these positions, and open up the teaching/learning questions for some rethinking" (Roy & Lee, 1999). With the sprouting of the Internet, it was created what we today understand as the web-based teaching, a method that brings innumerable benefits and challenges for the educators. This chapter describes the development of Applets in a General Technological Chemistry (QTG) course covering the topics of the discipline and joining them in a main "host" Applet, creating a virtual chemistry laboratory, which would be available for the students in the discipline's website.

Chapter 10. *Competitive Design of Web-Based Courses in Engineering Education* by Stelian Brad. Developing engineering study programs of high quality, able to satisfy customized needs, with flexible paths of study, with easy and rapid access to the most appropriate educational facilities and lecturers is a critical and challenging issue for the future of engineering education. The latest developments in communication and information technologies facilitate the creation of reliable solutions in this respect. Provision of web-based courses in engineering education represents one of these solutions. However, the absence of physical interactions with the training facilities and the specificity of remote collaboration with lecturers rise up additional challenges in designing a high-quality web-based engineering course. In order to define superior solutions to the complex set of requirements expressed by several stakeholders (e.g. students, lecturers, educational institutions and companies), a comprehensive planning of quality and an innovative approach of potential conflicting problems are required during the design process of web-based engineering courses. In this context, the present chapter introduces a generic roadmap for optimizing the design process of web-based engineering courses when a multitude of requirements and constrains are brought into equation. Advanced tools of quality planning and innovation are considered to handle the complexity of this process. The application of this methodology demonstrates that no unique, best-of-the-world solution exists in developing a web-based engineering course; therefore customized approaches should be considered for each course category to maximize the impact of the web-based educational process.

Chapter 11. *WIRE: A Highly Interactive Blended Learning for Engineering Education* by Yih-Ruey Juang. Many researchers have shown that blended learning can more effectively enhance motivation, communication skills, and learning achievement compared to single-form teaching methods. However, the crucial issue that needs to be addressed in blended learning is the question of how to integrate the selected blended format, technology, and teaching strategy into a coherent learning model, and while maintaining interaction between the teacher and students both inside and outside the classroom. Most useful functions of e-learning tools have not been meaningfully integrated into teaching and learning strategies, or into the course management systems that have been used in most campuses. This chapter introduces a highly interactive strategy, the WIRE model, for blended learning that incorporates web-based and face-to-face learning environments into a course by linking the Warm-up before class, Interactive teaching in class, and Review and Exercise after class, creating a sustained learning experience through meaningful use of technology in engineering curricula.

Chapter 12. *Sights Inside Virtual Engineering Education* by Giancarlo Anzelotti & Masuomeh Valizadeh. The critical tenet of engineering education reform is the integral role of virtual environment capabilities that is provided by fabulous advances in information technology. Current technological progresses combined with changes in engineering content and instructional method require engineering instructors to be able to design intensive and concentrated lessons for exploration and discovery of the engineering concepts through appropriate computer applications. In actual practice, however, most computer applications provided for engineering education consist of software designed for a specific educational purpose. Furthermore, economical constraints often stand in the way of incorporating such special purpose software into an instructional setting. This chapter discusses an alternative to the traditional approach that shifts the instructional focus from specific computer applications to more sophisticated uses of general purpose software. In particular, educational uses of purpose-oriented small software that can be implemented in multi purpose software are exampled as an introduction to this approach

Chapter 13. *Effective Design and Delivery of Learning Materials in Learning Management Systems* by Mehregan Mahdavi , John Shepherd, & Mohammad H. Khoobkar. Learning Management Systems (LMS) enable effective design and delivery of learning materials. They are Web-based software applications used to plan, implement, and assess a specific learning process. LMSs allow learners to connect to and interact with the educational material through the Internet. They enable tools for authors (instructors) to design learning materials that include text, html, audio, video, etc. They also enable learner activity management in the learning process. Moreover, they provide tools for effective and efficient assessment of the learners. This chapter explores learning management systems and their key components that enable instructors organize and monitor learning activities of the learners. It also introduces the authoring features provided by such systems for preparing learning material. Moreover, it presents assessment methods and tools that enable evaluation of the learners in the learning process. Furthermore, existing challenges and issues in this field are explored.

Chapter 14. *Web-Based Training: An Applicable Tool for Engineering Education* by Masoumeh Valizadeh, Giancarlo Anzelotti & Salesi Sedigheh. Due to its singular capabilities, web-based learning has entered and is widely used in every field of science and technology. It is not only warmly welcomed at schools and universities, but also in factories and houses. However utilization of web-based learning technique requires tender and comprehensive attentions in designing, applying and assessing configurations, directed by when, where and for which purpose it is being employed. Among different branches of human knowledge and sciences, engineering as well as medicine is more involved in practical and daily-life aspects, where the virtual utilities and educational software can be utilized to consummate the practical features of engineering education. Furthermore the virtual environment of e-learning courses can provide cheaper, safer, more comprehensive and more inclusive approaches to engineering educational material. What usually the students need to learn in laboratories and workshops and it is costive and demanding for the universities and the schools. The aim of this chapter is to count the requirements of engineering education and to accord the facilities and inadequacies of e-learning as training technique in engineering instruction.

Chapter 15. *A Diagnostic System Created for Evaluation and Maintenance of Building Constructions* by Attila Koppány. The successful diagnostic activity has an important role in the changes of the repair costs and the efficient elimination of the damages. The aim of the general building diagnostics is to determine the various visible or instrumentally observable alterations, to qualify the constructions from the suitability and personal safety (accidence) points of view. Our diagnostic system is primarily based on a visual examination on the spot; its method is suitable for the examination of almost all-important structures and structure changes of the buildings. During the operation of the diagnostic system a large number of data – valuable for the professional practice – was collected and will be collected also in the future, the analysis of which data set is specially suitable for revaluing construction and the practical application of the experiences later during the building maintenance and reconstruction work. For using the system a so-called "morphological box" has been created, that contains the hierarchic system of constructions, which is connected with the construction components' thesaurus appointed by the correct structure codes of these constructions' place in the hierarchy. The thesaurus was not only necessary because of the easy surveillance of the system, but to exclude the usage of structure-name synonyms in the interest of unified handling. The analysis of which data set is specially suitable for revaluing earlier built constructions and which data can help to create knowledge based new constructions for the future is the topic of this chapter.

SUMMARY

This book evaluates the usefulness of advanced learning systems in delivering instructions in a virtual academic environment for different engineering sectors. The learning process discussed in each chapter will include a walk-through of a case problem solving exercise in the virtual company environment. This will enable the reader to adjust to conditions of the world of work. This volume will demonstrate how to enter the virtual company module, select a company and a case problem of their choice. Each chapter plays a key role in designing and producing workable solutions. Each chapter provides the reader with opportunities to think critically and approach problems analytically. Hence, the reader will be able to link diverse skills and knowledge to tackle problems and maximize productivity in the design and implementation of web-based engineering education. Finally, this volume aims at providing a deep probe into the most relevant issues in engineering education and digital learning and offers a comprehensive survey of how digital engineering education has developed, where it stands now, how research in this area has progressed, and what the prospects are for the future. As a result this volume will be of significance to those interested in digital learning and teaching and training focusing on the university and industrial context. This book is also a productive resource to industrial engineering professionals who would like to see how digital learning works in practice. In addition, the book will be highly valuable to those researchers in the field interested in keeping abreast of current developments in the confluence of their fields of expertise and technological settings.

ACKNOWLEDGMENT

I would like to acknowledge the USA National Science Foundation for funding the ARROWS: Achieving Recruitment, Retention & Outreach With STEP grant and the collaboration and support of the professors and staff at the School of Computing and Engineering at the University of Missouri-Kansas City. I would like to specifically thank Dr. Koshrow Shoraby and Dr. Mark Hieber for their collaborative efforts in developing the ARROWS program to inform, engage and excite urban high school students about their potential to work in engineering and computer science careers.

REFERENCES

Anderson, J., Reder, L., & Simon, H. (1997). Situative versus cognitive perspectives: Form versus substance. Educational Researcher, 26(1), 18-21.

Anderson, L. & Gilbride, K.(2002). Discover Engineering: Assessing the impact of outreach programs. Retrieved from http://www.ccwest.org/cnu5news/images/Anderson%20and%20

Astin, H.. & Kent, L. (1983). Gender Roles in Transition: Research and Policy Implications for Higher Education, Journal of Higher Education, 54, 309-324.

Austin, A. (1985). Involvement: The Cornerstone of Excellence. Change, July/August.

Besterfield-Sacre, M., Atman, C. & Shuman, L., (1997). Characteristics of Freshman Engineering Students: Models for Determining Student Attrition in Engineering, Journal of Engineering Education, 139 – 149.

Colbeck, C. , Cabrera, A.,Terenzini, P. (2001). Learning professional confidence: linking teaching practices, students' self-perceptions, and gender. Review of Higher Education, 42, 324-352.

Delisio, E. (2002). Education World: The Educator's Best Friend. High Schools Introduce Engineering. Retrieved from http://www.educationworld.com/a_curr/curr396.shtml

Fortier, P.J., Fowler, E., Laoulache, R.N., Pendergrass, N. A., Sims-Knight, J., & Upchurch, R., (2002). Technology Supported Learning Applied to an Innovative, Integrated Curriculum for First-Year Engineering Majors. In Proceedings of the 35th Hawaii International Conference on System Sciences, IEEE Computer Society. Retrieved from http://csdl.computer.org/comp/proceedings/hicss/2002/1435/01/14350033.pdf

Fortier, P., Fowler, E., Laoulache, R., Pendergrass, N., Sims-Knight, J., & Upchurch, R., (2002). Technology Supported Learning Applied to an Innovative, Integrated Curriculum for First-Year Engineering Majors. In Proceedings of the 35th Hawaii International Conference on System Sciences. IEEE Computer Society

Hirsch, L., Gibbons, S., Kimmel, H., Rockland, R. & Bloom, J. (2003). High school students' attitudes to and knowledge About Engineering. 33rd ASEE/IEEE Frontiers in Education Conference, Nov 5-8, 2003, Boulder, CO (pp. F2A-7 – F2A 12). Retrieved from http://fie.engrng.pitt.edu/fie2003/papers/1145.pdf

Hirsch, L.S., Gibbons, S.J., Kimmel, H., Rockland, R. & Bloom, J. (2003). High School Students' Attitudes to and Knowledge About Engineering. 33rd ASEE/IEEE Frontiers in Education Conference, Nov 5-8, 2003, Boulder, CO (pp. F2A-7 – F2A-12).

Hoit, M. & Ohland, M.,(1998). The Impact of a Discipline-Based Introduction to Engineering Course on Improving Retention. Journal of Engineering Education, ASEE, 79–85. Retrieved from http://csdl.computer.org/comp/proceedings/hicss/2002/1435/01/14350033.pdf

Savery, J. R., & Duffy, T. M. (1996). Problem based learning: An instructional model and its constructivist framework. Educational Technology, 35(5), 31-38. Retrieved from http://fie.engrng.pitt.edu/fie2003/papers/1145.pdf

Jain, A. Ross, & Prabhakar, S. (2004). An Introduction to Biometric Recognition. IEEE Transactions on Circuits and Systems for Video Technology, Special Issue on Image- and Video-Based Biometrics, 14/1, 4-20.

Jonassen, D. H. (2000). Toward a design theory of problem solving. Educational Technology Research and Development, 48(4), 63-85.

Lave, J., & Wenger, E. (1991). Situated learning: Legitimate peripheral participation. Cambridge, UK: Cambridge University Press.

May, G. & Chubin, D. (2003). A Retrospective on Undergraduate Engineering Success for Underrepresented Minority Students. Journal of Engineering Education, ASEE, 27-39.

Rosebery, A., Warren, B. & Conant, F (1992). Appropriating scientific discourse: Findings from language minority classrooms. The Journal of the Learning Sciences 2, 61-94.

Taylor, P.C., Fraser, B.J. & White, L.R. (1994). The revised CLES: A questionnaire for educators interested in the constructivist reform of school science and mathematics. Paper presented at the annual meeting of the American Educational Research Association, Atlanta, GA

Yates, J.K., Vos, M, & Tsai, Kuei-wu, (downloaded, August 2002). Creating Awareness about Engineering Careers:Innovative Recruitment and Retention Initiatives. 29th ASEE/IEEE Frontiers in Education Conference, Nov. 10-13, 1999, San Juan, Puerto Rico, 13d7 9 – 13d7 14.

Donna L. Russell
University of Missouri, USA

Acknowledgment

We would like to express our deep appreciation to all the authors for their outstanding contribution to this book and to express our sincere gratitude for their generosity. All the authors eagerly shared their experiences and expertise in this new book. Special thanks go to the reviewers for their valuable works.

Donna L. Russell
University of Missouri, USA

A.K. Haghi
University of Guilan, Iran

Chapter 1
Designing of E-Learning for Engineering Education in Developing Countries:
Key Issues and Success Factors

B. Noroozi
University of Guilan, Iran & University of Cincinnati, USA

M. Valizadeh
University of Guilan, Iran

G. A. Sorial
University of Cincinnati, USA

ABSTRACT

Traditional education for engineers has shifted towards new methods of teaching and learning through the proliferation of Information and Communication Technologies (ICT). The continuous advances in technology enable the realization of a more distributed structure of knowledge transfer. This becomes critically important for developing countries that lack the resources and infrastructure for implementing engineering education practices. The two main themes of technology in designing e-Learning for engineering education in developing countries focus either on aspects of technological support for traditional methods and localized processes, or on the investigation of how such technologies may assist distance learning. Commonly such efforts are threefold, relating to content delivery, assessment and provision of feedback. This chapter is based on the authors '10 years' experience in e-Learning, and reviews the most important key issues and success factors regarding the design of e-Learning for engineering education in developing countries.

DOI: 10.4018/978-1-61520-659-9.ch001

1. INTRODUCTION AND BACKGROUND

Recent years have seen dramatic changes in engineering education in terms of increased access to lifelong learning, increased choice in areas of study and the personalization of learning. To advance across all domains seems to necessitate incompatible changes to the learning process, as practitioners offer individualized learning to a larger, more diverse engineering student base. To achieve this cost effectively and without overwhelming practitioners requires new approaches to teaching and learning coupled with access to a wide range of resources: practitioners need to be able to source and share engineering materials, adapt and contextualize them to suit individual needs, and use them across a variety of engineering educational models (Littlejohn et al, 2008).

Hence a great deal of effort has focused on the integration of new technologies such as multimedia video, audio, animation, and computers, with associated software, to achieve the improvement of traditional engineering education. The internet technologies have also been popularly applied to web-based learning (Hung et al, 2007). The growth of the information society provides a way for fast data access and information exchange all over the world. Computer technologies have been significantly changing the content and practice of engineering education (Gladun et al, 2008). Information and communication technologies (ICT) are rightly recognized as tools that are radically transforming the process of learning. Universities, institutions and industries are investing increasing resources to advance researches for providing better and more effective learning solutions (Campanella et al, 2007).

Figure 1. The most important and meaningful characteristics for engineering instruction

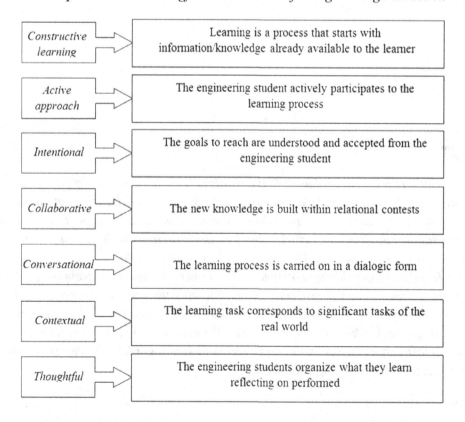

Constructive learning	→	Learning is a process that starts with information/knowledge already available to the learner
Active approach	→	The engineering student actively participates to the learning process
Intentional	→	The goals to reach are understood and accepted from the engineering student
Collaborative	→	The new knowledge is built within relational contests
Conversational	→	The learning process is carried on in a dialogic form
Contextual	→	The learning task corresponds to significant tasks of the real world
Thoughtful	→	The engineering students organize what they learn reflecting on performed

The most important aspects of a focused learning for engineering students is shown in Figure 1.

The final objective of a learning object is to realize three fundamental learning goals:

- To inform the engineering students to be responsible of their learning, capable to manage processes to reach aims and to understand their learning needs;
- To promote real and meaningful learning environments and contexts, enabling the engineering students to retrieve information and build knowledge by using different learning ways;
- To create stimulating situations and learning dynamics that prelude to wider learning tasks

Nowadays, the use of ICT has improved learning, especially when coupled with more learner-centered instruction, or convenience, where learning and exchange with the instructor can take place asynchronously at the learners own pace or on as-needed basis (Motiwalla, 2007). The consequent applications of all multimedia and simulation technologies, computer-mediated communication and communities, and Internet-based support for individual and distance learning have the potential for revolutionary improvements in education (Gladun et al, 2008).

Hence, electronic collaboration (e-collaboration) technologies for engineers are technologies that support e-collaboration. An operational definition of e-collaboration is collaboration among different individuals using electronic technologies to accomplish a common task. These e-collaboration technologies include several Internet-based technologies, such as e-mail, forums, chats, and document repositories (Padilla et al, 2008).

However the first computer-supported collaboration system emerged in 1984 from the need of sharing interests among product developers and researchers in diverse fields. This revolutionary approach was called computer-supported collab-

orative work and it was used to support learning by means of desktop and video conferencing systems. Consequently, a new paradigm arose around educational institutions which were defined as computer-supported collaborative learning (CSCL). This emerging system was based on the contributions of constructivist learning theories about the term collaborative learning, which focus on social interdependence and maintain that engineering students consolidate their learning by teaching one another. CSCL environments were created for using technology as a mediation tool within collaborative learning methods of instruction. Since then, and thanks to the great evolution of network technologies, engineering education is moving out of traditional classrooms. These collaborative e-Learning environments have caused a revolution in the academic community, providing a great amount of advantages for using both the Internet and technologies for 'any-time, any-place' collaborative learning (Jara et al, 2008).

It should be noted that e-Learning is becoming one of the most popular solutions to meet new needs especially in technical courses. In e-Learning course development and management for engineers the emphasis is often on technical aspects, whereas the relevance of learning products for the actual process of learning is not considered in depth. Indeed the most important aspect of a learning product is its aptitude to provide knowledge and skill by stimulating in dept study, further researches and close investigations (Campanella et al, 2007).

The present chapter is a study on the growth of e-Learning in engineering education as the most important objective area of science apart from the medical science. Using the definitions and aptitudes of e-Learning we tried to find a more specialized way to develop e-Learning in engineering education. To this aim, we have investigated the progresses being made on engineering e-Learning, and the benefits and difficulties of implementing e-Learning in engineering education. The study has focused on the importance of e-Learning design for engineering education; therefore, the concept of instructional design has been reviewed and oriented for engineering e-Learning. In this regard, application of ISS and ADDIE models parameters in engineering education has been introduced and analyzed. This study also presents an overview on International potential for designing e-Learning, with selected case studies of developing countries.

2. E-LEARNING, DEFINITIONS, APPLICATIONS AND PERFORMANCE

The definition of e-Learning for engineers may vary significantly, but perhaps "… learning the engineering concept is aided by information and communication technologies…" is one of the definitions, which is very close to reality. However, some authors and users define e-Learning only as "… the delivery of content via all electronic media, including the internet, intranets, extranets, satellite, broadcast, video, interactive TV, and CD ROM…". In this case the emphasis is only at the delivery and many unworthy engineering courses have been "developed" by this way –just delivery to students of existing files with handouts. This has the only advantage of reduced cost, but the educational results do not have significant value. Many authors support the first view that e-Learning should develop materials, which will increase the pedagogical effectiveness, and then deliver these to engineering students. Only in this case the full power of e-Learning can be utilized (Tabakov, 2008).

E-Learning for engineering students, at its best, is the kind of learning that complements traditional methods and gives a more effective experience to the learner (Magoha & Andrew, 2004). E-Learning refers to the use of electronic devices for learning, including the delivery of content via electronic media such as Internet/Intranet/Extranet, audio or video tape, satellite broadcast, interactive TV, CD-ROM, and so on

(Shee & Wang, 2008). Simply, e-Learning for engineers is the use of technology to support the learning process. Fundamentally, it is about putting the learner first by placing resources at the learner's fingertips. The engineering e-Learner is able to dictate the pace and balance of learning activities in a way that suits him/her. E-Learners can absorb and develop knowledge and skills in an environment that has been tailored to suit them – and at their own pace. As opposed to online courses in their strictest sense, e-Learning does not necessarily lead to an engineering certification or an engineering degree programmer but may be tailored, for example, to suit the needs of a specific company (Magoha & Andrew, 2004).

It should be noted that technology has proved its value in engineering education and in applied areas such as engineering management. For example, in Europe digital literacy is emerging as a new key competence required by workers and citizens in the new knowledge society. The integration of IT supported learning helps workers acquires the necessary skills and knowledge for their job. IT use can also improve the effectiveness of the learning process. Consequently, if the engineering student learns to use technology before starting his/her job; this could be an advantage for both the future profession and the employer. Moreover, the use of the Web as an educational delivery medium (e-Learning) provides the engineering students with the opportunity to develop an additional set of communication, technical, teamwork and interpersonal skills that mirror the business environment in which they will work. Meanwhile, the statistics on e-Learning show a considerable use of these tools in recent years. Universities are combining interactive technology and active ways of learning, which require students to develop or hone their computing skills and to take more responsibility for their own learning. Nevertheless, engineering students, contrary to the general idea that they can be considered digital natives, do not all react positively to IT learning; some prefer the traditional process. Engineering students may react differently to the online learning environment, depending on their own level and attitude. This is similar to the finding about teachers working in public centers who have shown some resistance towards IT implementation in engineering education. There is also the need to investigate the engineering students' acceptance of an Internet-based learning medium in order to understand the various drivers influencing acceptance (Padilla et al, 2008).

E-Learning for engineering students is important for building a technologically literate workforce as well as for meeting societies continuous need for rapid life-long learning delivered in increasingly more convenient forms (Buzzetto-More, 2008). In spite of all efforts, in the last years e-Learning has experienced slow user growth involvement and high dropout rates in many organizations: users become easily frustrated or unenthusiastic about the material and do not complete learning activities (Campanella et al, 2007). The development of e-Learning materials can be presented as a multi-layered process, including the following stages:

- Programming specific simulations;
- Building of e-Learning modules;
- Development of e-Learning programs.

These stages most often exist as separate entities, but the programs will include modules, and simulations can be applied at various stages (Tabakov, 2008) to provide five approaches for using technology in learning (Motiwalla, 2007):

a. Intelligent tutoring systems that have attempted to replace the teacher; these have never been successful due to their limited knowledge domains

b. Simulation and modeling tools that serve as learner's assistants or pedagogical agents embedded in applications that act as mentors providing advice;

c. Dictionaries, concept maps, learning organizers, planners and other resource aids that help learners to learn or organize knowledge with system tools and resources;

d. Personalized communication aids that can present materials depending on user abilities and experience with the system;

e. Simulated classrooms and labs that engage teachers and learners in an interaction similar to the real classrooms.

The effect on e-Learning is measured with an ISS model because it is also one of the information systems. The e-Learning success model evaluates e-Learning effectiveness based on the ISS model, constructivism and self-regulatory efficacy. In 1992 one information systems success (ISS) model has been suggested that is measured through six dimensions as was presented in Figure 2.

Figure 2. ISS model parameters

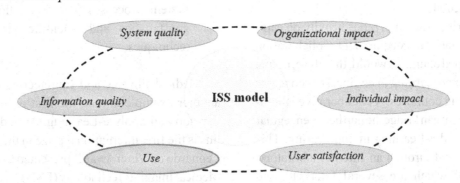

System quality implies an information process system quality based on production of produced information. Information quality is defined as the quality of system product outputs, and the usage and user satisfaction is defined as the recipients' interaction of information and information system product. Also, individual and organizational effects are measured by the information system which affects the users and the affiliated organization of the users (Lee & Lee, 2008).

3. E-LEARNING FOR ENGINEERING EDUCATIONS

Despite the dramatic expansion in e-Learning and distance education, e-Learning in engineering education still faces a number of setbacks that prevent an equivalent expansion rate. For effective and complete learning in engineering, science and technology, engineering education requires a mixture of theoretical and practical sessions. In order to understand how theoretical knowledge can apply to real world problems, practical exercises are essential. While it is relatively easy to simulate experiments, performing practical experiments online has continued to be a challenge. Coupled with this, engineering software is often very expensive and may not be easily affordable by the ordinary e-Learner. Although low cost alternatives that utilize freeware have been successfully developed and tested, practical laboratories that support engineering education are still difficult to implement online (Magoha & Andrew, 2004).

Engineering education relies heavily on capital-intensive laboratory equipment. Collaboration with developed countries would, therefore, be one path to enhance learning in engineering concepts for developing countries. A collaborative laboratory component can bridge the gap between regular e-Learning and e-Learning in engineering. This can be achieved through an Internet laboratory, examples of which are several Internet-based

laboratories have already been developed and have been described elaborately. Most of these laboratories are based on standard hardware systems, such as those provided by National Instruments (NI), and are commercially available. However they are costly and often require specialized training. The solution is to develop low cost systems that rely partially on expensive hardware at the server and open software at the client (Magoha & Andrew, 2004).

Numerous digital applications have been developed to demonstrate laboratory experiments, real-world landscape processes or microscopic objects. Engineering students learned faster and liked their classes more than students in traditional on-campus classes. Cognitive scientists believe that to learn, the material must have meaning to the learner. Cognitive science may be defined as a multi-disciplinary approach to studying how mental representations enable an organism to produce adaptive behavior and cognition. The following criteria have been considered for certain virtual field laboratory (Ramasundaram et al, 2005):

- Global access, i.e., web-based implementation.
- Stimulation of a variety of learning mechanisms.
- Interactivity to engage engineering students.
- Compartmentalization and hierarchical organizational structure.
- Abstraction of 2D and 3D geographic objects (e.g. soils, terrain) and dynamic ecosystem processes (e.g. water flow) using geostatistics and scientific visualization techniques.

Medical Physics and Engineering (MEP) is another example among the first professions to develop and apply e-Learning. An indicator for this is the first international prize in the field (EU Leonardo da Vinci Award) presented to European Medical Imaging Technology (EMIT) Consortium in 2004. During the last 15 years, a number of activities and publications addressed the questions of MEP Education and Training. These led to rapid development of the profession worldwide and now the next stage of e- Learning is to be addressed. A special issue on the subject was published by the Journal of Medical Engineering and Physics in 2005. Based on the paper and on the authors' 12 years' experience in e-Learning, the following key elements can be specified for the introduction and use of e-Learning in MEP:

- e-Learning is imperative for engineers because it offers quick and easy update of teaching materials – a very important function for this dynamic profession. This, combined with the fast delivery of the content through Internet, makes e-Learning materials the first choice for many engineering lecturers;

- e-Learning proposes an elegant way to solve the engineering problem through the understanding of complex physics models. Using interactive simulations, computer diagrams or images leads to increased effectiveness of the engineering learning process.

- Images are specifically important for engineers and e-Learning provides a cheap and effective means for publishing large number of images (either on CD or through the Web). Additionally e-Learning can offer a means of image manipulation, which has no analogue in other means of publications.

- The search function offered by various e-Learning materials is another important advantage. This is also imperative for engineers.

- Finally, the fact that many engineering students from around the world can use the materials through the guidance of most renowned specialists has no analogue in the other educational methods and media (Tabakov, 2008).

4. E-LEARNING BENEFITS AND DIFFICULTIES FOR ENGINEERING EDUCATION

E-Learning system for engineering education is able to be interpreted in various ways such as "computer based, education delivery system which is provided through the Internet", or "an educational method that is able to provide opportunities for the needed people, at the right place, with the right contents, and the right time" (Lee & Lee, 2008).

Some difficulties have been noted in using technology for professional development and four suggestions offered for product refinement and use. The learner must: (a) understand the navigation system, (b) spend time focusing on domain knowledge, rather than on navigation, (c) be prepared in domain area knowledge [e.g., mathematics] or be experienced in learning in technology environments, or (d) all three aspects. These attributes and concerns suggest the need for continued investigations about outcomes of the use of these programs in education and discussions about further product design (Pryor & Bitter, 2008).

E-Learning for engineering education offers unique pedagogical opportunities to enhance student learning: In the realm of e-education, there are clear benefits that can be derived from e-Learning as follows:

- It promotes exploratory and interactive modes of inquiry.

- It supports and facilitates team-orientated collaborations and expands the ease of access to engineering education across institutional, geographical and cultural boundaries, among others.

- Class notes and materials are posted on the Internet and students can access the sites from anywhere in the world.

- This is quite unlike distance learning, where an engineering student is given course materials and reads solely on his/her own until examination time.

- E-Learning is interactive; the software permits the engineering student to communicate, not only with the lecturer, but also with fellow classmates. It enriches and supplements the classroom experience by engaging the Web.

- E-Learning has the ability to communicate consistently to learners by providing the same concepts and engineering information – unlike classroom learning, where different instructors may not follow the same curriculum or teach different things within the curriculum.

- E-Learning is cost effective in terms of learners per instructor. In addition, it saves classroom time and this is very significant for learners who are employed on a full-time basis.

- Engineering students, instructors and evaluators can track learning outcomes more easily (Magoha & Andrew, 2004).

It has been found that all of the respondents considered the online course materials beneficial to their overall learning experience. Obtained results have shown a positive effect on engineering student learning, problem-solving skills, and critical thinking skills, with females responding more positively than males. Comparing undergraduates who completed quizzes online with engineering students who took traditional paper-based quizzes leads to reveal that students who took the quizzes online significantly outperformed engineering students who took the pencil-and-paper quizzes. It has been asserted that web-enhanced learning improves instruction and course management and offers numerous pedagogical benefits for learners. Engineering students in Web-enabled learning environments become more active and self directed learners, who are exposed to enhanced learning materials. Course Websites have proved to be an effective means of delivering learning materials,

with students responding positively to the quality resources they make available (Buzzetto-More, 2008).

Although the significant advantage of digital resources is that they can offer flexibility of format, and ease of storage and retrieval, however, digitization, on its own, is not enough to ensure flexibility and use. A major difficulty with early digital 'multimedia' resources was their inflexibility; they were not designed for reuse across a range of contexts. Consequently resources were often based around a single educational model and made available in a set format. Ease of finding suitable resources is another important factor in their effectiveness, as suggested by studies showing that the current generation of students often choose to source digital resources in preference to print based materials and their habitual use of 'Google' as a primary resource search engine. Among the reasons given for these preferences are that the computer terminal provides a 'one stop shop' for resources, and that while Google may not provide the best quality information or most efficient search, it is familiar and has a track record of producing results that are adequate. Similar criteria of unified access, familiarity and adequacy seem likely to apply to teachers' strategies in sourcing resources (Littlejohn et al, 2008).

For all its potential in dealing with learners individually, online learning does have its own invisibility problems. The other side of online learning can seem far away and isolated. For the learner, the price of empowerment is the responsibility of wielding that power. As with all distance learning, e-Learning relies on self-motivation. With no enforced discipline or deadlines, it is easy for the learner to be distracted and put off work for a distant tomorrow. With no human presence, it is also impossible for a learner with a problem to obtain help easily. Thus, personal interaction between the instructor and engineering students is either absent or else very different from traditional face-to face learning. Other subtle disadvantages of e-Learning include the ability to read text from the computer screen. Research has shown that linear text is often difficult for people to read from the computer. Hence, learners often have to reformat the text and print it out for reading, necessitating the need for a printer. The problems of invisibility, anonymity and isolation can be dealt with in several ways. Once more, the key solution is communication and there are several channels available to learners. The first is a national help line where learners can talk to someone trained to give them advice on their course. Many e-Learners are also assigned an online tutor who checks on their progress and can be contacted by e-mail with any queries or worries. Online message boards and chat rooms provide learners with the means to contact and talk to other people doing the same course, exchanging advice, ideas and encouragement. For those without access to computer at home or who require additional support there is a network of centers across countries. Surely, no engineering educator would argue against a legitimate teaching method that highly motivates students to learn their subject matter and promotes student interaction in the process. To witness students achieve an increased mastery of ideas, while also setting greater expectations for themselves, would be fulfilling educators' dreams. Table 1 summarizes a comparison between e-Learning and traditional learning (Magoha & Andrew, 2004).

5. INSTRUCTIONAL DESIGN AND ENGINEERING EDUCATION

We believe that the Design and implementation of e-Learning functionalities in the form of Web services provides mechanisms to reuse e-Learning functionalities and to create flexible platforms that can easily be adapted to the individual needs of learners. By application of already existing standards from the field of Web services, functionalities can easily be integrated.

It should be noted that Instructional Design (ID) is a multi-disciplinary domain that helps

people learns better. ID has unwavering focus on learning engineering goals, content types, learners, and technology. Therefore, the use of ID in e-Learning ensures effective learning design.

Our past experiences show that ID is the systematic development of instructional specifications using learning and instructional theory to ensure the quality of instruction. It is the entire process of analysis of learning needs and engineering goals and the development of a delivery system to meet those needs. It includes development of instructional materials and activities; and tryout and evaluation of all instruction and learner activities. As a discipline, Instructional Design is that branch of knowledge concerned with engineering research and theory about instructional strategies and the process for developing and implementing those strategies. Nevertheless, this is the science of creating detailed specifications for the development, implementation, evaluation, and maintenance of situations that facilitate the learning of both large and small units of subject matter at all levels of complexity (Siemens, 2002). Based on our teaching experiences, Implementation of this

Table 1. Comparison of e-learning versus conventional engineering learning methods

E-Learning	Traditional engineering learning mthods
It can relies on learners' and it is self-motivation	Lecturer always plays a leading role in motivating and directing the engineering students
Assessment of examinations conduced at learners' place	Assessment and examinations time does not depend on learners
Greater achievement is expected in number of students going through engineering courses	Learner restricted to those attending university or college
Innovative methods required to reach practical assignments and experiments	Laboratories readily available for practical assignments and experiments
Duration of course normally decided by the engineering student	College of engineering has calendars and set durations for courses

multi-disciplinary science in e-Learning and its design, and specially the design of e-Learning for engineering courses seem unavoidable. Hence in this session we try to make a clear scheme of ID and relate it to engineering e-Learning.

There are different models for designing instruction, as there are different learning theories. Martin Ryder of Denver School of Education at University of Colorado has divided the instructional design models in to two groups:

1. Modern models (including several models in three categories: Behaviorist, Cognitivist and Prescriptive Models) and
2. postmodern phenomenological models (including several other models in Constructivist Models category) (Ryder, 2008).

Among all presented models in ID, ADDIE which is a prescriptive design model has been selected here to be developed according to engineering e-Learning demands. ADDIE is one of the first Design Models. There has been much discussion about its effectiveness and appropriateness. ADDIE has been selected for its simplicity, ease of application, and cyclic nature.

In 1975, Florida State University developed the ADDIE 5-steps model of (see Figure 3).

1. Analysis (Assess and analyze needs)
2. Design (Develop materials)
3. Development (Develop materials)
4. Implementation (Implement activities and courses)
5. Evaluation (Evaluate participant progress and instructional materials effectiveness)

ADDIE was selected by the Armed Services as the primary means for developing training. At the time, the term "ADDIE" was not used, but rather "SAT" (Systems Approach to Training) or "ISD" (Instructional Systems Development). As a general rule, the military used SAT, while their civilian counterparts used ISD. The "D" in "ISD" first stood for "Development" but now normally means "Design" (Boot, 2008; Clark, 2004 ; Dick & Carey, 1996).

Analysis (Assess)

In analysis as the first step of ADDIE, the designer identifies the learning difficulties, the aims and objectives, the learners' needs, knowledge, and any other relevant characteristics. Analysis also considers the learning situation, any constraints, the delivery options, and the timeline for the project (Colston, 2008). It should be noted that in engineering education, this concept needs very special attention.

Design

Design is concerned with subject matter analysis, lesson planning, and media selection. The choice of media is determined by contingencies of the participant's needs and available resources. This is a systematic process of specifying learning objectives. Detailed storyboards and prototypes are often made, and the look and feel, graphic design, user-interface and content are determined here (Colston, 2008). Malachowski from city college of San Francisco, believes that Lesson planning requires that you determine your (Malachowski, 2002):

Figure 3. Five-steps of ADDIE model

- Objectives defined in terms of specific measurable objectives or learning outcomes.
- Skills, knowledge and attitudes to be developed.
- Resources and strategies to be utilized.
- Structuring, sequencing, presentation, and reinforcement of the content.
- Assessment methods matched to the learning objectives to ensure agreement between intended outcomes and assessment measurements.

Joyce and Flowers list seven instructional functions To show how to incorporate available technology into your presentations (Colston, 2008; Malachowski, 2002).

- Informing the learner of the aims,
- Presenting motivations,
- Increasing learner interest,
- Helping the learner review what they have previously learned,
- Providing conditions that will irritate presentation,
- Determining order of learning,
- Prompting and guiding the training.

Development

The actual creation (production) of the content and learning materials based on the Design phase. Development is a process of creation and testing of learning experiences and we tried to seek and answer questions such as (Malachowski, 2002):

- Have the learning needs and characteristics of the participants been accurately analyzed?
- Were the problem statement, the instructional goals and the instructional objectives appropriate for the learning needs of the participants?
- To what extent are the teaching resources, instructional strategies and the participant learning experiences successful in effec-

tively meeting the instructional goals and objectives of the target learner?
- Is it possible to accurately assess participant learning with the proposed course of instruction?

Unlike other branches of educations, e-Learning in engineering education is a highly complicated concept. Here are some important points that the designer has to pay special attention:

- Boring and weak instructional content
- A technology which its application is difficult or unconfident
- A culture which is not informed of e-Learning or developing sorts of e-Learning designs which are not consistent with the culture. This point illustrates more in developing countries or the countries with fewer internet users.

Implementation

Negative responses indicate a need for revision. Implementation is the presentation of the learning experiences to the participants utilizing the appropriate media. Learning, skills or understanding, are "demonstrated" to the participants, who practice initially in a "safe" setting and then in the targeted workspace. It may involve showing participants how to make the best use of interactive learning materials, presenting classroom instruction, or coordinating and managing a distance-learning program. The progress of the learning frequently follows cyclic patterns based on motivation and intention. Curriculum should be organized in a spiral manner such that the participant continually builds upon what they have already learned.

During implementation, the plan is put into action and a procedure for training the learner and teacher is developed. Materials are delivered or distributed to the student group. After delivery, the effectiveness of the training materials is evaluated.

Evaluation

One purpose of evaluation is to determine if the instructional methods and materials are accomplishing the established goals and objectives. Implementation of instruction represents the first real test of what has been developed. Try to pre-test instruction on a small scale prior to implementation. If this is not possible, the first actual use will also serve as the "field test" for determining effectiveness. This step consists of two objectives (Willis, 2008):

1. *Develop an evaluation strategy*: Plan how and when to evaluate the effectiveness of the instruction.

Formative evaluation can be used to revise instruction as the course is being developed and implemented. For example, the distance educator can give students pre-addressed and stamped postcards to complete and mail after each session. These "mini-evaluations" might focus on course strengths and weaknesses, technical or delivery concerns, and content areas in need of further coverage.

Within the context of formative and summative evaluation, data are collected through quantitative and qualitative methods. *Quantitative evaluation* relies on a breadth of response and is patterned after experimental research focused on the collection and manipulation of statistically relevant quantities of data.

In contrast, *qualitative evaluation* focuses on a depth of response, using more subjective methods such as interviews and observation to query a smaller number of respondents in greater depth. Qualitative approaches may be of special value because the diversity of distant learners may defy relevant statistical stratification and analysis. The best approach often combines quantitative measurement of student performance with open-ended interviewing and non-participant observation to collect and assess information about attitudes

toward the course's effectiveness and the delivery technology (Willis, 2008).

Summative evaluation consists of tests designed for criterion-related referenced items and providing opportunities for feedback from the users. Revisions are made as necessary (Colston, 2008). This is conducted after instruction is completed and provides a data base for course revision and future planning. Following course completion, consider a summative evaluation session in which students informally brainstorm ways to improve the course. Consider having a local facilitator run the evaluation session to encourage a more open discussion (Willis, 2008).

2. *Collect and analyze evaluation data*: Following implementation of the course/materials, collect the evaluation data. Careful analysis of these results will identify gaps or weaknesses in the instructional process. It is equally important to identify strengths and successes. Results of the evaluation analysis will provide a "springboard" from which to develop the revision plan (Willis, 2008).

The benefits of evaluation in engineering e-Learning are (Astrom, 2008):

* Justify and expand training funds.
* Selecting the best instruction strategies.
* Obtain and analyze learners' fulfillment feedback.
* Quantify performance improvement.

And the risks of not evaluating (Astrom, 2008):

* Waste money on ineffective programs.
* Terminate training program.
* Ineffective and Inefficient transfer of training.
* Training program does not support objectives.
* Learners' training needs not met or unknown.
* Presentation improvement unknown.
* Expectations are not set nor communicated.

6. INTERNATIONAL POTENTIAL FOR DESIGNING E-LEARNING: SELECTED CASE STUDIES IN DEVELOPING COUNTRIES

There are numerous e-Learning programs in developing countries. In Africa, for example, several distance-learning projects have been implemented successfully, the majority through collaboration with institutions in developed countries; a comprehensive list can be found from the African Distance Learning Association. Concerted efforts through outside collaborations and partnerships with developing countries and a number of African universities have culminated in various distance learning degree programs. The African Virtual University offers a range of degree programs, most of them in collaboration with Australian institutions. For developing nations, the realization of e-Learning, especially in engineering education, is yet to be thought of. There are several obstacles that developing countries have to overcome if e-Learning is to be successfully implemented. The first and major obstacle is the lack of technological resources. While the Internet has been around for over 15 years, it is only in the last few years that most developed countries have had access or even had networks on the Internet. While the Internet is dependent upon a reliable telecommunications system, most such networks in developing countries are either badly maintained or obsolete and often do not have the requisite bandwidth to support e-Learning activities. In many developing countries, such networks belong to government agencies and are usually underfunded. A second obstacle is the lack of skilled personnel who are capable of implementing and maintaining an e-Learning environment. According to ILO statistics, for instance, less than 1% of employed workers in Africa are professionals with an IT background. All these factors, together with the lack of financial resources, have contributed to the slow growth in e-Learning in engineering education in developing countries to date. While

the progress is slow, a framework for sustained growth could be created through collaborative efforts, especially among institutions in developing countries and partnerships with institutions in developed nations (Magoha & Andrew, 2004).

The difficulties of e-Learning evolution in developing countries can be counted as:

1. Lack of appropriate technological substructures
2. Financial problems
3. Lack of awareness on IT and information on the benefits of e-Learning
4. Problems in the way of communication, which are primarily the language skills, or Computer skills
5. Cultural consistency

However the developing countries have potential propensities motivating the growth of e-Learning:

1. The growing demand for knowledge and technology
2. The easiness and cost-effectiveness comparing to following the education in a developed country, and encountering with living costs and communication difficulties in these countries
3. The expanding growth of Internet utilization

In the following sections a brief review on the efforts and accomplishments of some Asian countries on e-Learning for engineers is presented.

E-Learning Progress for Engineers in Thailand; "A Brief Review"

E-Learning in Thailand like many other developing countries, deals with barriers such as culture, organization, limited technology and availability low household penetration of the ICT such as telephones, Satellites, PC's, Internet connections, broadband and etc. here again technologies available during the last decade did not appear to support the engineering education sector perfectly. They have been rare, limited and relatively expensive for engineering educational purposes. Multimedia presentation is a well-presented (and well tested) form of learning that requires a large bandwidth for the transmission of the contents to all destination users, certainly pushing the broadband diffusion too. It is important not to forget the localization as Thai students do not love studying non-Thai contents and this is why almost well known foreign (English) contents did not succeed in Thailand market. The majority of existing e-Learning Contents available are in English. Thus, e-Learning Contents in non-English are highly in demand (Sirinaruemitr, 2004; Vate-U-Lan, 2007; Witsawakiti et al, 2006).

The first traces of e-Learning for engineers in Thailand returns to 1993. When after several years of usage of internet in 1-2 academic centers in Thailand, i.e. Asian Institute of Technology and Chulalongkorn University, they felt that Internet should be made available to the whole Assumption University (AU). The outcome was implementing a project, called AuNet. Purposing (Chorpothong & Charmonman, 2004):

1. To educate the engineering students, faculty and staff members on the concepts of local and international networking.
2. To prepare the engineering students to enter into information society where networking will be the norm rather than the exception.
3. To provide full Internet access to all engineering students, faculty and staff members for their personal and educational usage.

Concerning to the rapid expansion of knowledge and widespread demand of engineering education, in 2002 this program proposed a project on e-Learning with the aim of (Chorpothong & Charmonman, 2004):

1. To serve the country by allowing those interested in engineering education the opportunity to continue their studies conveniently, no matter from where or when.
2. To promote Life-Long Learning by using the Internet.
3. To expand Assumption University from traditional classroom-based engineering education to Internet-based distance education.
4. To increase the number of students at Assumption University from about 18,000 persons in 2002 to 100,000 persons later.

Establishment of the College of Internet Distance Education on April 25, 2002, with over 2000 Internet terminal and 1000 phone lines, and mirror sites for degree programs from other countries has been the first step of this project which looks for degree-level e-Learning in Thailand. The College of Internet Distance Education of Assumption University has entered into cooperation with many universities such as UKeU, Middlesex and UNITAR, hoping that by the year 2010, the target of 100,000 students, will be reached. Now the attempts on developing e-Learning seem to be more on e-Learning designing and overcoming infrastructure problems. Local case studies are more emphasizing on the role of design of hypertext or hyperlink and the flexibility of presentation (Morse & Suktrisul, 2006; Siritongthaworn & Krairit, 2004; Vate-U-Lan, 2007; Witsawakiti et al, 2006).

Furthermore, Thailand is quite slow in deploying the services including the necessary (broadband) infrastructure and transforming the ways of learning and the (digital) contents are very rare too. In some open universities such as Rhamkhamhaeng University, and Sukhothai Thammatirat University, it's true that they are working on e-Learning projects but they are still familiar with one way broadcasting or two-way video conferencing through the satellite and Fiber Optics. This is because the digital contents available may not deliver to the learners or students by means of using effective broadband ADSL for lack of availability. As a step ahead, ICT ministry of Thailand has launched a program to address provincial home market and enabled people in the rural areas to have home computers. In addition, low-cost software has been introduced to support the hardware program and in turn Thailand can now see quite a low cost of Microsoft licenses and freeware such as Linux and Pladao being deployed. Ministry sponsored low cost software's and cheap computers as a result of high competitions and reduced production cost have for the first time allowed Thai people to be able to afford reasonable computers at home in larger and growing quantities. However, people still perceive that Internet access cost is still too high. This is the barrier to the use of Internet for education. Broadband access fees are also still relatively high and are considered as a premium. Affordability is still an important consideration as far as broadband was concerned (Sirinaruemitr, 2004).

E-Learning Progress for Engineers in India: "A Brief Review"

E-Learning or electronic learning in India is gaining prominence slowly, but indeed steadily. This is due to the fact that more than half the population of India today is below 25 years of age and the number of Internet users is growing continuously. According to UNESCO reports at 2006, 65.2% of adults and 81.3% of youth are literate (UNESCO, 2006). The tremendous growth of the economy in the recent past has also helped in the growth of online education in India. E-Learning in India is especially popular with the young professionals who have joined the work force quite early but still would like to continue their education that may help them move up their career ladder quickly and safely. They find online education in India very convenient, as the nature of the course work does not require them to attend regular classes. Moreover reputed institutes like Indian Institute of Management, Indian Institute of Technology, Indian Institute of Foreign Trade are today offering e-Learning courses. The scope of online educa-

tion in India is actually much wider. Apart from proper course works, some E-Learning portals in India are also conducting mock tests for various competitive examinations like engineering, medical, management etc.

In India, the engineering education has got both government and private players in the market. It consists of arts, science, and management, technical and professional education. Since the Indian knowledge industry is entering into the take off stage, the strategy of survival is high. The foreign players are also trying to join the competition. And hence the less effective educational institutions for engineers are forced to merge themselves with others or they are forced to go out of market. However there is a paradox in e-Learning among various institutes in India. Few institutions join the race, while the rest suffer from lack of knowledge or from lack of realization of the importance of e-Learning. Institutes like IITs are adopting all latest technologies and are keeping their engineering students enlightened from various parts of the world. E-Learning has vast potential in India.

The IT service sector export of India has grown up from US $ 754 million in 1995-95 to US $ 12,000 million in 2004-05. The annual growth based on the trend analysis is US $ 573 in India and US $ 1038 in Tamil Nadu. The Indian IT sector especially Tamilnadu IT sector is growing at a faster rate. This same speed of growth is not replicated in e-Learning application. If all these efforts, are directed properly the ups and downs in knowledge growth can be removed.

In engineering education, virtual classroom or a teacher free classroom has got bright future in India. A virtual classroom is one where the virtual reality is enhanced. It is a totally technology enabled class room environment. Virtual Reality (V.R.) is a 3D learning environment where the learner can explore the learning concept. Learning is experienced through games or simulated situation. It brings a real environment while wearing a headset and data glove in an immersive virtual reality environment. In the areas of Medicine, Engineering, Astronomy and other skill trainings, virtual e-Learning will become indispensable (Nelasco et al, 2007).

E-Learning Progress for Engineers in China: "A Brief Review"

Cyber-education has been growing rapidly in China since the late 1990s, especially in the fields of higher education and basic education. Technical and engineering institutions in China are currently encountering a high pressure onto their schooling capacity due to an increasing population of education pursuers. Cyber-education is considered as a fast and economic approach to ease the pressure. In September 1998, the MOE started to grant special licenses to Tsinghua University, Beijing Post and Telecommunication University, Zhejiang University, and Hunan University as the first set of higher educational institutions pioneering cyber-education. In 1999, Beijing University and the Central Broadcast and Television University were added to the pioneer list. Simulated by favorable policies, a considerable number of Chinese universities started to invest in Cyber-education since then. At the end of 2002, up to 67 universities in China have received cyber-education licenses. It is estimated that over 1.6 millions of students are enrolled in these cyber-education institutions, involved in 140 specialties from 10 academic disciplines. Besides, the Central TV University which has over 2 million of students, enrollment is moving to cyber-education (Zhiting, 2004).

E-Learning Progress for Engineers in Iran: "A Brief Review"

The first traces of e-Learning returns to 25 years ago with distance learning. Iran is a country with a high interest in education. This country has an expanded network of private, public and state affiliated universities offering degrees in higher education. Reports of UNESCO show that up to

2006, 84.0% of adults and 97.6% of youth are literate and this is well balanced between male and female population. Furthermore a significant majority of the youth population is at or approaching collegiate levels (UNESCO, 2006). However the tendency of getting higher levels graduates is obviously growing between the youth population and particularly in graduated students.

Therefore the development of e-Learning in Iran glances in a brief view. Several universities in Iran has started e-Learning programs for distance education, specially Sharif University of Technology which is the most prestigious technical university in Iran and middle east, with different on-line courses on mechanic, electronic, and advanced modeling methods. However, many other authentic universities such as University of Tehran, Amirkabir University of Technology, University of Science and Industry and University of Shiraz have developed several on-line courses with the main concentration on engineering education.

Despite of all the efforts being made to expand e-Learning methods in educational structure of Iran, the process of transferring the traditional education in the Iranian society involves many pressing difficulties which according to the recent studies can be abridged as (Dilmaghani, 2003; Giveki, 2003; Hejazi, 2007; Noori, 2003):

- Lack of realistic comprehension concerning the process of learning
- Ambiguous understanding about students' educational needs in different levels
- Defective implementation of computer hardware and software
- Limited experienced IT professionals
- Incompatible educational resources for e-Leaning

SUMMARY

This chapter is based on the authors '10 years' experience in e-Learning in engineering faculty, and reviews the most important key issues and success factors regarding the design of e-Learning for engineering education in developing countries. It weighs the potential challenges and benefits of implementing e-Learning for engineers in developing countries. It provides a broad perspective or e-Learning design in engineering education. The main themes of technology in engineering education for developing countries focused either on aspects of technological support for traditional methods and localized processes, or on the investigation of how such technologies may assist e-Learning design. Commonly such efforts are threefold, relating to content delivery, assessment and provision of feedback. This chapter also focused and reviewed key issues regarding the implementation of e-Learning for engineering students in developing countries that could be tailored to satisfy the needs of a limited educational infrastructure. Moreover, for e-Learning design to succeed in the developing world, it needs to build one important pillar: the existence of infrastructure, along with some degree of connectivity. Nevertheless a key challenge could be technological requirements which must be kept to a minimum in order to increase the participation of developing countries in e-Learning design for engineering students.

REFERENCES

Astrom, A. (2008). *E-learning quality Aspects and criteria for evaluation of e-learning in higher education*. Stockholm: National Agency's Department of Evaluation.

Boot, E. W., van Merrienboer, J. J. G., & Theunissen, N. C. M. (2008). Improving the development of instructional software: Three building-block solutions to interrelate design and production. *Computers in Human Behavior, 24*, 1275–1292. doi:10.1016/j.chb.2007.05.002

Buzzetto-More, N. A. (2008). Student Perceptions of Various E-Learning Components. *Interdisciplinary Journal of E-Learning and Learning Objects, 4*, 113–135.

Campanella, S., Dimauro, G., Ferrante, A., Impedovo, D., Impedovo, S., & Lucchese, M. G. (2007). Engineering e-learning surveys: a new approach. *International Journal of Education and Information Technologies, 1*(2), 105–113.

Chorpothong, N., & Charmonman, S. (2004). An eLearning project for 100,000 students per year in Thailand. *International Journal of The Computer, the Internet and Management, 12*, 111-118.

Clark, D. (2004). ADDIE - 1975. from http://www.nwlink.com/%7Edonclark/history_isd/addie.html

Colston, R. (2008). *ADDIE model*. Retrieved from http://www.learning-theories.com/addie-model.html

Dick, W., & Carey, L. (1996). *The systematic design of instruction* (4th ed.). New York: Harper Collins.

Dilmaghani, M. (2003). *National providence and virtual education development capabilities in higher education*. Paper presented at the Virtual University Conference Kashan, Iran.

E-learning. (2006). *Evaluation management*. Retrieved from http://www.elearning-engineering.com/evaluation.htm

Giveki, F. (2003). *Learning New Methods in Distance Higher Education*. Paper presented at the Virtual University Conference, Kashan.

Gladun, A., Rogushina, J., Garcıa-Sanchez, F., Martınez-Bejar, R., & Fernandez-Breis, J. T. (2009). An application of intelligent techniques and semantic web technologies in e-learning environments. *Expert Systems with Applications, 36*(2), 1922–1931. doi:10.1016/j.eswa.2007.12.019

Hejazi, A. (2007). *New wine-old bottles*. Retrieved from http://ictarticles.blogspot.com/2007/05/new-wine-old-bottles.html

Hung, T., Liu, C., Hung, C., Ku, H., & Lin, Y. (2007, September 3-7). *The Establishment of an Interactive E-learning System for Engineering Fluid Flow and Heat Transfer*. Paper presented at the International Conference on Engineering Education – ICEE 2007, Coimbra, Portugal.

Jara, C. A., Candelas, F. A., Torres, F., Dormido, S., Esquembre, F., & Reinoso, O. (2009). Real-time collaboration of virtual laboratories through the Internet. *Computers & Education, 52*(1), 126–140. doi:10.1016/j.compedu.2008.07.007

Lee, J., & Lee, W. (2008). The relationship of e-Learner's self-regulatory efficacy and perception of e-Learning environmental quality. *Computers in Human Behavior, 24*, 32–47. doi:10.1016/j.chb.2006.12.001

Littlejohn, A., Falconer, I., & Mcgill, L. (2008). Characterising effective eLearning resources. *Computers & Education, 50*, 757–771. doi:10.1016/j.compedu.2006.08.004

Magoha, P. W., & Andrew, W. O. (2004). The global perspectives of transitioning to e-learning in engineering education. *World Transactions on Engineering and Technology Education, 3*(2), 205–210.

Malachowski, M. J. (2002). *ADDIE based five-step method towards instructional design*. Retrieved from http://fog.ccsf.cc.ca.us/~mmalacho/OnLine/ADDIE.html

Morse, A., & Suktrisul, S. (2006). Introducing eLearning into Secondary Schools in Thailand. *Special Issue of the International Journal of the Computer, the Internet and Management, 14*, 81-85.

Motiwalla, L. F. (2007). Mobile learning: A framework and evaluation. *Computers & Education, 49*, 581–596. doi:10.1016/j.compedu.2005.10.011

Nelasco, S., Arputtharaj, A. N., & Alwinson, E. G. (2007). ELearning for Higher Studies of India. *Special Issue of the International Journal of the Computer . The Internet and Management, 15*, 161–167.

Noori, M. (2003). *Traditional education or learning with computer*. Paper presented at the Virtual University Conference at Kashan Payam-e Noor College, Kashan.

Padilla-Melendez, A., Garrido-Moreno, A., & Aguila-Obra, A. R. (2008). Factors affecting e-collaboration technology use among management students. *Computers & Education, 51*, 609–623. doi:10.1016/j.compedu.2007.06.013

Pryor, C. R., & Bitter, G. G. (2008). Using multimedia to teach inservice teachers: Impacts on learning, application, and retention. *Computers in Human Behavior, 24*, 2668–2681. doi:10.1016/j.chb.2008.03.007

Ramasundaram, V., Grunwald, S., Mangeot, A., Comerford, N. B., & Bliss, C. M. (2005). Development of an environmental virtual field laboratory. *Computers & Education, 45*, 21–34. doi:10.1016/j.compedu.2004.03.002

Ryder, M. (2008). Instructional design models. Retrieved from http://carbon.cudenver.edu/~mryder/itc_data/idmodels.html#modern

Shee, D. Y., & Wang, Y. (2008). Multi-criteria evaluation of the web-based e-learning system: A methodology based on learner satisfaction and its applications. *Computers & Education, 50*, 894–905. doi:10.1016/j.compedu.2006.09.005

Siemens, G. (2002). *Instructional design in e-learning*. Retrieved from http://www.elearnspace.org/ Articles/InstructionalDesign.htm

Sirinaruemitr, P. (2004). Trends and forces for eLearning in Thailand. *International Journal of The Computer, the Internet and Management, 12,* 132-137.

Siritongthaworn, S., & Krairit, D. (2004). Use of Interactions in E-learning: A Study of Undergraduate Courses in Thailand. *International Journal of The Computer . The Internet and Management, 12,* 162–170.

Sun, P., Cheng, H. K., Lin, T., & Wang, F. (2008). A design to promote group learning in e-learning: Experiences from the field. *Computers & Education, 50,* 661–677. doi:10.1016/j.compedu.2006.07.008

Tabakov, S. (2008). e-Learning development in medical physics and engineering. *Biomedical Imaging and Intervention Journal, 4*(1), e27. doi:10.2349/biij.4.1.e27

UNESCO. (2006a). *UIS statics in brief-India.* Retrieved from http://stats.uis.unesco.org/unesco/ TableViewer/ document.aspx?ReportId=121&IF_ Language=eng&BR_Country=3560

UNESCO. (2006b). *UIS statics in brief-Iran.* Retrieved from http://stats.uis.unesco.org/unesco/ TableViewer/ document.aspx?ReportId=121&IF_ Language=eng&BR_Country=3640

Vate-U-Lan, P. (2007). Readiness of eLearning connectivity in Thailand. *Special Issue of the International Journal of the Computer, the Internet and Management, 15,* 21-27.

Willis, B. (2008). *Instance education in glance.* Retrieved from http://www.uiweb.uidaho.edu/ eo/dist3.html

Witsawakiti, N., Suchato, A., & Punyabukkana, P. (2006). Thai language E-training for the hard of hearing. *Special Issue of the International Journal of the Computer . The Internet and Management, 14,* 41–46.

Zhiting, Z. (2004). The development and applications of elearning technology standards in China. *International Journal of The Computer, the Internet and Management, 12,* 100-104.

Chapter 2
Architectural Web Portal and Interactive CAD Learning in Hungary

Attila Somfai
"Széchenyi István" University, Hungary

ABSTRACT

It is opportune to show the teaching web portal of the Faculty of Architecture at "Széchenyi István" University (www.arc.sze.hu/indexen.html), its conformation and use. Nowadays, the Internet helps to look into Hungarian and foreign study aids, architectural websites, and novelties. The Internet has created potential new and effective ways of cooperation between lecturers and students of the university and other institutions of higher education. The teaching web portal mentioned above realizes the diversity and complexity of architecture with efficient grouping of information, and is attentive to high professional standards. Computer Aided Architectural Modeling (www.arc.sze.hu/cad) is one of the new types of online lecture notes, where many narrated screen capture videos show the proper usage of CAD software instead of text and figures. This interactive type of learning helps students become more independent learners. This type of teaching modality provides the opportunity for students who need more time to acquire subject matter by viewing video examples again. Success of our departments' common web initiations can be measured through Internet statistics and feedback of the students and external professionals.

MAIN GOALS AND PHILIOSOPHY

It is high time to speak of the importance of the electronic knowledge-bases in the teaching of Hungarian architects and students, as well its multidirectional possibilities for use. The pro-

fessional web portal of the "Széchenyi István" University in Győr is introduced, together with some of its subject-matters of instruction. In our time the world has been widely opened, internet gives support in getting acquainted with teaching packages from domestic and foreign sources, professional portals and novel topics, and moreover, it makes possible to effectively cooperate among

DOI: 10.4018/978-1-61520-659-9.ch002

lecturers, professors and students, as well as among institutions of higher education. Actuality of this initiative from Győr is best shown by the dynamic increase in the number of visitors since the beginning, and now achieves a rate of as high as 2300 visitors in a month. URL-address of the portal, and links to actual works are incorporated onto the homepages of numerous institutions dealing with the profession or the education in general, however, real acknowledgement is proved by responses from the users – both students and industrial experts.

Our "Database for Architecture" (www.arc.sze. hu/indexen.html) started in 2001 with the target to provide *up-to-date and easily useable, editable* knowledge materials *that can be easily accessed at any time and from any place*, to the education of architects at any level in Győr. A more effective education at a higher level can be achieved with the help of electronic subject-matters following the changes in the profession in a flexible way. Knowledge on the fields of related branches, the most recent aims of science and technique, and examples realized in practical life are dynamically coupled with basic knowledge. Through the step-by-step building up of the complexity-aimed database-concept of several special branches, *versatility and richness of our profession* has also been outlined, and at the same time it was an important point of view to fulfill proper selection by *correct technical content, its perfection levels and actuality of information.*

Realization of the above targets till now means – on the example of the subject Building Constructions – that lecture notes, practical guides and study aids with references, electronic stores of drawings, and planning and construction sheets and brochures of leading companies manufacturing building materials – all that can be achieved at the same place. In addition to that, further homepages of professional and scientific character, journals, reviews and periodicals from Hungary and abroad, as well as a selection of independent articles can be found here.

Successful operation of the Database for Architecture is maintained – beside its logical structural arrangement, good selection of contents and favourable appearance – by *numerous special services* (search possibilities, integrated dictionaries, forum, students' administration, news bar).

In our days, it is a general experience that the county borders are getting more and more permeable within Europe of the regions – thereby making wide-ranging collaborate and cooperation very easy. An important precursor and catalyst of this process is the world-wide web where results of international research and practical achievements can be published, as well. Internet is, however, much more than simply one of the media because cooperation can be established and practised on a daily basis among different institutions of higher education owing to its *interactivity.* There is a possibility to more reasonable sharing of the activities, or publishing the results in a common knowledge base. The electronic type knowledge bases serve – owing to the interactivity they provide, too – as *educational tools of extreme efficiency.*

In the meanwhile it must always be taken into consideration that education of future architects has the characteristic feature of often being highly individual regarding the duties given and solutions obtained therefore a personal master-and-follower contact is and shall be of uppermost importance. The knowledge base provides a very useful background for these works, too, for example in sensible and impressing performing of the variety of good solutions.

ELECTRONIC TEACHING KNOWLEDGE BASES AS ARCHETYPES

Through existing domestic professional databases and ideas there were favourable impulses given for a new viable concept suitable for the above detailed wide-ranging demands at our university. These were then complemented a continuous "evolutionary" development–after the starting period–taking into consideration the users' feedbacks.

We were familiarized with one of the considerable initiatives as early as in 1999, on the XXIV. Conference on Building Construction where the concept of the educational computer program package on building construction (called "FAL", meaning "Wall") developed at the Budapest Technical University was introduced us by Nándor Bártol (see: Figure 1). The author wished to compromise the electronic lecture aids from the university department and links from outside of the university – representing the professional life – into a common knowledge base. The ideas on the structure of the database, the professional requirements and the way of actualizing were formulated with extreme good sense.

Another ingenious and since lots of years successfully used example for the professional knowledge transfer is the motion of Dr. József Orbán from Pécs, a catalogue on building materials published by the company Orisoft in Pécs (see: Figure 2). Professional knowledge and skill offered with excellently suited measure can be accessed in a clearly and distinctly detailed hierarchy. This compilation is perfectly passed and utilized both

Figure 1. "FAL" – Conceptual scheme of an Educational Program Package on Building Construction developed at the Budapest Technical University (© 1999, Nándor Bártol. Used with permission).

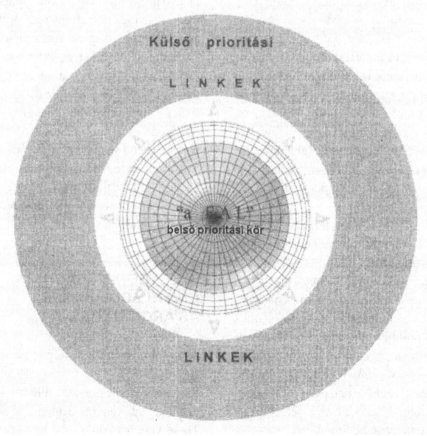

in the practical life and education, owing to its encyclopaedia-like structure.

After having selected the main theme desired, the subgroups in the Orisoft Catalogue are provided in *matrix-array* thereby making it possible to choose from the secondary and tertiary points of view of information collecting. So it can be decided whether to choose from different company information within one single product group, or to choose a product group within an actual company.

Another advantage of the matrix principle is compactness, i.e. being unexpanded. Activities of a great number of companies become easily perspicuous and comparable. However, in the small grid field there is only place enough for one link therefore information attached can be seen when the mouse cursor on top of the link moves and displays an information windows with variable content besides the matrix. In case more information should be needed you can click onto the link in the grid field and navigating away onto a new page.

STRUCTURE AND USE OF THE KNOWLEDGE BASE FROM GYŐR

At the time of establishing the Architectural Database in Győr (Figures 3 and 4) our target was not only to utilize already available experiences but to implement a lot of new solutions.

Visitors are greeted on the main page showing a mosaic in details from the most beautiful sceneries of the town then there comes a picture dissolution directing onto a webpage with professional information. Catching the eye of our visually susceptible students, their interest is going to be increased by a header toolbar in individual design showing animation over the main themes.

In the database, university lecture notes, additional supporting materials, and selected outside homepages, documents and videos (of informative, research, company information, etc. character) are available. It has a distinctly theme-oriented structure, and even more, *it is quite unique among domestic (Hungarian) web portals since its main structure orientation is subject to that of*

Figure 2. Scheme of a catalogue on building materials (company: Orisoft in Pécs) organized by main themes, and one selected of the subgroup matrices.

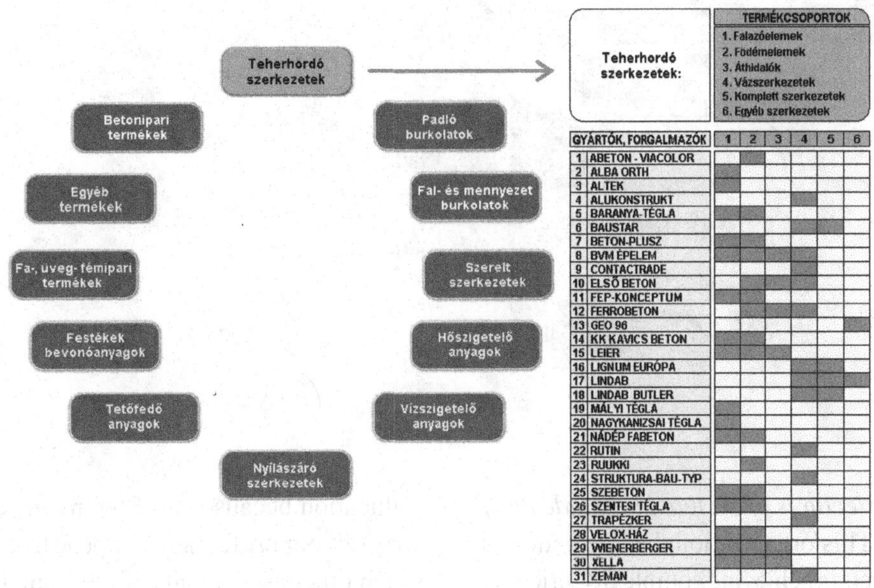

Figure 3. Architectural database for students (designed in Győr)

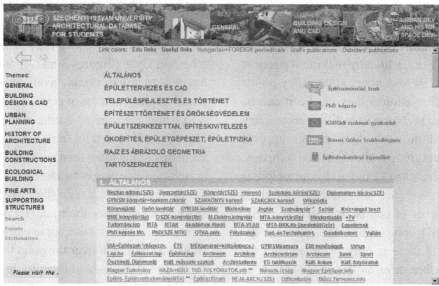

Figure 4. On-line lecture notes and practical aids for each semester on building construction and building effectuation

the university lectures of the teaching students of architecture. This organizational construction is one of our targets in achieving complexity-oriented education because all of the main segments and parts of our profession compose true unity beside each other as elements of the same importance.

At the starting location of the professional surface, one of the following main themes can be selected:

- General
- Building Design & Construction, and CAD
- Urban & Settlement Development, and History
- History of Architecture, and Protection of Heritage
- Building Construction Science, and Building Execution
- Ecological (Green) Buildings, Building Installation, and Building Physics
- Design Drawing and Descriptive Geometry
- Supporting Structures

If one may change the main theme selected before, it is not necessary to go back to the top of the homepage because there are direct links for the main themes on the side bar, too. At the very beginning of the professional surface, but distinctly separated in graphical way from the main themes, certain emphasis is given to the access to some of the most important homepages (e.g. Department of Architect Engineering, PhD Education, Professional Practice Abroad, College "Baross Gábor" for Advanced Technical Studies, or a link to the homepage of the Association of Building Construction Science Branch in Győr).

The structure of the database is *similar to a site-map* (a kind of link-maps) which results in choice able *to demonstrate nearly at the same time* from all the main theme blocks down to the links representing in this meaning the smallest building elements. A novel kind of solution used that links are not listed featureless but are organized into five subgroups kept apart by *spectacular colour markings*.

The following subgroups can be found within all of the main themes:

- Education material (lilac);
- Useful professional links (blue);
- Domestic and foreign periodicals, journals, etc. (red);
- Articles written by university lecturers (green);
- Certain articles of higher importance written by authors "from the outside" (brown).

Owing to this multi-level grouping, it is much easier to search for something, and at the same time, a requirement set by the lecturers' community is also fulfilled, i.e. *engineering ability and sense of our students in making things more reasonable and rationalized must be developed, and the way from great agglomeration (clusters) to the smaller sections of elements must be observed and realized.*

This site-map principle makes – similarly to the matrix-principle – several information gaining strategies possible. It is also possible to decide whether to choose from the five subgroups within each main theme, or the subgroups with the same colour (i.e. periodicals, journals, etc.) represent the primary point of view for the search and the selection of a certain main theme (i.e. a given field of profession) remains at the second level (secondary point of view).

Since there are several links in the cluster selected on the basis of both the primary and the secondary selection point of view (i.e. several journals on building construction), thereby a third level of differentiation should be derived.

Through modification of the link names, a quaternary differentiation may partly also be realized:

- Separation of lecture notes from education supports and aids with the help of determining additional words;
- Discrimination of the most important document owners with supplemental words (authors, universities, countries, etc.);
- Emphasized display of the names of periodicals and journals from foreign countries by capital letter;
- Separation of periodicals and journals with on-line full-text from those with only con-

tent information (these latter-mentioned are distinguished with the supplementary word "info").

Since these spectacularly coloured site-map markings can be easily overview through a simple mouse scrolling, therefore the last mentioned particular points of view can even be chosen as primary selection feature (and thereby it is possible to browse only e.g. the cluster of foreign periodicals, or that of the university lecture notes, etc.).

This *multi-differentiated site-map* as shown can flexibly be developed; however, it is our important target to support our students only with well pre-selected mass of information that can be reviewed at a glance. Sometimes – when an extension of the material happens – certain other elements are removed, or links that are close to or in strong relationship with each other (e.g. practical guides and helping supplements for building construction) are compiled onto a single independent page to be opened by a single click.

SPECIAL SERVICES OBTAINABLE FROM THE KNOWLEDGE BASE

Through these special services, efficiency of use and up-to-date information transfer are significantly increased.

Previously, possibility of multiple-purpose selection of the line for the points of view for access to information has been mentioned. It is then associated with the feature that the links already visited are displayed in and distinguished with a slightly changed background shade thereby systemic (or on the contrary, "on the random basis") browsing of the database items, and the return onto a former item are greatly facilitated.

Not only "search what you see" but some kinds of *"computer-aided search"* can also be chosen. A simple quick search (CTRL+F) is used for search among link names, and an individually modified purpose-oriented Google-service (under "Search"

of the side bar) helps in finding something within the full content of the university's web-documents (when choosing the so-called "browse in saved content" the exactly matched places of occurrence of the search words can be seen within the document brought into eye-catching coloured limelight). This menu point is also suitable not only for search on the internet but for use with the Google Scholar search machine developed for educational and scientific purposes.

From the side menu it is possible to enter the recently developed *"Forum"* page. This is a kind of interactive discussion surface where students and teachers, lecturers and "outside" participants working in the line of business can communicate with each other. Themes of this "virtual round-table" are determined by the user, and anybody can add a comment to them. At the very beginning, some themes were designated as start-up topics by the lecturer in order to facilitate deeper discussions and formation of professional directions (e.g. Designers' Forum, Forum for Urbanists and Heritage Protection, Forum for Building Constructors). This is also the very forum that can create a community, a kind of society among our students who stay in different foreign countries under the work with "Leonardo" or "Erasmus" programs.

A prompt help can be achieved during browsing homepages in foreign languages when the *on-line dictionaries* (developed by SZTAKI) in five languages and in both translation directions are turned on through clicking on the bottom part of the side bar menu.

On the lowest part of the side bar menu, *actual and latest news* are temporarily running (e.g. about certain lectures, exhibitions, conferences, "Day of Open Gates", admission information).

From the homepage *"Neptun"* (*an administrative service for students*) can be accessed where information regarding different subjects and time schedules are published. This is one of the ways for the students to register themselves for learning a subject and later for exams, and here they can also check their graduation notes. This kind of information on documents and graduation is subject to change only with password of the university department.

Scrolling down onto the bottom of the database, the Database for Architecture can be set with a single click as start page in the internet browser; and there is the possibility to jump onto a download page of free software for the most commonly used file types (extensions .pdf, .doc, .ppt, .xls).

HAVE A TASTE FROM THE CURRICULAR SCHOOL-WORKS OF THE KNOWLEDGE BASE

In case of certain subjects – for example, building construction, building execution, history of architecture – curricular school-works were published on a CD good ten years ago. Owing their format HTML they could easily be inserted functionally into the internet knowledge base, too; their menu system should have undergone a modernization process. Through the recently used side bar menu and read-windows the notes became easier to overview, and their use is also more convenient. Other school-works can be accessed not in HTML but the also most popular PDF format (e.g. Organization of Building Activities, Planning of Buildings).

Our newest (and also on CD accessible) school-work is "Computer-Based Architectural Modelling" in order to support ArchiCAD teaching. Important roles are here given – besides the texts and pictorial information – to *educational videos* and *interactive practicing examples*. This material was prepared in 2006 within the framework of HEFOP-program, and its realization was also supported by the company Hyperionics with its "HyperSnap" screen capture software and "HyperCam" screen recorder software.

In the videos, example solutions can exactly be documented, the oral explanation of the lecturer can be heard, and the places of mouse clicks are also marked with colours. This educational film can be run by the students parallel to the CAD-software; it can be stopped and re-started, etc. These films are never-stopping educational supports to learn step-by-step the examples. The students can thereby more efficiently learn the special knowledge than earlier. When using ArchiCAD, preparation of technical drawings is similar to a certain extent to the classical and traditional drawing on tracing-paper with china-ink but spatial modelling requires a specific way of ArchiCAD-Thinking, therefore greater emphasis has been laid upon this (see: Figure 5).

In our days, numerous architect offices prepare own homepages, and even accept orders not only for architectural planning but web design-type works, thereby turning visual art abilities to business. In accordance with the challenges of our time, in the above mentioned lecture note on CAD there is a short summary given about publishing of vision schemes on an internet webpage, with interactive web design-examples (when modifying the software lines written in HTML in the practical examples, different results as reactions can be displayed).

MAINTENANCE OF THE KNOWLEDGE BASE; FURTHER DEVELOPMENTS

Construction and editing of the internet knowledge base is based in general on html-knowledge, but is accompanied with numerous Java Script applications (animations, interactive VR-panorama, on-line dictionary request, etc.). Both editing and automated looking for link-failure are realized with free softwares. During steady maintenance and actualization, ideas, proposals and opinions of the users are taken into account. Our students give reports on the great help this knowledge base provides them in learning and being informed, and the visitor-statistics give also high *access figures as feedback* (see: Figure 6).

Figure 5. Computer aided architectural modelling (online lecture notes)

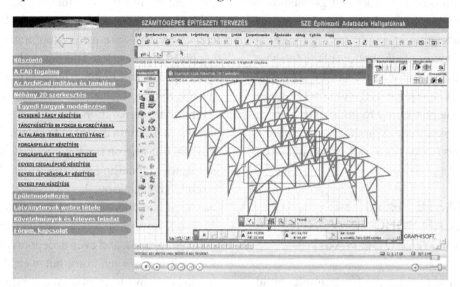

The Database for Architecture has been and is continuously developed and its content actualized. It is our plan to prepare further school-works, and we try to utilize the new possibilities created by computer science. The database is already pre-edited for incorporating later additional educational materials of the English-language teaching.

Our professional webportal is truly integrated both into the homepage of "Széchenyi István"

University, Department of Architecture and into those of the related special departments. So this Knowledge Base provides *new opportunities for both the students and their lecturers for thinking together*. Our portal can, maybe, give motivation to others who wish to learn out of their professional feeling, and may, as we do hope, help them after obtaining a diploma, as well.

Figure. 6. Visitor-statistics of arc.sze.hu

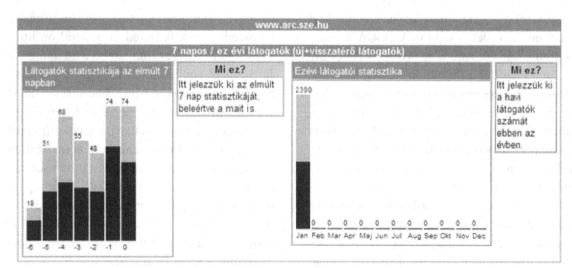

REFERENCES

Bártol, N. (1999). *Új segédanyag az épületszerkezettan hatékonyabb oktatásáért.* (New Education Aids for a More Effective Teaching of Building Construction Theory.) In *XXIV. Conference on Building Construction, "Széchenyi István" University, Győr* (pp. 54-62).

Dr. Fátrai, G. (1998). *Építéskivitelezés online-jegyzet* (On-Line Lecture Notes on Building Effectuation). Retrieved March 24, 2009, from www.arc.sze.hu/kivitelea

Dr. Koppány, Attila (1998 and 2006). *Épületszerkezettan online-jegyzet* (On-Line Lecture Notes for Building Construction). Retrieved March 24, 2009, from www.arc.sze.hu/epszerkea

Dr. Orbán, J. (2000-2007). *Orisoft Építőanyagipari Katalógus* (Orisoft Catalogue of Building Materials). Retrieved March 24, 2009, from www.orisoft.pmmf.hu

Dr. Somfai, A. (2006). *Számítógépes Építészeti Modellezés online-jegyzet* (On-Line Lecture Notes on Computer Aided Architectural Modelling). Retrieved March 24, 2009, from www.arc.sze.hu/cad

Molnárka, G. (2006). *Hallgatói mobilitási portál* (Mobility Portal for Students). Retrieved March 24, 2009, from http://www.sze.hu/leonardo

Chapter 3
Adapting Engineering Education to the New Century

A.K. Haghi
University of Guilan, Iran

B. Noroozi
University of Guilan, Iran & University of Cincinnati, USA

ABSTRACT

In this book chapter, the authors summarize their retrospections as engineering educators for more than 20 years. Consideration is given to a number of educational developments to which the authors have contributed during their career in academia and the contribution made to engineering and technological education. Increasing emphasis is being placed on establishing teaching and learning centers at the institutional level with the stated objective of improving the quality of teaching and education. The results of this study provide information for the revision of engineering curricula, the pedagogical training of engineering faculty and the preparation of engineering students for the academic challenges of higher education in the field. The book chapter provides an in-depth review of a range of critical factors liable to have a significant effect and impact on the sustainability of engineering as a discipline. Issues such as learning and teaching methodologies and the effect of E-development; and the importance of communications are discussed.

INTRODUCTION

Ernest Boyer (Boyer, 1990) states that:

"...scholarship means engaging in original research, it also means stepping back from one's investigation, looking for connections, building bridges between theory and practice, and communicating one's knowledge effectively to students".

Therefore, the scholarship of teaching engineering, seeks to find effective ways to communicate knowledge to students. The realization that traditional instructional methods will not be adequate to equip engineering graduates with the knowledge, skills, and attitudes they will need to

DOI: 10.4018/978-1-61520-659-9.ch003

meet the demands likely to be placed on them in the coming decades, while alternative methods that have been extensively tested offer good prospects of doing so (Rugarcia, et al., 2000).

Engineering is the profession in which knowledge of the mathematical and natural sciences gained by study, experience and practices are applied with judgment to develop ways to utilize economically the materials and forces of nature for the benefit of mankind. Engineering is a unique profession since it is inherently connected to providing solutions to some expressed demand of society with heavy emphasis on exploiting scientific knowledge. In the real world, engineers must respond to sudden changes. The engineers of today, and in the decades ahead, also must be able to function in a team environment, often international, and be able to relate their technical expertise to societal needs and impacts. Yet we start at making transformative changes in our educational system. Our educational challenge is itself a design challenge—making the "right" engineers for our nation's future. The basis for the reform of engineering education is made up of unique experiences, traditions and everlasting values of specialist training at universities. Engineering educators have to focus on market demand and stop defending the obsolescent and obsolete programs. In order to prepare engineers to meet these new challenges, engineering training and education must be revised and modernized. Today's engineer cannot be merely a technician who is able to design the perfect bridge or the sleek skyscraper.

Today's engineer must not only have a breadth and depth of expertise, but must be able to communicate effectively, provide creative solutions with vision, and adapt to ever-changing demands. Today's engineer, like any other modern professional, must be someone who can see the big picture

1. INTEGRATING "WHAT" INTO "WHY" IN ENGINEERING EDUCATION

A professional needs to recognize the "why" dimension as well as the "what" in order to provide a wisdom and understanding. Also, for the profession to attract students there needs to be an enhanced community respect for engineering. This can be assisted if we integrate a person-centered and nature-respecting ethic into engineering education (Hinchcliff, 2000).

The urgent need to change the teaching method of the current engineering education system was the reason for which the author launched to a new plan. The new plan envisaged changes in the curriculum to meet the demands of the industry, now facing strong competition as a consequence of the recent technological changes. With this aim, the authors developed the courses considering following issues:

1.   **"Why"** not try replacing one quarter of the lectures with an online resource? As part of online resources, lecture-based courses are taught at many institutions using videotaped lectures, live compressed video, television broadcasts or radio broadcasts (CSU, 2004; Forks, 2009; UI, 2004). The student can have an easier time communicating online as opposed to in a full classroom (Purcell-Roberston & Purcell, 2000). In addition, student is at the center of his or her information resources. The content information is not delivered as a lecture for the students to hear but rather as information for the students to use. Students are free to explore and learn through their successes and sometimes failures Instead of the lecture, students could spend time over a month working through some online materials complete with self-tests, interactions, mini-project or whatever.

We can use the replaced lecture time to manage the online course and deal with queries. The presentation would be varied without too much change at once.

2. **"Why"** not replace half the lectures with a simple online resource, and run smaller seminar groups in the time the lectures would have used (obviously not practical if you lecture to 200 students, but if you online lecture to 40, this might give you a chance to do something different with them face to face. When all of a course's lectures, readings, and assignments are placed online, anyone with a computer and Internet access can use online resources in the course from any location, at any time of day or night. One institutional goal of this movement toward computer-aided education based on online resources is to make higher education more economical in the long run through an "economy of scale." If all of lectures, syllabi, and assignments are digitized and put online, tutors could spend less of their time teaching a larger number of students, and fewer of those students would be on campus using the university's resources (Frances, et al., 1999). According to the survey results conducted in university of Wisconsin-Madison, the vast majority of the students took advantage of the fact that lectures were online to view material in ways that are not possible with live lectures. When asked if it was easier to take notes to understand the material when viewing lectures on electronic resources than it would have been attending the same lecture live, almost two thirds (64%) of students agreed, either strongly (27%) or somewhat (37%) (Foertsch, et al., 2002).

Online education of engineering students requires more than "what" the transfer of the knowledge and skills needed by the profession because should always keep in mind that:

- A classical lecture is not the best way to present materials any more.
- We have to demonstrate something visual this is difficult for everyone to see in a lecture.
- We have to work through a simulation or case study which would be better done at a student's own speed.
- We have to ask regular questions during the presentation of content and we have to monitor responses.

2. FUNDAMENTAL ASPECTS

There are a number of technologies whose integration into modern society has made dramatic changes in social organization. These include the Internet and its attendant promises of more democratic information access and distribution, including distance learning. A totally effective education and training environment, when applied to information technology instructional strategies that are enhanced by the world-wide web, will include factors that have long been identified as contributing to an optimal and multi-dimensional learning context - a personalized system of instruction (Wild, et al., 2002). The ingredients of such a system have long been known to contribute to an optimal learning environment for the individual student ("Overview of e3L," 2005).

Electronic based distance learning as a potential lever play a extraordinary role in the creation and distribution of organizational knowledge through the online delivery of information, communication, education, and training (Wild, et al., 2002). Numerous studies have been conducted regarding the effectiveness of e-learning. To date, there are only a few that argue that learning in the online environment is not equal to or better than traditional classroom instruction Web-based also learning provides education access to many non-traditional students, but if applied to replace existing educational spaces, the potential effects of this replacement must be evaluated and assessed (Coleman, 1998; Kolmos, 1996; "Overview of e3L," 2005). However, e-learning is not meant to replace the classroom setting, but to enhance it, taking advantage of new content and delivery technologies to enable learning. Key characteristics of online learning compared to traditional learning is shown in Table 1 (Wild, et al., 2002).

A. Interactivity

E-learning should be interactive. There are opportunities for both learners and knowledge holders to build on the information being conveyed in the online environment. Opportunities include threaded discussions, chat areas, and exercises that invite learners to interact with the content and respond. Most importantly, learners and instructors should have a means to contact the content expert or others in the actual learning environment. E-mail, discussion bulletin boards, groupware environments such as Lotus Notes (TM), etc., can be used to support interactivity.

Table 1. Knowledge presentation consideration for e-learning

Characteristics of traditional learning	Characteristics of online learning
Engages learners fully Promote the development of cognitive skills Use learners previous experience and existing knowledge Use problems as the stimulus for learning Provide learning activities that encourage Cooperation among them members	E-learning should be interactive E-learning should provide the means for repletion and practice E-learning should provide a selection of presentation style E-learning content should be relevant and practical Information shared through E-learning should be accurate and appropriate

B. Repetition and practice

E-learning should include a means for repetition and practice. In other words, the courses should engage and challenge the learners to evaluate, select, and use the information in their everyday lives. The content should be relevant to the learner's frame of reference (i.e. content that is practical and understandable to the user). Case studies, simulations, and "what would you do" exercises help learners grasp the content and find ways to use the new information creatively in their lives. The Web is an excellent resource for establishing a frame of reference for exercises.

C. Presentation Styles

E-learning should provide a selection of presentation styles. The most beneficial courses offer several ways for learners to absorb the material. Written content is fine, but more learners grasp concepts with illustrations that accompany the content. Video, audio, and other multi-media choices within an online environment contribute to the richness of presentation choices. Instructors can inquire in the online classroom about other ways to present the material if alternative delivery methods are available or preferred by members of the learning group.

D. Content

E-learning content should be relevant and practical. Adults learn better when the objectives of the course are directly linked to issues, theories, case studies, research, and knowledge that is practical. Simply putting material online does not make e-learning successful. Learners require some amount of integration of all of the information being provided in the learning environment so that it makes sense and has meaning and utility in their lives. Company intranets permit employees to access issues facing the organization that can enrich the content of learning activities.

E. Accuracy and Relevance

Information shared through e-learning should be accurate and appropriate. Instructors should employ measures (usually by direct contact with learners, but also by assessments such as surveys) to ensure that the content provided in the course is appropriate to learning needs. The course content should be reviewed regularly to ensure accuracy and relevance.

Several considerations must be taken into account for e-learning to be a beneficial investment and an effective knowledge management tool. The elements of the e-learning planning process include assessing and preparing organizational readiness (factors to consider before going online), determining the appropriate content (content that ties into the goals of knowledge management), determining the appropriate presentation modes (considering factors contributing to effective e-learning), and implementing e-learning (content and technology infrastructure considerations).

In academic institutions, ideally, both students and faculty should be provided with an e-learning environment that is optimal for each to be inspired to do the best job possible. For the student, the objective is to accumulate and learn as much knowledge as their personal make up will allow. The faculty must be stimulated and inspired to be outstanding faculty members for the students and to simultaneously grow professionally as rapidly as their personal abilities will allow. This is a complex environment to build as each individual involved will have different needs and hence different emphasis need to be placed on various components that make up the environment. No doubt it is impossible to provide the optimal e-learning environment for all individuals among the students and the faculty. It is important however to give considerable attention to this problem and to build the best e-learning environment possible with the financial resources available to the academic institution. Planning and faculty discussions make it possible to build

e-learning and growing environments that are far better than would otherwise occur. E-Learning although is a growing trend in the education community Given all its merits, however, institutions considering this option should be aware of the particulars of Web-based training. With proper planning, implementation, and maintenance, you can create an optimal e-Learning environment at your institution that benefits both the academic institution and students.

In the following sections, different components and influences that impact on the magnitude of students and faculty's accumulation of and handling of knowledge is discussed.

3. TEACHING CULTURE

The development of a suitable teaching culture is a prerequisite of the academic world, once the principle to *"have learned to learn"* has been promoted. On the other hand, the development of the concept of sustainable employability determines that the university trains people to become future professionals with basic competences: cognitive, social and affective, allowing them to achieve creative and effective professional performances in quick-change work environment (Oprean, 2006). Boyer states that:

"...without the teaching function, the continuity of knowledge will be broken and the store of human knowledge dangerously diminished" (Boyer, 1990).

Teaching is not a process of transmitting knowledge to the student, but must be recognized as a process of continuous learning for both the lecturer and student. The old adage, if you want to know something teach it, certainly applies. But it needs to be extended (Al-Jumaily & Stonyer, 2000).

The most obvious and instantaneous effect of development of suitable culture can be considered as improving the classroom environment and also classroom effectiveness. Through describing some important concepts the effective strategy for mutual relationship between learner and lecturer will be addressed.

A. Practical Applications

Completing a thriving course involves connecting the program of study to existent life problems and to present events. This is often made promising by including realistic applications in the classroom, and the instructor can carry out this by drawing on personal experience or by using student examples as sources of practical problems (Finelli, et al., 2001). Students who contribute in certified work programs, research activity while in college can be a precious source of such information. Since there are many situations where bring students into contact with realistic problems, all that is necessary is to take advantage of those activities in teaching situations. Following paragraph give one example related to this partnership between industry and academic institute.

Faculty of engineering in Guilan University has a compulsory cooperative education program. Students are placed with an employer in their field of study (currently the school has partnerships with over 100 companies), and they alternate semesters at their worksite and on campus beginning. Under this program, undergraduate students have some financial support from the university, and they get practical experience by working for the industry, for governmental agencies, or for public or private non-profit organizations. The connection with industry begins in the first year when students are teamed with alumni to assist the students with their semester writing projects. Through all of these experiences, students are often able to provide practical insight to classroom activities.

B. Learning Styles and Class Participation

It is extremely attracting and challenging area if one strikes a balance between lecturing and engaging in alternative teaching techniques to stimulate students with various learning styles. Learning style is a biologically and developmentally imposed set of personal characteristics that make some teaching (and learning) methods effective for certain students but ineffective for others. These include the Myers-Briggs Type Indicator, (Myers & McCaulley, 1985) Kolb's Learning Style Model, (Felder, 1993; Kolb, 1981) the Felder-Silverman Learning Style Model, (Dunn, 1990) and the Dunn and Dunn Learning Style Model (Dunn, et al., 1989). The following statements has been offered based on current research on learning styles to assure that every person has the opportunity to learn (Dunn, 1992).

A. Each person is unique, can learn, and has an individual learning style.
B. Individual learning styles should be acknowledged and respected.
C. Learning style is a function of heredity and experience, including strengths and limitations, and it develops individually over the life span.
D. Learning style is a combination of affective, cognitive, environmental, and physiological responses that characterize how a person learns.
E. Individual information processing is fundamental to a learning style and can be strengthened over time with intervention.
F. Learners are empowered by knowledge of their own and others' learning styles.
G. Effective curriculum and instruction are learning-style based and personalized to address and honor diversity.

H. Effective teachers continually monitor activities to ensure compatibility of instruction and evaluation with each individual's learning style strengths.

I. Teaching individuals through their learning style strengths improves their achievement, self-esteem, and attitude toward learning.

J. Every individual is entitled to counseling and instruction that responds to his/her style of learning.

K. A viable learning style model must be grounded in theoretical and applied research, periodically evaluated, and adapted to reflect the developing knowledge base.

L. Implementation of learning style practices must adhere to accepted standards of ethics.

An instructor who strives to understand his/her own learning style may also gain skill in the classroom. *"Consider the question, how does the way you learn influence the way that you teach?"*

A. Most instructors tend to think that others see the world the way they do, but viewing things from a different learning perspective can be useful. It is good practice to specifically consider approaches to accommodate different learners, and this is often easiest after an instructor learns about his/her own learning style. An instructor with some understanding of differences in student learning styles has taken steps toward making teaching more productive (Finelli, et al., 2001).

B. It is also important for the instructor to encourage class participation, but one must keep in mind the differences in student learning style when doing so. Research has shown that there are dominant learning characteristics involved in the perception of information through concrete versus abstract experience (Kolb, 1981; Kolb, 1984).

C. Some learners need to express their feelings, they seek personal meaning as they learn, and they desire personal interaction with the instructors as well as with other students. *"A "characteristic*

question of this learning type is why?" This student desires and requests active verbal participation in the classroom. Other individuals, though, best obtain information through abstract conceptualization and active experimentation. This learner tends to respond well both to active learning opportunities with well-defined tasks and to trial-and-error learning in an environment that allows them to fail safely. These individuals like to test information, try things, take things apart, see how things work, and learn by doing. *A characteristic question of this learning type is how.* Thus, this student also desires active participation; however, hands-on activity is preferred over verbal interaction.

D. The instructor must have a sincere interest in the students. However there is no best single way to encourage participation. Individual student differences in willingness to participate by asking questions often surface (Baron, 1998; Jung, 1971) Still, although the number of times an individual speaks up strongly depends on student personality qualities, a class where all are encouraged to enter into dialog is preferable, and opening the lecture to questions benefits all students.

C. Active Learning

An ancient proverb states:

"Tell me, and I forget; Show me, and I remember; Involve me, and I understand."

This is the basis for active learning in the classroom, (Mamchur, 1990) and extensive research indicates that what people tend to remember is highly correlated with their level of involvement. It has been shown that students tend to remember only 20% of what they hear and 30% of what they see. However, by participating in a discussion or other active experience, retention may be increased to up to 90% (Dale, 1969).

4. TEACHING METHODOLOGY

Cooperative learning is a formalized active learning structure involves students working together in small groups to accomplish shared learning goals and to maximize their own and each other's learning. Their work indicates that students exhibit a higher level of individual achievement, develop more positive interpersonal relationships, and achieve greater levels of academic self-esteem when participating in a successful cooperative learning environment (Johnson, et al., 1991; Johnson, et al., 1992).

However, cooperation is more than being physically near other students, discussing material with other students, helping others, or sharing materials amongst a group, and instructors must be careful when implementing cooperative learning in the classroom. For a cooperative learning experience to be successful, it is essential that the following five elements be integrated into the activity (Johnson, et al., 1991; Johnson, et al., 1992)

1. **Positive Interdependence:** Students perceive that they need each other in order to complete the group task.
2. **Face-to-Face Interaction:** Students promote each others' learning by helping, sharing, and encouraging efforts to learn. Students explain, discuss, and teach what they know to classmates. Groups are physically structured (i.e., around a small table) so that students sit and talk through each aspect of the assignment.
3. **Individual Accountability:** Each student's performance is frequently assessed, and the results are given to the group and the individual. Giving an individual test to each student or randomly selecting one group member to give the answer accomplishes individual accountability.
4. **Interpersonal and Small Group Skills:** Groups cannot function effectively if students do not have and use the required social skills. Collaborative skills include leadership, decision making, communication, trust building, and conflict management.
5. **Group Processing:** Groups need specific time to discuss how well they are achieving their goals and maintaining effective working relationships among members, Group processing can be accomplished by asking students to complete such tasks as: (a) List at least three member actions that helped the group be successful, or (b) List one action that could be added to make the group even more successful tomorrow. Instructors also monitor the groups and give feedback on how well the groups are working together to the groups and the class as a whole.

When including cooperative learning in the classroom, the instructor should do so after careful planning. Also, the students may be more receptive to the experience if the instructor shares some thoughts about cooperative learning and the benefits to be gained by the activity.

In our classes, the students should be able to do group work; in fact much of the assessment takes the form of performance presentations in which two students from each group act as directors in a given week. The remaining group members are just helping, so it is important that they have reliable access to the directions in advance of each week's performance. The student's on-line collaboration can work on their joint presentations at a distance from each other. I have to oversee the group interactions to check that groups weren't copying from each other and to ensure that directions were posted on time, so I helped to set that up.

It should be noted that there is a big difference from one module to another. So in modules where the tutor has told the students that they will log-on to the on-line discussion at such and such a time and they give replies and responses to the student's exchange - then you see the majority of students participating.

There is an interesting balance - the students want to know you are there even if you are not directly participating in the discussion. We think the students want validation that the opinions they're expressing are appropriate - so if they feel that the tutor is in the background and will correct any major mistakes they seem to be more open to learn from each other.

In particular, the objective was to assess the students' ability to engage with the material in a certain way, rather than to have them reiterate a series of facts about the material. There is an acceptance across the whole School that e-learning is part of teaching now. Some students are happy to embrace it whilst others are still a bit anxious and have workload concerns. If a School does want a wide-scale adoption of e-learning there does need to be a School based individual who can coordinate and provide support for students? I could not correlate between team work and e-learning in this paragraph).

We think that our graduates should also be provided with an international platform as a foundation for their careers. With this in mind, language skills are essential for engineers who are operating in the increasingly global environment. Therefore, it is necessary to speak one or two foreign languages in order to be able to compete internationally. It is also necessary to include international subjects in an institution's programs of study; to make programs attractive to international students.

Nevertheless, engineering education institutions still grapple with the fundamental concepts and ideas related to the internationalization of their activities and courses. Comprehensive studies concerning curriculum development and its methodology are essential in order to ensure that the main stream of academic activities is not completely lost in the process of globalization. Research should be undertaken on a global engineering education curriculum, in order to identify fundamental issues and concerns in an attempt to devise and develop a proper methodology, which would be used in curriculum development in an era of globalization.

5. TEACHING INSTRUMENTS

It is widely believed that using traditional methods like writing of teaching material on whiteboard or blackboards take lots of time and cause tutor contact diminished to lowest level. Now a days, fortunately, majority of universities are equipped with Audio- video learning devices to communicate with students. Using electronic devices like opaque, overhead and video projectors take the lowest time for write up of material in the class, and attract student's attention to completely focus on matter without wasting time. Also there is a direct eye contact between tutor and learner which is necessary to oblige student to be in class without any absence of mind. It is necessary that all material would be transferred as hard copy in advance to student to completely use an interactive environment and don't coerce to write teaching material in class. Student goes ahead with tutor and main part of understanding process is completed in class.

6. DISCUSSION

The economic future within Europe and worldwide increasingly depends on our ability to continue to provide improved living standards, which in turn depends on our ability to add value to products and services continuously. To achieve this, employers increasingly need to look to engineers throughout their organizations to come forward with innovative and creative ideas for improving the way business performs and equally to take greater responsibility within their work areas through personal development on a life-long basis.

To compete in the rapidly changing global markets it has become essential for organizations to recognize that they can no longer depend on educated elite but, increasingly, must maximize the potential of the workforce within the organization by harnessing all available brain power to assure economic survival and success. Life-long learning will provide for the development of a more cohesive society much more able to operate within a cross-cultural diversity.

We expect steady growth in global engineering e-learning programs. The main drivers will be the strong industry interest in recruiting students who have some understanding and experience of global industry, government funding agencies who are slowly developing support for global engineering education, rising faculty awareness, and high levels of student interest. It is not, however, easy to do, and resource issues will slow the growth rate. The following are the reasons why we think global engineering e-learning programs are good;

1. Preparing students for the global economy. This is necessary and it will happen.
2. Everyone learning from the comparative method: education, research, and service global collaborations make everyone smarter.
3. 3. There are good research prospects through the education activities such as optimizing virtual global teams, and through research collaborations that are a byproduct of the educational collaborations.
4. It builds international and cross-cultural tolerance and understanding

7- CONCLUDING REMARKS

One key finding was that the students wanted a greater sense of community; they wanted more interaction with lecturers and the university. Many professors post a course syllabus, homework assignments, and study guides on the Web, and some ambitious faculty with large classes may even give exams online. But educators aren't ready to plunge completely into the electronic learning environment and fully adopt simulated lab experiments or self-guided online instruction. One reason is the time it would take for faculty to learn how to use new computer programs and to develop online materials to replace their current versions. Another reason is that laboratory science requires intuitive observations and a set of skills learned by hands-on experience, none of which can be fully imitated in the digital realm. Simulations are a bridge from the abstract to the real, combining old tech with new tech to connect theory in the classroom with real-world experience in the lab. The purpose is to provide a creative environment to reinforce or enhance traditional learning, not to replace it. As students progress through the theory section, they can quiz themselves on what they have learned. As an example, in a virtual lab section, students learn how to construct a data table by using chemical shifts and coupling patterns from spectra of common compounds. Errors entered by the students are automatically corrected and highlighted in red. Once the table is complete, the student is directed to select the molecular fragment associated with an NMR signal, and then assemble the fragments to form the molecule. E-learning has great potential for engineering education. Broader use of e-learning will be driven by the next generation of students who will have had exposure to e-learning programs in high school and will start to ask for similar systems at the undergraduate level. E-learning also will likely be adopted more quickly for distance-learning courses or courses for which lab costs or lack of lab facilities may be a factor, such as those for nonscience majors or those offered by community colleges or high schools. It should be noted that the e-learning model adopted in one university cannot be the best model to follow in another college. The providers of distance learning may have to accept that there are limitations in all models of distance education. The best opportunity lies in identifying and offering the mode that suits the most students in a particular cultural and regional context at a particular point in time. There is still work to be done on distance learning, but initial signs are positive. For all modes of delivery, learning is the active descriptor, distance is secondary. Nevertheless, the Website appears to be an effective supplementary tool for students with all learning style modalities. The correlations between course grade and the Website usage were weakest for Active and Sensing students. It may be an issue that needs to be addressed through instructional design to make the materials more engaging for these particular modalities. Also, the relatively small sample of styles may have affected the results.

REFERENCES

Al-Jumaily, A., & Stonyer, H. (2000). Beyond Teaching and Research . *Changing Engineering Academic Work Global Journal of Engineering Education*, *4*(1), 89–97.

Baron, R. (1998). *What Type Am I? Discover Who You Really Are*. New York: NY Penguin Putnam Inc.

Boyer, E. L. (1990). *Scholarship Reconsidered: Priorities of the Professoriate*. Stanford: Carnegie Foundation for the Advancement of Teaching.

Coleman, D. J. (1998). *Applied and academic geomatics into the 21st Century*. Paper presented at the FIG Commission 2, XXI Inter. FIG Congress, Brighton, England.

CSU. (2004). *Chico Distance & Online Education*. Retrieved from http://rce.csuchico.edu/online/site.asp

Dale, E. (1969). *Audiovisual Methods in Teaching* (3rd ed.). New York, NY: Dryden Press.

Dunn, R. (1990). Understanding the Dunn and Dunn Learning Styles Model and the Need for Individual Diagnosis and Prescription. *Reading . Writing and Learning Disabilities*, *6*, 223–247.

Dunn, R. (1992). Learning Styles Network Mission and Belief Statements Adopted. *Learning Styles Network Newsletter*, *13*(2), 1.

Dunn, R., Beaudry, J. S., & Klavas, A. (1989). Survey of Research on Learning Styles. *Educational Leadership*, *46*(6), 50–58.

Felder, R. M. (1993). Reaching the Second Tier: Learning and Teaching Styles in College Science Education. *Journal of College Science Teaching*, *23*(5), 286–290.

Finelli, C. J., Klinger, A., & Budny, D. D. (2001). Strategies for improving the classroom environment . *Journal of Engineering Education*, *90*(4), 491–501.

Foertsch, J., Moses, G., Strikwerda, J., & Litzkow, M. (2002). Reversing the lecture / homework paradigm using e-TEACH Web-based streaming video software. *Journal of Engineering Education*, *91*(3), 267–275.

Forks, G. (2009). *UND Online & Distance Education*. Retrieved from http://distance.und.edu/

Frances, C., Pumerantz, R., & Caplan, J. (1999). Planning for instructional technology. What you thought you knew could lead you astray. *Change*, *31*(4), 24–33. doi:10.1080/00091389909602697

Hinchcliff, J. (2000). Overcoming the Anachronistic Divide: Integrating the Why into the What in Engineering Education. *Global Journal of Engineering Education*, *4*(1), 13–18.

Johnson, D. W., Johnson, R. T., & Smith, K. A. (1991). *Active Learning: Cooperation in the College Classroom*. Edina, MN: Interaction Book Company.

Johnson, D. W., Johnson, R. T., & Smith, K. A. (1992). *Cooperative Learning: Increasing College Faculty Instructional Productivity. ERIC Digest*. Washington, D.C: The George Washington University, School of Education and Human Development.

Jung, C. G. (1971). *Psychological Types*. Princeton, NJ: UA University Press.

Kolb, D. A. (1981). *Learning styles and disciplinary differences*. San Francisco, CA: Jossey-Bass.

Kolb, D. A. (1984). *Experiential Learning: Experience as the Source of Learning and Development* (1st ed.). Englewood Cliffs, NJ: Prentice-Hall.

Kolmos, A. (1996). Reflections on project work and Problem-Based Learning. *European Journal of Engineering Education, 21*(2), 141–148. doi:10.1080/03043799608923397

Mamchur, C. (1990). *Cognitive Type Theory and Learning Style, Association for Supervision and Curriculum Development. A teacher's guide to cognitive type theory and learning style*. Alexandria, VA: Association for Supervision and Curriculum Development.

Myers, L. B., & McCaulley, M. H. (1985). *Manual-Guide to the Development and Use of the Myers-Briggs Indicator*. Palo Alto, CA: Consulting Psychologists Press.

Oprean, C. (2006). The Romanian contribution to the development of Engineering Education. *Global Journal of Engineering Education, 10*, 45–50.

Overview of e3L. (2005). Retrieved from http://e3learning.edc.polyu.edu.hk/main.htm

Purcell-Roberston, R. M., & Purcell, D. F. (2000). *Interactive distance learning. Distance Learning Technologies: Issues, Trends and Opportunities*. Hershey, PA: IGI Global.

Rugarcia, A., Felder, R. M., Woods, D. R., & Stice, J. E. (2000). The future of engineering education I: a vision for a new century. *Chemical Engineering Education, 349*(1), 16–25.

UI. (2004). *Engineering Outreach Distance Education*. Retrieved from http://www.uidaho.edu/evo

Wild, R. H., Griggs, K. A., & Downing, T. (2002). A framework for e-learning as a tool for knowledge management. *Industrial Management & Data Systems, 102*(7), 371–381. doi:10.1108/02635570210439463

Chapter 4
3D Virtual Learning Environment for Engineering Students

M. Valizadeh
University of Guilan, Iran

B. Noroozi
University of Guilan, Iran & University of Cincinnati, USA

G. A. Sorial
University of Cincinnati, USA

ABSTRACT

Virtual Reality and Virtual Learning Environments have become increasingly ambiguous terms in recent years because of essential elements facilitating a consistent environment for learners. Three-dimensional (3D) environments have the potential to position the learner within a meaningful context to a much greater extent than traditional interactive multimedia environments. The term 3D environment has been chosen to focus on a particular type of virtual environment that makes use of a 3D model. 3D models are very useful to make acquainted students with features of different shapes and objects, and can be particularly useful in teaching younger students different procedures and mechanisms for carrying out specific tasks. This chapter explains that 3D Virtual Reality is mature enough to be used for enhancing communication of ideas and concepts and stimulate the interest of students compared to 2D education.

INTRODUCTION

Distance learning is a re-invented method of education, rather than a new one. Distance learning is broader than e-Learning, as it covers both non-electronic (e.g. written correspondence) and technology-based delivering of learning. Technology-based learning is delivered via any

technology, so it entails distance learning, too. Resource-based learning is the broadest term because any technology could be used as a resource in the learning process, where learners are active. In its early days, distance learning consisted of correspondence education, televised courses, collections of video tapes, and cassette recordings. Figure 1 shows a brief history of distance learning (Hamza-Lup & Stefan, 2007; Harper, et al., 2004).

DOI: 10.4018/978-1-61520-659-9.ch004

Figure 1. The history of distance learning

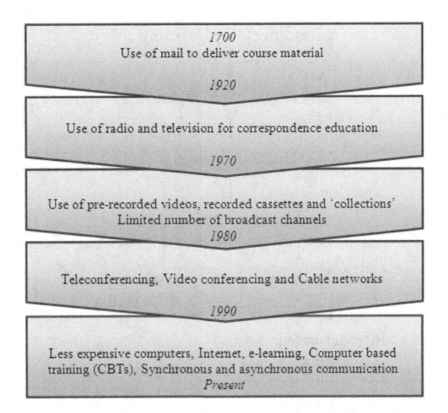

The concept of Internet-based learning is broader than Web-based learning (see Figure 2). The Web is only one of the Internet services that uses a unified document format (HTML), browsers, hypertext, and unified resource locator (URL) and is based on the HTTP protocol. The Internet is the biggest network in the world that is composed of thousands of interconnected computer networks (national, regional, commercial, and organizational). It offers many services not only Web, but also e-mail, file transfer facilities, etc. Hence, learning could be organized not only on the Web basis, but also for example, as a correspondence via email. Furthermore the Internet is based not only on the HTTP protocol, but on other proprietary protocols as well (Anohiina, 2005; Hamza-Lup & Stefan, 2007).

Of particular interest is the growing number of students from developing or transitional economies studying Western university degrees. They enroll either as a foreign student at a Western university, or join an internationally accredited and qualified educational institution in their home country which collaborates with a Western university (Van Raaij & Schepers, 2008).

Virtual Reality and Virtual Learning Environment have become increasingly ambiguous terms in recent years. The powerful 3D graphics hardware available in desktop computers provides an attractive opportunity for enhancing interaction. It may be possible to leverage human spatial capabilities by providing computer generated 3D scenes that better reflect the way we perceive our natural environment. 3D environments have the potential to position the learner within a meaningful context to a much greater extent than traditional interactive multimedia environments (Cockburn & Mckenzie, 2004; Dalgarno & Hedberg, 2001).

In the last two decades collaborative virtual environments (CVEs) have been largely adopted

Figure 2. Different levels of resource-based learning

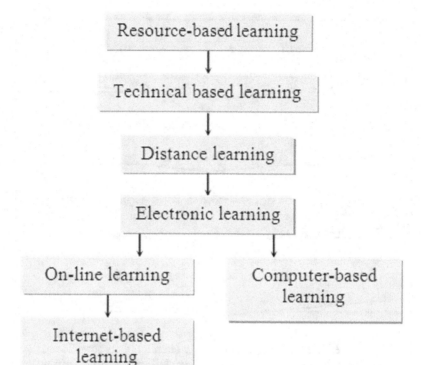

to favor social interaction and learning. They offer the possibility to simulate the real world as it is or to create new worlds. Interacting with these environments, people can actively experiment situations useful for understanding concepts as well as learning to accomplish specific tasks. CVEs support independent viewpoints for users: they share the virtual environment to do highly synchronous collaborative work, manipulating the same virtual objects. Nevertheless, it is not an easy task to create such an environment.

Virtual reality has been very popular and successful in other areas including entertainment and urban planning. It has also been extensively used within manufacturing industries and military bodies. In addition, the benefits of 3D graphics for education have been explored. Many 3D re-

sources have already been developed in this area. 3D models are very useful to familiarize students with features of different shapes and objects, and can be particularly useful in teaching younger students. Many games have been developed using 3D images that the user must interact with in order to learn a certain lesson. Interactive models increase a user's interest and make learning more fun. 3D animations can be used to teach students different procedures and mechanisms for carrying out specific tasks. Virtual reality has also been used extensively for simulations and visualization of complex data. For example, medical disciplines use virtual reality to represent complex structures and increasingly scientists are using this technology for visualization and in particular as a teaching aid (Monahan, et al., 2008).

It is necessary to produce a world model, to animate avatars which have to temporally and spatially share the environment, and to implement communication facilities. Thus, the delivery of educational 3D environments based on virtual reality technologies can be very expensive, and, as a consequence, such solutions are not widely accessible to learners (De Lucia, et al., 2008; Ling, et al., 2003).

BACKGROUND

The term 3D environment has been chosen to focus on a particular type of virtual environment that makes use of a 3D model. Currently several trends of research are looking to progress further towards more interactive and immersive three-dimensional user interfaces based on languages like Virtual Reality Modeling Language (VRML). Many of the early 3D environment research focused on physically immersive environments, which require expensive hardware such as head-mounted displays, rather than desktop environments, which use standard computer hardware. Intelligent user interfaces are seen by many developers as the next step in the evolution of user interfaces.

The single most prominent goal in the development of intelligent user interfaces is to offer a customized interface and interaction mechanism for each individual user (Dalgarno & Hedberg, 2001; Mangina & Kilbride, 2008).

There are two main principles of exploring 3D models (Sperka, 2004):

1. Browser displays sequence of images (photographs or computer generated), representing panorama or the pictures of 3D object viewed from the different angles.
2. Browser calculates 2D projections of 3D object (rendering) seen from the different viewing angles.

Open systems allow everybody to create and publish 3D worlds with own tools (VRML). Closed systems require proprietary authoring tools. Open Web3D system is the cheapest form of desk top virtual reality systems. It allow to create virtual worlds and publish them on Internet for all, who have Internet connection (can be used also off-line) and simple computer with the Internet browser and any text editor. The most used are the following Web3D solutions: VRML, X3D, JAVA 3D, MACRO MEDIA SHOCK WAVE 3D, ADOBE ATMOSPHERE, VIEWPOINT, SUPER SCAPE, CULT, EON Reality etc (Sperka, 2004).

The spectrum of Web3D applications is very broad. The list of the main existing and future application groups is presented in Table 1 (Sperka, 2004).

The use of Virtual Reality and 3D graphics for e-learning is now being further extended by the

Table 1. Web 3D applications

E-Business	– Sales and marketing-more effective showcase product offering to customers – Shops on the web and product presentation with the virtual touch and test – Real estate presentation and virtual tours – Internet hotel rooms presentations – Virtual panoramas of the recreation areas, ski centers, swimming pools, etc.
Education and training	– 3D virtual laboratories and instruments – 3D e-library – 3D classroom – Distant education
Services	– Location based services-link 3D city maps with GIS/GPS – Virtual cities – Hyper-markets – Galleries and museums and archeological parks – Airports – Hospitals, etc
Tele maintenance	– Field maintenance procedures to remote locations
Data visualization	– Distributed CAD/CAM systems – Distributed 3D scientific and medical visualization systems – e-Science
Entertainment	– Internet 3D games

provision of entire Virtual Reality environments where learning takes place. This highlights a shift in e-learning from the conventional text-based on-line learning environment to a more immersive and intuitive one. Since Virtual Reality is a computer simulation of a natural environment, interaction with a 3D model is more natural than browsing through 2D WebPages looking for information. These Virtual Reality environments can support multiple users, further promoting the notion of collaborative learning where students learn together and often from each other. CLEV-R, a Collaborative Learning Environment with Virtual Reality, is a web-based multi-user 3D environment that provides a virtual reality university where students go to learn collaborate and socialize online. As with a real university, students are aware of each other within the environment and they can partake in lectures, group meetings and informal chats. Virtual Reality can bring a great deal to an e-learning experience in these ways (Monahan, et al., 2008).

Communication methods provided in the system can also act as a means of social interaction for students and their peers. The system consists of a series of WebPages where prospective students can register to use CLEV-R and returning students can login to the 3D environment. Once a student provides their username and password, they are presented with a personalized webpage with information on the courses they are registered for. From this page, users can access the 3D learning environment and begin to take part in their course. The 3D environment is presented through a webpage which is split into two distinct sections. The upper section consists of the actual Virtual Reality environment while the lower section provides a Graphical User Interface (GUI) with tools for communication (Monahan, et al., 2008).

The Virtual Reality Environment

The Virtual Reality world is modeled to contain many of the features found in a traditional university. Virtual Reality is expected to enable universities and industry to make experiments in a safe and economic way, in a virtual environment rather than in the real plant where such experiments cannot be performed because of the costs and risks involved. Such experiments can provide the needed measurements and qualitative evaluations of human factors and serve as test beds of different hypothesis, scenarios and methods for reviewing the human organization and work processes, improving design quality for better safety in new or existing plants and retrospectively examining past accidents to gain hints for the future procedures (Zio, 2009).

Martins (2004) believe that Virtual Learning Environment is a web-based communications platform that allows students, without limitation of time and place, to access different learning tools, such as program information, course content, teacher assistance, discussion boards, document sharing systems, and learning resources (Martins & Kellermanns, 2004).

The processes of collaboration and communication between learners and teachers are increasingly computer-mediated, such as via the Internet. From the learner's perspective, perhaps the most significant and detrimental factors to the success of a virtual learning environment are stress, association with technology use, and dissatisfaction towards the technology itself. It is suggested, conceivably, that the success of any virtual learning environment depends on the adequate skills and attitudes of learners. This proposition is evidenced by the popularity of online course delivery at postgraduate level when compared with undergraduate degree courses; as it is commonly believed that postgraduate students are mature and motivated to undertake self-study as required in most virtual learning environments. The authors present a study with the purpose to access preparation of learners (Lee, et al., 2002).

Virtual Reality Environment consists of essential elements facilitating a consistent environment for learners. These elements are *virtual classrooms, virtual library, meeting rooms (chat rooms),* social areas and a graphical user interface. Each room or area is equipped with tools and features to match their purpose (Bouras, et al., 2001; Monahan, et al., 2008).

The lecture room or classroom is at the heart of the Virtual Reality environment. To create a virtual classroom, one must plan for the following tasks: advising, curriculum development, content development, articulation and credentialing, learning delivery, hardware choice, and assessment. This scenario mainly concerns the lectures given throughout the virtual environments, anticipating for virtual classrooms where appropriate teaching material could be displayed and presented by a teacher to the students. In a virtual Reality Environment much of the structured learning takes place in a 3D classroom. The room is designed for use by tutors to address students synchronously in a live lecture. The room contains a large presentation board where the teacher can upload files in an intuitive fashion. When a tutor clicks an upload button on their virtual lectern, a webpage is presented, where they can upload a new file to the presentation board or select from a list of previously uploaded files. The presentation board can currently display PowerPoint presentations, word files and a number of image formats. The tutor can use the communication tools described above to accompany these lecture slides. Furthermore they can use audio communication to comment on the lecture slides or a web-cam broadcast to demonstrate certain points associated with the lecture. The lecture room also contains a video board, which facilitates the tutor to upload movie files to the Virtual Reality environment (Bouras, et al., 2001; Harper, et al., 2004; Monahan, et al., 2008).

An important advantage in this system is the possibility of integration of virtual laboratories with Virtual Reality environments. The use of Virtual Reality tools does not only enhance the interaction of scientists with the experiment by offering an easy way of virtually exploring its

components (human virtual reality) but also offers a solid base for rationalizing what is happening at microscopic level by providing a Virtual Reality representation of its elementary components (molecular virtual reality, MVR). Such a combination of HVR and MVR is also of invaluable help in designing innovative e-learning approaches. This scope is absolutely important in applicable sciences such as engineering, chemistry, physics and medicine (Gervasi, et al., 2004).

As another element of Virtual Reality environment, Meeting Rooms have been designed to facilitate group meetings and discussions. They allow students to work together on projects and other group tasks. To this end, these rooms are equipped with similar features to those found in the lecture room. A presentation board and video board are available for students to upload their own files for others to see and discuss. When a student wishes to speak with others in the meeting room, they can use the text-chat or audio chat facilities discussed previously. One of the main differences between the meeting rooms and the lecture room is the level of restrictions which apply. Upload facilities in the lecture room are reserved for use by tutors only, thus if a student tries to upload files their actions are refused. Such restrictions do not apply in the meeting rooms on the other hand, Industry generally supports collaboration and interaction features which may include around the clock mentoring, expert led chats, peer-to-peer chats, seminars, threaded discussions, mentor ed exercises, discussion boards, workshops, study groups, and online meetings. Many of these activities are a simulation by technology of face-to-face interactions (Harper, et al., 2004; Monahan, et al., 2008).

With regard to the characteristics of communication task, media richness theory states that the purpose of communication is to reduce uncertainty and Equivocality in order to promote communication efficiency. Uncertainty is associated with the lack of information. Organization creates structures such as formal information systems, task forces, and liaison roles that facilitate the flow of information to reduce the uncertainty. The role of media in uncertainty reduction is its ability to transmit the sufficient amount of correct information. Equivocality is associated with negotiating meanings for ambiguous situations. To deal with equivocality, people in an organization must find structures that enable rapid information cycles among them so that meaning can emerge (Ramasundaram, et al., 2005).

While the virtual meeting rooms and classrooms discussed above are intended for group learning and for replicating a real-life lecture scenario, a Collaborative Learning Environment with Virtual Reality also caters for individual learning with facilities for students to review and acquire the lecture notes. Obtaining the learning material is achieved in a natural and intuitive way through a virtual library. The library contains a bookcase and a number of desks. When a tutor uploads notes to the system, they are automatically represented by a book on the library shelf. Students can then enter the library and browse the catalogue of lecture notes available. When a student clicks on a book, the notes associated with that particular book are placed on the nearest free table. The user can then review the notes, flicking back and forth through them and also download them to their own computer for review at a later date. At present the library contains eight desks. If students are viewing notes at all eight desks, then the next student attempting to view a set of lecture notes is instructed to download them directly to their own computer for viewing. Personal notes taken by the student during a learning session can also be accessed via the library. In addition, a number of links to external websites are available. For example, clicking on the dictionary opens a webpage for the online version of an English dictionary, and likewise clicking on the encyclopedia opens a webpage for an online encyclopedia that students can use (Monahan, et al., 2008).

As a summary the specifications of a Virtual Reality Environment for E-learning can be according to Table 2 (Bouras, et al., 2001).

GUI, the Graphical User Interface

The main purpose of the Graphical User Interface (GUI) is to host a suite of communication controls for the system. It is made up of a series of panels. The initial panel details personal user information such as their name and status.

This is a messaging service where users can type a message and send it to all those connected or alternatively send it privately to individual users. The GUI also provides a panel with tools for users to broadcast their voice. Once a user connects to the communication server, they can select an area or virtual room in the 3D environment to broadcast to. For example, they can choose to broadcast to a meeting room allowing anyone in this room to hear the voice broadcast. Furthermore users can broadcast a live stream from their web-cam to a media board in the Virtual Reality environment. The web cast is automatically displayed in the 3D environment of all users currently connected. This is actually a Face-to-face conversation tool

Table 2. General concepts for collaborative Virtual Reality based e-learning

Scalability	The ability to support a maximum number of simultaneous users, which could vary according to the specific settings of each virtual world.
Persistence	This is realized by distributing and synchronizing user input as well as user independent behavior in order to achieve the impression of a single shared world.
Extensibility	It should be possible to customize an existing VE, before Presenting it to the users.
Openness	It should be possible to interface a Virtual Environment to external applications.
Communication	The users in a Learning Virtual Environment should be able to communicate.
Compatibility	A Web-based application implemented with international accepted standards and technologies (HTTP, HTML, VRML, and JAVA).
Support of variable content	Several forms of data should be supported and embedded in the Learning Virtual Environment.

which is considered the richest medium because it provides immediate feedback. Face-to-face also provides multiple cues via body language and tone of voice, and message content is expressed in natural language (Monahan, et al., 2008; Ramasundaram, et al., 2005).

A Comparison between 2D and 3D Interfaces in Engineering E-Learning

It is clear that the 3D virtual environment adds something new to e-learning and teaching. It opens up a number of new possibilities, for example for group work, presentations, meetings and socialization. The results of the pilot project suggest that a 3D learning environment is a feasible possibility for parallel use with more traditional asynchronous learning environments. Until very recently the 3D was exclusively used by professionals in other fields (e.g. movie makers, web designers, etc.). It would be interesting for money traders or stock market advisers to use 3D graphics easily without spending much time learning the details of the application. A first step in this direction was taken by the 3DStock software (Hamza-Lup & Stefan, 2007; Jonasson, 2005).

Real opportunities exist for the development of novel educational and training materials, particularly for science applications where 3D visualization is critical for understanding concepts. A 3D virtual space brings advantages such as increased motivation on behalf of the student and increased efficiency in explaining difficult concepts. There are fields, such as medicine, where the Web 3D-based applications have proved their utility already (Hamza-Lup & Stefan, 2007).

The "digital classroom" provided by 2D tools does not resemble the reality of the conventional classroom, since it relies only on classical desktop windows metaphors rather than emulating real world as in 3D environments. The key feature of 3D virtual environments is the ability for lecturer/students to visualize the presence and location of other participants to the action, and, in general,

to increase the presence and awareness sensation, as better described in the next section(De Lucia, 2008).

Dalgarno and Hedberg (2001) stated that the main specifications of a 3D environment could be as follows (Anohiina, 2005; Chen & Hsiang, 2007; Dalgarno & Hedberg, 2001):

- The environment is modeled using 3D vector geometry with objects represented in x, y and z coordinates describing their shape and position in 3D space.
- The learning process is based on some technology partly or entirely replacing a human teacher.
- The user's view of the environment is rendered dynamically according to their current position in 3D space.
- The user has the ability to move freely through the environment and their view is updated as they move.
- Creating an online virtual classroom and providing specific learning time and space within the company may reduce the issue of inefficient use of online resources.
- At least some of the objects within the environment respond to user action, for example doors might open when approached and information may be displayed when an object is clicked on.

Modern multimedia technologies are matured enough allowing educators to have more options in designing the instructions. For the development of computer-based tutorials for spatial visualization skills, the comparison of 2D and 3D representations is considered necessary and informative in informing future design decision and the literature of multimedia learning (Hung, et al., 2007).

The potential of the Virtual Reality and 3D graphical interface as tools for delivering education is promising at all levels of educational hierarchy. However, despite all the assertions of lower opportunity cost, greater convenience, and expanded accessibility for all, it is generally accepted that the preparation of both the university faculty members and students for web-based virtual learning will need to be further enhanced. Some researchers have suggested that demands on students' time to maintain the hardware, software, and connectivity for their web-based courses are expected to be high and students might not have the maturity or skill to allocate their time productively to make progress in web-based distance education courses. Furthermore, as collaboration and communication between learners and teachers are increasingly being mediated by the computer via Internet, the likely stress and dissatisfaction about the technology itself may be the most significant and foremost detrimental factor to the success of such a virtual learning environment (Lee, et al., 2002).

3D in Engineering Education

Using three dimensional graphics for more realistic and detailed representations of topics, offering more viewpoints and more inspection possibilities compared to 2D education. For example the WebTOP system helps in learning about waves and optics by visually presenting various kinds of physical phenomena, may are not available in the real world, but have invaluable potential for education (Chittaro & Ranon, 2007).

In improvement of engineer education, some authors use CFD (Computational Fluid Dynamics) program to run the case and observe the physical phenomena on single-plane software rather than via internet. Frederick et al. (Hung, et al., 2007) used the commercial software, FlowLab, to create a CFD education interface for engineering course and laboratories, had been proven that is an effective and efficient tool to help students learning. Li et al. (Olivier, 2003) used the web-base to develop an on-line mass transfer course system. They also indicated that learning via internet will be the trend in the future.

The concept of thermal-hydraulics e-learning system was initiated by Hung et al. (Hung, et al., 2007). They merged existing computational fluid dynamics capability into the e-learning concept to improve traditional engineer education in fluid flow and heat transfer. In the paper, they referred to HTML (Hyper text markup language) and ASP (Active server pages) where an interactive user interface as the core of the system and system structure has been preliminarily planned. The prototype of a postprocessor program was also established. But the ability of post-processor in drawing for flow field, stream lines, and isothermal contours has to be further enhanced. Liu et al. (Liu, et al., 2006) quoted the concept of Hung et al (Hung, et al., 2007) to implement and strengthen the architecture of this thermal hydraulics e-learning system. In order to establish a cross platform for e-learning, they used an HTML embedded scripting language, PHP (Hypertext Preprocessor), and integrated with MySQL database management system to manage all teaching materials (Hung, et al., 2007).

In general, virtual 3-D models can be used in engineering education to stimulate processes to enhance the understanding of the concepts used in a particular process. Students could experience the 3-D structure of a specific process and the means of controlling direction and movement within a system.

CONCLUSION

Web3D is a good and available platform for experimenting with creation of new tools as well as applications for tele-presence in form of 3D worlds models. 3D Virtual Reality, software tools and associated Web technologies are mature enough to be used in conjunction with advanced e-Learning systems. 3D based content can enhance communication of ideas and concepts and stimulate the interest of students.

REFERENCES

Anohiina, A. (2005). Analysis of the terminology used in the field of virtual learning. *Educational Technology & Society, 8*, 91–102.

Bouras, C., Philopoulos, A., & Tsiatsos, T. (2001). e-Learning through distributed virtual environments. *Journal of Network and Computer Applications, 24*, 175–199. doi:10.1006/jnca.2001.0131

Chen, R., & Hsiang, C. (2007). A study on the critical success factors for corporations embarking on knowledge community-based e-learning. *Information Sciences, 177*, 570–586. doi:10.1016/j.ins.2006.06.005

Chittaro, L., & Ranon, R. (2007). Web3D technologies in learning, education and training: Motivations, issues, opportunities. *Computers & Education, 49*, 3–18. doi:10.1016/j.compedu.2005.06.002

Cockburn, A., & Mckenzie, B. (2004). Evaluating spatial memory in two and three dimensions. *International Journal of Human-Computer Studies, 61*, 359–373. doi:10.1016/j.ijhcs.2004.01.005

Dalgarno, B., & Hedberg, J. (2001). *3D Learning environments in tertiary education*. Paper presented at the 18th annual conference of the Australasian Society for Computers in Learning in Tertiary Education, Melbourne, Australia.

De Lucia, A., Francese, R., Passero, I., & Tortora, G. (2008). *Development and evaluation of a virtual campus on Second Life: The case of SecondDMI*. Computers & Education.

Gervasi, O., Riganelli, A., Pacifici, L., & Laganà, A. (2004). VMSLab-G: a virtual laboratory prototype for molecular science on the Grid. *Future Generation Computer Systems, 20*, 717–726. doi:10.1016/j.future.2003.11.015

Hamza-Lup, F. G., & Stefan, V. (2007). *Web 3D & Virtual Reality - Based Applications for Simulation and e-Learning*. Paper presented at the 2nd International Conference on Virtual Learning, ICVL, Constanta, Romania.

Harper, K. C., Chen, K., & Yen, D. C. (2004). Distance learning, virtual classrooms and teaching pedagogy in the Internet environment. *Technology in Society, 26*, 585–598. doi:10.1016/S0160-791X(04)00054-5

Hung, T. C., Wang, S. K., Tai, S. W., & Hung, C. T. (2007). An innovative improvement of engineering learning system using computational fluid dynamics concept. *Computers & Education, 48*, 44–58. doi:10.1016/j.compedu.2004.11.003

Jonasson, J. (2005). *3D learning environment. Will it add the 3rd dimension to e-learning and teaching?* Paper presented at the 3rd International Conference on Multimedia and ICTs in Education, Cáceres, Spain.

Lee, J., Hong, N. L., & Ling, N. L. (2002). An analysis of students' preparation for the virtual learning environment. *The Internet and Higher Education, 4*, 231–242. doi:10.1016/S1096-7516(01)00063-X

Ling, C., Gen-Cai, C., Chen-Guang, Y., & Chuen, C. (2003). *International Conference on Communication Technology proceeding, 2*, 1655-1661.

Liu, C. C., Hung, C. T., Hung, T. C., Pei, B. S., & Zhang, L. (2006). *The development of an innovative interactive e-learning system in computational thermal-hydraulics for engineers*. Paper presented at the EDMEDIA.

Mangina, E., & Kilbride, J. (2008). Utilizing vector space models for user modeling within e-learning environments. *Computers & Education, 51*, 493–505. doi:10.1016/j.compedu.2007.06.008

Martins, L. L., & Kellermanns, F. W. (2004). A model of business school students' acceptance of a web-based course management system. *Academy of Management Learning & Education, 3*, 7–26.

Monahan, T., McArdle, G., & Bertolotto, M. (2008). Virtual reality for collaborative e-learning. *Computers & Education, 50*, 1339–1353. doi:10.1016/j.compedu.2006.12.008

Olivier, B. A. (2003). Learning content interoperability standards . In Littlejohn, A. (Ed.), *Reusing Online Resources: A Sustainable Approach to eLearning. London: Kogan Page*. Liber, O.

Ramasundaram, V., Grunwald, S., Mangeot, A., Comerford, N. B., & Bliss, C. M. (2005). Development of an environmental virtual field laboratory. *Computers & Education, 45*, 21–34. doi:10.1016/j.compedu.2004.03.002

Sperka, M. (2004). *Web 3D and new forms of human computer interaction*. Paper presented at the 2ed International Symposium of Interactive Media Design, Istanbul, Turkey.

van Raaij, E. M., & Schepers, J. J. L. (2008). The acceptance and use of a virtual learning environment in China. *Computers & Education, 50*, 838–852. doi:10.1016/j.compedu.2006.09.001

Zio, E. (2009). Reliability engineering: Old problems and new challenges. *Reliability Engineering & System Safety, 94*, 125–141. doi:10.1016/j.ress.2008.06.002

Chapter 5

Online Automated Essay Grading System as a Web Based Learning (WBL) Tool in Engineering Education

Siddhartha Ghosh
G. Narayanamma Institute of Technology and Science, India

ABSTRACT

Automated Essay Grading (AEG) or Scoring (AES) systems are not more a myth they are reality. As on today, the human written (not hand written) essays are corrected not only by examiners / teachers also by machines. The TOEFL exam is one of the best examples of this application. The students' essays are evaluated both by human & web based automated essay grading system. Then the average is taken. Many researchers consider essays as the most useful tool to assess learning outcomes, implying the ability to recall, organize and integrate ideas, the ability to supply merely than identify interpretation and application of data. Automated Writing Evaluation Systems, also known as Automated Essay Assessors, might provide precisely the platform we need to explicate many of the features those characterize good and bad writing and many of the linguistic, cognitive and other skills those underline the human capability for both reading and writing. They can also provide time-to-time feedback to the writers/students by using that the people can improve their writing skill. A meticulous research of last couple of years has helped us to understand the existing systems which are based on AI & Machine Learning techniques, NLP (Natural Language Processing) techniques and finding the loopholes and at the end to propose a system, which will work under Indian context, presently for English language influenced by local languages. Currently most of the essay grading systems is used for grading pure English essays or essays written in pure European languages. No one in today's world can ignore the use of English in Engineering education. Better to tell in professional courses. All the Engineering branches or streams are normally supported with modern English and sometimes known as English-for-Engineers. This write-up focuses on the existing automated essay grading systems, basic technologies behind them and proposes a new framework to show that how best these AEG systems can be used for Engineering Education. E-learning has created the path of alternate education. Whereas the Web-based-learning (WBL) has made the path much easier. Use of AEG systems in a web based learning environment helps the students to know, use,

DOI: 10.4018/978-1-61520-659-9.ch005

and understand English much better than they used to do in normal classroom based study. Such kinds of AEG systems are very useful mainly for non-English spoken students, better to say – students whose mother tongue is not English. Normally found that English used by such students are influenced by local languages. Use of a AEG system will not only help students to write better English essay, score better in English and others subjects written in English.

1. INTRODUCTION

Evaluation and Grading considered playing a central role in the educational process. The interest in the development and in use of *Computer-based Assessment Systems* (CbAS) has grown exponentially in the last few years, due both to the increase of the number of students attending universities and to the possibilities provided by e-learning approaches to asynchronous and ubiquitous education. Presently more than forty commercial CbAS are currently available on the market. Most of those tools are based on the use of the so-called objective-type questions: i.e. multiple choice, multiple answer, short answer, selection/association, hot spot and visual identification. Most researchers in this field agree on the notion that some aspects of complex achievement are difficult to measure using objective-type questions. Learning outcomes implying the ability to recall, organize and integrate ideas, the ability to express oneself in writing and the ability to supply merely than identify interpretation and application of data, require less structuring of response than that imposed by objective test items (Gronlund, 1985). It is in the measurement of such outcomes, corresponding to the higher levels of the Bloom's (1956) taxonomy (namely evaluation and synthesis) that the essay question serves its most useful purpose. One of the difficulties of grading essays is the subjectivity, or at least the perceived subjectivity, of the grading process. Many researchers claim that the subjective nature of essay assessment leads to variation in grades awarded by different human assessors, which is perceived by students as a great source of unfairness.

Furthermore essay grading is a time consuming activity. It is found that about 30% of teachers' time is devoted to marking. A system for automated assessment would at least be consistent in the way it scores essays, and enormous cost and time savings could be achieved if the system can be shown to grade essays within the range of those awarded by human assessor. Furthermore using computers to increase our understanding of the textual features and cognitive skills involved in the creation and in the comprehension of written texts, provide a number of benefits to the educational community.

Purpose of this paper is to present a new concept over the existing ones, through which we can overcome the problem of influence of local Indian languages in English essays. The system can do the grading of English essays as well as it can also provide sufficient feedback so that the students/user can understand what are the basic errors (spelling, grammar, sentence formation etc.) made by them and whether there essay is influenced by local language or not and how to overcome all these problems. The paper also discusses the current approaches to the automated assessment of essays (English Essays) and utilizes this as a foundation for the new framework. Thus, in the next section, research of some of the following important automated grading systems will be discussed: Project Essay Grade (PEG), Intelligent Essay Assessor (IEA), Educational Testing service I, Electronic Essay Rater (ERater), C-Rater, BETSY, Intelligent Essay Marking System, SEAR, Paperless School free text Marking Engine and Automark. All these systems are currently available either as commercial systems or as the result of research in this field. In the later chapters the concept of the new system is described.

Advantages of Web-Based Learning

The general benefits of Web-based training when compared to traditional instructor-led training include all those shared by other types of technology-based training. These benefits are that the training is usually self-paced, highly interactive, results in increased retention rates, and has reduced costs associated with student travel to an instructor-led workshop.

When compared to CD-ROM training, the benefits of Web-based training stem from the fact that access to the content is easy and requires no distribution of physical materials. This means that Web-based training yields additional benefits, among them.

Access is available anytime, anywhere, around the globe. Students always have access to a potentially huge library of training and information whether they are working from home, in the office, or from a hotel room. As cellular modems become more popular, students will even be able to access training in a place that doesn't have a traditional phone line or network connection.

Per-student equipment costs are affordable. Almost any computer today equipped with a modem and free browser software can access the Internet or a private Intranet. The cost of setup is relatively low.

Student tracking is made easy. Because students complete their training while they are connected to the network, it is easy to implement powerful student-tracking systems. Unlike with CD-ROMs that require students to print reports or save scores to disk, WBT enables the data to be automatically tracked on the server-computer. This information can be as simple as who has accessed the courseware and what are their assessment scores, to detailed information including how they answered individual test questions and how much time they spent in each module.

Possible "learning object" architecture supports on demand, personalized learning. With CD-ROM training, students have access only to the information that can be held by one CD-ROM. The instructional design for this type of delivery, therefore, has been to create entire modules and distinct lessons. But with WBT, there is virtually no storage limitation and content can be held on one or more servers. The best WBT is designed so that content is "chunked" into discrete knowledge objects to provide greater flexibility. Students can access these objects through pre-defined learning paths, use skill assessments to generate personal study plans, or employ search engines to find exact topics.

Content is easily updated. This is perhaps the single biggest benefit to WBT. In today's fast-paced business environment, training programs frequently change. With CD-ROM and other forms of training, the media must be reduplicated and distributed again to all the students. With WBT it is a simple matter of copying the updated files from a local developer's computer onto the server-computer. The next time students connect to the Web page for training, they will automatically have the latest version.

2. VARIOUS AUTOMATED ESSAY-GRADING SYSTEMS

Automated scoring capabilities are especially important in the realm of essay writing. Essay tests are a classic example of a constructed-response task where students are given a particular topic (also called a prompt) to write about1. The essays are generally evaluated for their writing quality. Surprisingly for many, automated essay scoring (AES) has been a real and viable alternative and complement to human scoring for many years. As early as 1966, Page showed that an automated "rater" is indistinguishable from human raters (Page, 1966). In the 1990's more systems were developed; the most prominent systems are the

Intelligent Essay Assessor (Landauer, Foltz, & Laham, 1998), Intellimetric (Elliot, 2001), a new version of the Project Essay Grade (PEG, Page, 1994), and e-rater (Burstein et al., 1998).

Ellis Page set the stage for automated writing evaluation (see the timeline in Figure 1). Recognizing the heavy demand placed on teachers and large-scale testing programs in evaluating student essays, Page developed an automated essay-grading system called Project Essay Grader (PEG). He started with a set of student essays that teachers had already graded. He then experimented with a variety of automatically extractable textual features and applied multiple linear regressions to determine an optimal combination of weighted features that best predicted the teachers' grades. His system could then score other essays using the same set of weighted features. In the 1960s, the kinds of features someone could automatically extract from text were limited to surface features. Some of the most predictive features Page found included average word length, essay length in words, number of commas, number of prepositions, and number of uncommon words—the latter being negatively correlated with essay scores.

In the early 1980s, the Writer's Workbench tool (WWB) set took a first step toward this goal. WWB was not an essay-scoring system. Instead, it aimed to provide helpful feedback to writers about spelling, diction, and readability. In addition

to its spelling program—one of the first spelling checkers - WWB included a diction program that automatically flagged commonly misused and pretentious words, such as regardless and utilize. It also included programs for computing some standard readability measures based on word, syllable, and sentence counts, so in the process it flagged lengthy sentences as potentially problematic. Although WWB programs barely scratched the surface of text, they were a step in the right direction for the automated analysis of writing quality.

In February 1999, **E-rater** became fully operational within ETS's Online Scoring Network for scoring GMAT essays. For low-stakes writing-evaluation applications, such as a Web-based practice essay system, a single reading by an automated system is often acceptable and economically preferable. The new version of e-rater (V.2) is different from other automated essay scoring systems in several important respects. The main innovations of e-rater V.2 are a small, intuitive, and meaningful set of features used for scoring; a single scoring model and standards can be used across all prompts of an assessment; modeling procedures that are transparent and flexible, and can be based entirely on expert judgment.

Figure 2 shows a popular common frame work of the automated essay grading systems. Most of the modern systems train the system

Figure 1. A timeline of research developments in writing evaluation

Pioneering Writing-evaluation research		Recent Essay grading research		Operational systems	Current ETS research	Future research & application		
			Computer Analysis of Essay Content Burstein et al.	**PEG** Page	**Wiriting diagnosis**	**Short-answer scoring**	**Questioning – answering system**	
				e-rater ETS	Chodorow & Leacock	Leacock & Chodorrow	Light et al.	
			Intelligent Essay Assessor Landauer et al.	**Latent semantic analysis**			**Verbal test creation tools**	
				Knowledge Analysis Technologies	Mitsakaki & Kulich	Hirchman et al		
	Writer's Workbench MacDonald et al.		**PEG** Page & Petersen		Burstien & Marcu	Breck at al.	**Students-centered instructional systems Erater – V.2**	
PEG Page		**PEG** Page		**Criterion** ETS Technologies				
1966-1968	1982	1994 - 1995	1997	1998 - 2000	2000	2000 - 2006		

with all most thousands of pre-assessed essays (corpus). Then, once the essay input is given, it gives the grade and as well as a proper feedback to improve. Hence some of these systems can be used for self-learning by students as well as by the teachers or institutes for grading huge amount of essays. Today (from 2007) the internationally recognized TOEFL exam gives the grade to the students' essays as a combination of human & machine assessment.

3. HOW THE AEG SYSTEMS WORK?

AEG systems are a combination of any two, three or all the techniques mentioned here - NLP (Natural Language Processing), Statistics, Artificial Intelligence (Machine Learning), Linguistics and Web Technologies, Text Categorization, annotated large corpora etc. It must be noted that seven out of ten most popular systems are based on the use of Natural Language Processing tools, which in some

cases are complemented with statistical based approaches. How come it comes under Artificial Intelligence? The time, machine can grade human written essays, which requires some expertise, we can tell that this is Artificial Intelligence. As because, the commonly available systems cannot perform that task. Text categorization is the problem of assigning predefined categories to free text document. The idea of automated essay grading, based on text categorization techniques, text complexity features and linear regression methods was first explored by Larkey (1998). The underlying idea of this approach relies on training of binary classifiers to distinguish "good" from "bad" essays and on using the scores produced by the classifiers to rank essays and assign grades to them. Several standard text categorization techniques are used to fulfill this goal: first, independent Bayesian classifiers allow assigning probabilities to documents estimating the likelihood that they belong to specific classes; then, an analysis of the occurrence of certain words in the documents is carried

Figure 2. A common framework for the existing automated essay grading systems

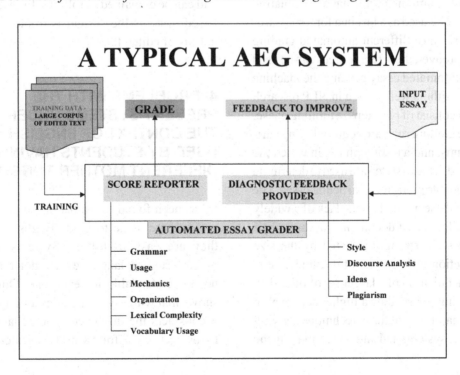

out and a k-nearest neighbor technique is used to find those essays closest to a sample of human graded essays; finally, eleven text complexity features are used to assess the style of the essays. Larkey conducted a number of regression trials, using different combinations of components. She also used a number of essay sets, including essays on social studies, where content was the primary interest and essay on general opinion where style was the main criteria for assessment.

A growing number of statistical learning methods have been applied to solve the problem of automated text categorization in the last few years, including regression models, nearest neighbor classifiers, Bayes belief networks, decision trees, rule learning algorithms, neural networks and inductive learning systems (Ying, 1997). This growing number of available methods is raising the need for cross method evaluation.

But the most relevant problem in the field of automated essay grading is the difficulty of obtaining a large corpus of essays (Christie, 2003; Larkey, 2003) each with its own grade on which experts agree. Such a collection, along with the definition of common performance evaluation criteria, could be used as a test bed for a standardized comparison of different automated grading systems. Moreover, these text sources can be used to apply to automated essay grading the machine learning algorithms well known in NLP research field, which consist of two steps: a training phase, in which the grading rules are acquired using various algorithms, and a testing phase, in which the rules gathered in the first step are used to determine the most probable grade for a particular essay. The weakness of these methods is the lack of a widely available collection of documents, because their performances are strongly affected by the size of the collection. A larger set of documents will enable the acquisition of a larger set of rules during the training phase, thus a higher accuracy in grading. A major part of these techniques, giving training to the systems and later stage, making the

systems to learn from new essays or experience is nothing but machine learning.

The feature set used with some modern AEG systems include measures of grammar, usage, mechanics, style, organization, development, lexical complexity, and prompt-specific vocabulary usage. This feature set is based in part on the NLP foundation that provides the instructional feedback to students who are writing essays. In some cases a web-based service evaluates a student's writing skill and provides instantaneous score reporting and diagnostic feedback. The score engine or score reporter (see Figure 2) provides score reporting. The diagnostic feedback is based on a suite of programs (writing analysis tools) that identify the essay's discourse structure, recognize undesirable stylistic features, and evaluate and provide feedback on errors in grammar, usage, and mechanics. The writing analysis tools identify five main types of grammar, usage, and mechanics errors – agreement errors, verb formation errors, wrong word use, missing punctuation, and typographical errors. The approach to detecting violations of general English grammar is corpus based and statistical, and can be explained as follows. In case of corpus based systems the system is trained on a large corpus of edited text.

4. PROBLEMS WITH THE PRESENT SYSTEMS UNDER THE CONTEXT OF ENGLISH USED BY STUDENTS HAVING DIFFERENT MOTHER TONGUE

It has been found that most of the popular AEG systems are made to grade English essays and they are easy to follow. Systems developed in non-English languages are not popular and not understandable for everyone. Our research shows that while system grades an English essay it considers the influence of local languages as Error. Hence the following two sentences (used in India, Hyderabad) will show error once they are evaluated by machine as well as by a English spoken man. Ex 1– Prime Minister Manmohan Sing Garu has visited Osmania University. Ex 2 – Hyderabad says albida to monsoon. Where the 'Garu' is a pure Telugu word and used in English newspapers published form Andhra Pradesh. 'Albida' is an Urdu word and very much used in English newspapers coming out form Luckhnow and Hyderabad. Local languages influence same as English used in Chiana, Japan, Korea or France and no one considers them as Error. Where as from English point of view they are wrong. Of course a good number of such words got chance to be included in Oxford dictionary. Research shows present AEG systems illustrate 10 - 15% lower score while using Indian English text as input. In a broader form it can be mentioned that the English spoken and written by non-native English people (i.e. - Asians) are very much influenced by local languages. India is a multilingual country with as many as 22 scheduled languages and only 5% (plz. read five percent!!) of the population is able to understand English and that's also not like USA or UK people. Hence a goal is to develop a framework for an AEG system (for those countries where English is not a common language), which can be used for correcting essays influenced by local languages, and also to teach how to write better English Essays. In this chapter I propose a standard framework to develop any Automated Essay Grading System under such context. This model can be executed or s/w can be build as per the requirement as for example in which part of world the system is going to be used. Because while writing English the students of Japan are not influence by Hindi or Tamil, it is Japanese. Hence a single system will not be able to solve the problem. But this framework can be used as a benchmark to develop the other AEG systems. The framework follows IEEE Std. 1471 –2000, which is about "IEEE Recommended Practice for Architectural Description of Software Intensive Systems".

5. PROPOSED FRAMEWORK

Under the above circumstances a need of a specialized AEG system was felt very much which can be used for engineering education and we can make it web based also. The system will have the capability of identifying the local languages (Ex - Indian) present in the submitted essay and it will also find out how much effect is there for these words. It will also help the students to resubmit the essay with corrections where the students will be asked to re-enter the similar words for those local languages. Their essays will be graded as it is they have entered the equivalent English words by their own. For the instructors or teachers it will also give a proper score-card mentioning that still how much the essay is influenced by the local languages and the no of local words present, number of times corrections made by students (i.e. they can be given two or three chances to enter equivalent English words for those local words (i.e. albida = good bye). This above-mentioned action is a part of the scoring engine. These functionalities are added as a new functional module in the scoring engine or score reporter.

The feedback module is also supported with a 'local language' engine which helps the students providing proper feedback and development notes along with English Grammatical mistakes improvement, fed back on use of too many weak or common words etc. This engine will be very much useful in the learning stage. At the very beginning this engine will identify the local languages present in the written essay. Then it will give a chance to the students to overcome this problem by providing equivalent English words by their own. Then it will show them the projected score with number of general (English) errors and presence of number of local languages and what are those. For the remaining local words in the essay the system will then suggest equivalent English words with similar English words of those. Now the students get a chance to substitute remaining local words, phrases with suggested English words.

Figure 3. Proposed framework of the AEG system with local language engines

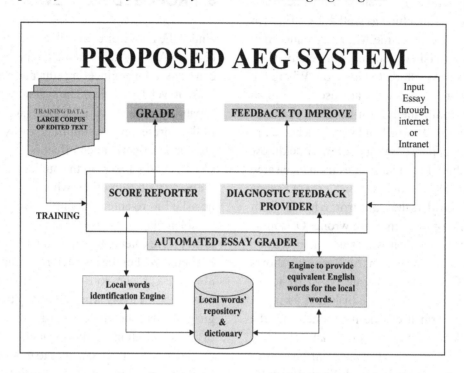

After submission they get the final projected score. Hence these engines help the students to learn better English. To make these engines effective the system I strained with a good number of local words, which are very much used in normal English (spoken English, news paper English). To make a proper collection of local words the local English news papers are used as a source. As for example – to make the engine working in Andhrapa Pradesh it is trained on collection of local words used in the news papers like Deccan Chronicle, Hindu (AP edition), Times of India (AP edition) etc., collected over last couple of years. It is found that this specific region's English is influenced by Telugu and Hyderabadi Hindi (a good mixing of Hindi and urdu).

6. CONCLUSION

In his paper 'Region Effects in AEG & human discrepancies of TOEFL score' Attali (2005) mentioned Asian Students show higher organization scores and poor grammar, usage and mechanics scores compare to other students. More over local languages influence them. Serious work in the area of AEG can bring significant changes in this direction and also can give a new shape to Indian NLP & Machine Learning research work.

Future plans - In near future the following things will be taken into consideration so that some solutions can be given as - Solution for machine translated essays (how to recognize them?), Capturing the mental status of the student writing essay (psychometric models will be considered). Detection of Anomalous Essays

REFERENCES

Bloom, B. S. (1956). *Taxonomy of educational objectives: The classification of educational goals. Handbook I, Cognitive domain.* New York: Longmans, Green.

Burstein, J., Kukich, K., Wolff, S., Chi, L., & Chodorow, M. (1998). Enriching automated essay scoring using discourse marking. In *Proceedings of the Workshop on Discourse Relations and Discourse Marking, Annual Meeting of the Association of Computational Linguistics, Montreal, Canada.*

Burstein, J., Leacock, C., & Swartz, R. (2001). Automated evaluation of essay and short answers. In M. Danson (Ed.), *Proceedings of the Sixth International Computer Assisted Assessment Conference, Loughborough University, Loughborough, UK.*

Christie, J. R. (1999). Automated essay marking-for both style and content. In M. Danson (Ed.), *Proceedings of the Third Annual Computer Assisted Assessment Conference, Loughborough University, Loughborough, UK.*

Christie, J. R. (2003). Email communication with author. 14th April.

Cucchiarelli, A., Faggioli, E., & Velardi, P. (2000). Will very large corpora play for semantic disambiguation the role that massive computing power is playing for other AI-hard problems? *2nd Conference on Language Resources and Evaluation (LREC), Athens, Greece.*

de Oliveira, P. C. F., Ahmad, K., & Gillam, L. (2002). A financial news summarization system based on lexical cohesion. In *Proceedings of the International Conference on Terminology and Knowledge Engineering, Nancy, France.* Breck, E.J. et al. (2000). How to Evaluate Your Question Answering System Every Day… and Still Get Real Work Done, In *Proc. LREC-2000, Linguistic Resources in Education Conf., Athens, Greece.*

Deerwester, S. C., Dumais, S. T., Landauer, T. K., Furnas, G. W., & Harshman, R. A. (1990). Indexing by latent semantic analysis. *Journal of the American Society for Information Science American Society for Information Science, 41*(6), 391–407. doi:10.1002/(SICI)1097-4571(199009)41:6<391::AID-ASI1>3.0.CO;2-9

Ghosh, S., & Fatima, S.S. (2007a). *Use of local languages in Indian portals. CSI Communication,* 4-12.

Ghosh, S., & Fatima, S. S. (2007b). *Retrieval of XML data to support NLP applications.* Paper presented at ICAI'07- The 2007 International Conference on Artificial Intelligence Monte Carlo Resort, Las Vegas, Nevada, USA, June 25-28, 2007.

Ghosh, S., & Fatima, S. S. (2007c). *A Web Based English to Bengali Text Converter.* Paper presented at the 3rd Indian International Conference on Artificial Intelligence (IICAI-07), Pune, India, December 17-19, 2007.

Grondlund, N. E. (1985). *Measurement and evaluation in teaching.* New York: Macmillan.

Hearst, M. (2000). The debate on automated essay grading. *IEEE Intelligent Systems, 15*(5), 22–37. doi:10.1109/5254.889104

Honan, W. (1999, January 27). High tech comes to the classroom: Machines that grade essay. *New York Times.*

Jerrams-Smith, J., Soh, V., & Callear, D. (2001). Bridging gaps in computerized assessment of texts. In *Proceedings of the International Conference on Advanced Learning Technologies* (pp. 139-140). IEEE.

Laham, D., & Foltz, P. W. (2000). The intelligent essay assessor . In Landauer, T. K. (Ed.), *IEEE Intelligent Systems.*

Landauer, T. K., Foltz, P. W., & Laham, D. (1998). An introduction to latent semantic analysis. *Discourse Processes, 25*. Retrieved from http://lsa.colorado.edu/ papers/dp1.LSAintro.pdf. doi:10.1080/01638539809545028

Larkey, L. S. (1998). Automatic essay grading using text categorization techniques. In *Proceedings of the 21st ACM/SIGIR (SIGIR-98)* (pp. 90-96). ACM.

Larkey, L. S. (2003). Email communication with author. 15th April. Mason, O. & Grove-Stephenson, I. (2002). Automated free text marking with paperless school. In M. Danson (Ed.), *Proceedings of the Sixth International Computer Assisted Assessment Conference, Loughborough University, Loughborough, UK.*

Ming, P. Y., Mikhailov, A. A., & Kuan, T. L. (2000). Intelligent essay marking system. In C. Cheers (Ed.), *Learners Together, Feb. 2000, NgeeANN Polytechnic, Singapore*. http://ipdweb.np.edu.sg/lt/feb00/ intelligent_essay_marking.pdf

Mitchell, T., Russel, T., Broomhead, P., & Aldridge, N. (2002). Towards robust computerized marking of free-text responses. In M. Danson (Ed.), *Proceedings of the Sixth International Computer Assisted Assessment Conference, Loughborough University, Loughborough, UK.*

Page, E. B. (1968). The Use of the Computer in Analyzing Student Essays. *International Review of Education, 14*, 210–225. doi:10.1007/BF01419938

Page, E. B. (1994). New computer grading of student prose, using modern concepts and software. *Journal of Experimental Education, 62*(2), 127–142.

Page, E. B. (1996). *Grading essay by computer: Why the controversy?* Handout for NCME Invited Symposium.

Palmer, J., Williams, R., & Dreher, H. (2002). Automated essay grading system applied to a first year university subject- How can we do it better. In *Proceedings of the Informing Science and IT Education (InSITE) Conference, Cork, Ireland* (pp. 1221-1229).

Rudner, L. M., & Liang, T. (2002). Automated essay scoring using Bayes' Theorem. *The Journal of Technology. Learning and Assessment, 1*(2), 3–21.

Thompson, C. (2001). Can computers understand the meaning of words? Maybe, in the new of latent semantic analysis. *ROB Magazine*. Retrieved from http://www.vector7.com/client_sites/ROB_preview/html /thompson.html

Valenti, S., Cucchiarelli, A., & Panti, M. (2000). Web based assessment of student learning. In Aggarwal, A. (Ed.), *Web-based Learning & Teaching Technologies, Opportunities and Challenges* (pp. 175–197). Idea Group Publishing.

Valenti, S., Cucchiarelli, A., & Panti, M. (2002). Computer based assessment systems evaluation via the ISO9126 quality model. *Journal of Information Technology Education, 1*(3), 157–175.

Chapter 6
Future Challenges of Mobile Learning in Web-Based Instruction

Rochelle Jones
Lockheed Martin Corporation, USA

Chandre Butler
University of Central Florida, USA

Pamela McCauley-Bush
University of Central Florida, USA

ABSTRACT

Mobile learning is becoming an extension of distance learning, providing a channel for students to learn, communicate, and access educational material outside of the traditional classroom environment. Because students are becoming more digitally mobile, understanding how mobile devices can be integrated into existing learning environments is advantageous however, the lack of social cues between professors and students may be an issue. Understanding metrics of usability that address the concern of student connectedness as well as defining and measuring human engagement in mobile learning students is needed to promote the use of mobile devices in educational environments.

INTRODUCTION

Distance education is the fastest growing educational modality because of the advances information technology has made over the past 25 years. Students, through technical tools and mobile devices such as Personal Digital Assistants (PDAs) and smartphones they use on the job, are becoming more digitally literate and mobile, making the ability to access class work on the go a necessity. Because many students already own mobile devices, understanding how they can be integrated into learning environments is advantageous however, the lack of social cues mobile devices and computer mediated communications may introduce remain a concern. Using mobile devices for educational purposes opens the lines of communication between professors and classmates without the need of being at a designated location, at a designated time. Understanding if

DOI: 10.4018/978-1-61520-659-9.ch006

these devices help facilitate some of the challenges of distance education such as learners having a sense of connectedness is valuable.

The term "connectedness" is used across various knowledge domains and is sometimes synonymous with the term "engagement." Interactive multimedia learning literature defines human engagement in terms of physiological arousal, mainly measures of heart rate, respiration, brain activation, and eye movement. The definition neglects the psychosocial aspect of the user experience because of the inherent complexity of interpreting both quantitative and qualitative data simultaneously within a common contextual frame of reference. Researchers are addressing this gap, where physiological measures are merged with psychosocial measures to define human engagement from a more inclusive perspective.

The purpose of this chapter is to introduce contemporary topics of applied mobile learning in distance education and the viability of mobile learning (m-learning) as an effective instructional approach. Metrics of usability that address the concern of user connectedness as well as defining and measuring human engagement in m-learning students is examined.

Application and Benefits of Mobile Learning

Mobile learning is learning that uses wireless, portable, mobile computing, and communication devices (namely smartphones, pocket personal computers (PCs), tablet PCs, PDAs, mobile phones, and iPods) to deliver content and learning support (Brown, 2005). Advances in mobile communication technologies including Wi-Fi networks, Third Generation (3G), and Worldwide Interoperability for Microwave Access (WiMAX) are enabling students to access class material without subjecting them to a physical classroom or in front of a computer at a set point in time. Despite the various functionality and capabilities of mobile devices, mobile learning is becoming

an extension of distance learning, providing a channel for students to learn, communicate, and access educational material outside the traditional classroom environment.

Thornton and Houser (2005) conducted a study of Japanese students learning English as a second language. Students were sent text and video lessons that defined new terms, story episodes that used target words, and English idioms on cell phones. The study resulted in students positively favoring the messages in their educational effectiveness. The students stated they felt comfortable reading the text and viewing the videos on the devices' small screens.

MP3 players and iPods are used to download class lectures and tutorial notes. Students can listen and view audio and video files (podcasts) to recap and review lectures (when preparing for tests or recovering from missed lectures) and take notes of class sessions (Guertin et al., 2007). Companies are taking advantage of podcasting as well by sending employees training material, company compliance videos, and videos introducing a new tool to be used on the job. This implementation benefits the company and employees because it cuts costs of employees in the field having to travel to a central training location, and it cuts the use of "impersonal manuals to read and digest" during training sessions.

Incorporating mobile devices into existing learning environments can benefit the university community. First, mobile devices are cheaper than desktop and laptop computers, thus making them more accessible to students. According to Informa Telecoms & Media (2008), the number of mobile devices active in the world is more than 60 percent of the world's total population. Second, the use of mobile devices for learning can help maximize the time students allocate for studying. Students can access course material whenever they have a free moment; during break hours or in transit. It allows students to customize and maximize their studying time around their busy schedules.

Mobile Device: Hardware

Mobile devices in m-learning environments have followed similar developmental and design trends of other electronic devices such as consumer electronics and personal computing. As microelectronics technology continues to mature, the quest for the ever smaller, faster, more power efficient device is still the focus of research and developmental efforts. Sources for hardware information such as technology journals and trade shows (i.e. the International Consumer Electronics Show [CES], produced by the Consumer Electronics Association [CEA], and IEEE Technical Committees), address the future of the technology not just the development of consumer electronic products. The following sections discuss hardware devices that have capabilities conducive for m-learning, and the direction for future technological development.

Handhelds

Handheld devices are the central devices of m-learning applications. Their functional mobility and wireless capabilities have evolved as advances in microelectronics technology have persisted. The handheld may be defined by its physical size. Typically the device may be held by one or both hands without excessive effort. The following sections discuss various types of handheld devices.

Personal Digital Assistant (PDA) and Smartphones

The Personal Digital Assistant (PDA) is a handheld device that has comparable functionality with a laptop, notebook, or desktop computer. The PDA is especially designed to increase the organizational capacity of the individual in a mobile environment untethered from the desktop. Screen sizes usually do not exceed 4 inches (100 millimeters) diagonally. In comparison, smartphones share much of the functionality of PDAs with the added function of wireless cellular communication. Names of

the top most popular (defined by the number of web inquires and searches for specific handheld devices) PDAs and smartphones include:

1. Palm Pre (released 2009)
2. Nokia E63 (released 2009)
3. RIM BlackBerry Curve 8330 (released 2008)
4. Apple iPhone 3G (released 2008)
5. Nokia 5800 Xpress Music (released 2008)
6. HTC Touch Pro2 (released 2009)
7. Samsung Alias 2 (released 2009)
8. HTC Touch Diamond2 (released 2009)
9. T-Mobile G1 (released 2008)
10. Nokia E71 (released 2008)

Handheld Portable Gaming Devices

Handheld portable gaming devices have evolved from purely entertainment-based user devices to Wi-Fi enabled m-learning capable hardware. The appeal as potential m-learning devices is hinged on the fact that devices such as the Nintendo DS family have supported software that targets learning and training among adult users. Another example of the growth of the handheld portable gaming device is Nokia's N-Gage QD, which includes an option of using the combination device (gaming and cellular phone) software application that can be downloaded to Nokia N-Gage ready devices. Finally, most noted for bringing a more true multimedia experience to portable gaming devices is the Sony PSP, another gaming device that can be used in a m-learning environment.

Wireless Portable Reading Devices (eReader Digital Book)

Wireless Portable Reading devices, also known as eReaders, occupy a small niche in m-learning environments. Much like a printed book or newspaper, eReaders are designed to make print electronic, thus minimizing the need for printed versions of text and informational documents. The focus of eReader development is the display functionality; the goal is to develop eInk technology that rivals the depth and extending viewing angle characteristics of printed text. Released products include Amazon's Kindle DX, the BEBOOK, iRex iLiad, and the Sony Reader PRS-700BC.

Portable Audio-Video Media Players
The portable media player, most commonly referred to as the mp3 player sometimes with video capability, has revolutionized the way audio and video is created and distributed. A device usually consists of a file storage mechanism (memory), power solution (rechargeable battery), a method to download files or synch with desktops and laptops (USB connection or removable memory card), and an audio output (headphone port). The portable media player's physical size may be as small as a postage stamp, to the size of a laptop. More advanced versions have video displays which enable encoded video file playback.

The Apple iPod has become the ubiquitous portable media player of choice, specifically among younger users. Apple's marketing strategy has been extremely successful, as the portable media player market is one of Apple's highest revenue generators. Other manufactures have tried to claim market share as they produce their versions of the portable media player. A few examples include Microsoft Zune, Creative Zen, and Samsung YP-P2.

Ultra-Mobile PC (UMPC/Netbook)
Ultra-Mobile PCs are the newest developments in personal computing. UMPCs retain the same functionality of laptops and notebooks, while shrinking the technology into a smaller package. Advances in microelectronics design and manufacturing have enhanced this smaller portable PC. The common display size of the UMPC is about 8 inches (203.2 millimeters) diagonally compared to 15.4 inches (391 millimeters) for the typical laptop. The future of the UMPC may depend on a single concern; battery life. Unfortunately existing battery technology is limited by the physical size of the cell; typically, the smaller the cell, the smaller the storage capacity. Besides battery life,

the screen size may pose ergonomic challenges for designers and users. Types of UMPCs include Acer ASPIRE One, ASUS EeePC, HP Mini-Note, Sony VAIO P500, Samsung Q1EX, and Viliv X70EX.

Mobile Device: Software

The advent of new mobile device hardware technology is the impetus for innovative software that optimizes the functionality of mobile devices in m-learning environments. As with personal computing, operating systems (OS) on mobile handheld devices have evolved. The most noticeable difference between the desktop PC market and the mobile computing market is a movement from proprietary based software to open source software design spearheaded by organizations such as the Symbian Foundation, the Linux Mobile (LiMo) Foundation, the Open Handset Alliance (OHA), and Open Source in Mobile (OSIM). It is projected that open source software will spur innovation, facilitate improved device performance, enable more user-friendly software design, and increase the level of positive user experiences.

Mobile Device Operating Systems

The Symbian OS, developed by Symbian Software Ltd., is the most used and developed mobile OS with licensed copies in more than 250 million shipped units. Though portions of the software code of the Symbian platform are open to the public, by 2010 it is expected that the entire Symbian platform will be open source and free to the public. This type of open source OS will enable continuity and increased compatibility of developed software in m-learning environments. Such open source initiatives are managed by the Symbian Foundation. The Symbian Foundation, which became operational in early 2009, is the premier lead organization in the move toward open source mobile OS implementation. The foundation is composed of the following mobile device industry leaders:

1. AT&T
2. LG
3. Motorola
4. Nokia
5. Samsung
6. Sony Ericsson

Other members exist, all dedicated to the mission of ensuring free open source mobile software to the public that enriches the user's mobile experience.

Linux, a software platform developed by founder Linus Torvalds, is one of the few platforms that was developed initially as an open source initiative and continues to remain open. Concerning mobile device support, the LiMo Foundation, founded in 2007, was established to spur innovation and the use of open source software by mobile device manufacturers. The LiMo Foundation's founding and core members include the following mobile device industry leaders:

1. NEC
2. NTT DOCOMO
3. Orange
4. Panasonic
5. Samsung
6. Vodafone
7. Access
8. Aplix
9. Azingo
10. Casio-Hitachi
11. LG Electronics
12. Myriad
13. SK Telecom
14. Telefonica
15. Verizon
16. Wind River

Other associate members include Sony Ericsson and Motorola. More than 30 different LiMo based mobile devices have been developed and are available to consumers. As the foundation's objectives are met, mobile device OS software development will benefit.

The Android OS developed by the Open Handset Alliance (OHA), has experienced exponential growth in usage since its inception in 2007. It is a true open source software OS, based on the Linux platform. Android differs from other open source initiatives in that it offers a packaged software solution that not only includes the OS but middleware and third-party applications that have equal access to a mobile device's resources.

The OHA consists of mobile device manufacturers, mobile operators, software developers, and semiconductor companies. The integrated cross-industry approach of the OHA is an innovative systems solution for software development. The OHA has a membership of more than 47 members including some of the following handset manufacturers:

1. Acer Inc.
2. ASUSTeK Computer Inc.
3. Garmin International Inc.
4. HTC Corporation
5. Huawei Technologies
6. LG Electronics Inc.
7. Motorola Inc.
8. Samsung Electronics
9. Sony Ericsson
10. Toshiba Corporation

The most notable handheld to date that uses the Android software platform is the Google G1 manufactured by HTC Corporation for T-Mobile.

Java Micro Edition (ME) from Sun Microsystems is an open source software solution that is used in billions of mobile and embedded applications. Sun Microsystems is committed to providing development and support activities by ensuring royalty free open source software to the larger mobile and embedded developer community. A major effort of Sun Microsystems is ensuring compatibility with other open source software platforms such as Linux.

Other proprietary mobile device OS types include:

1. BlackBerry RIM
2. Garnet OS (Palm OS)
3. Apple iPhone OS (OS X)
4. Microsoft Windows Mobile

The limitations of internally developed (proprietary) software have become evident. If developers cannot produce relevant, innovative, and abundant software applications that use the proprietary software platform, sales could potentially suffer. To avoid this, it is possible that proprietary operating systems developers will follow Symbian and Sun Microsystems' lead, by also providing open source software. The benefit of software applications support and development from a larger external global community should not be overlooked. The future challenges of mobile operating systems development for m-learning environments are being addressed by established open source software initiatives.

Specialized Third-Party Applications for Mobile Device Software

The open source movement is not only relegated to operating systems software but also to third-party applications development. Most m-learning third-party applications are scaled-down mini versions of desktop designed software. The list of third-party applications is extensive however, this section will discuss general categories of m-learning software and the challenges encountered.

Mobile Learning Software

Mobile Learning Software is software that enables m-learning activities and is especially designed for reading, editing, and publishing content on mobile devices. Acrobat Reader for Palm OS and Pocket PCs, Pocket Exam, eReader, Microsoft Reader for Pocket PCs, and Symex mTeacher are examples of third-party m-learning software. A function of some of the mentioned software includes enabling the creation and presentation of tests, quizzes, and exams on mobile devices. Third-party software may also enable a user to view content with enriched interactivity. Other examples include feedback and instruction assessment functions, which are made possible by administrative and management type third-party applications designed for m-learning.

Mobile Media Player Software

Mobile Media Player Software allows users to download, listen to, and view content. The software is usually limited to one-way communication. Apple has acquired a sizeable share of the media player market with its iPod brand. Podcasts, Apple's version of broadcast content, have become a major component of m-learning. Third-party applications support includes iPodSoft and Apple's iTunes. Preventing full implementation of media players in m-learning environments is the higher cost of the device and the apparent advantage of access the tech-savvy have over the tech-illiterate (Corbeil & Valdes-Corbeil, 2007). Other mobile media player software include RealOne Player for Mobile Devices, Flash Player for Pocket PC, and Windows Media Player for Pocket PC, which all allow multimedia content to be viewed on mobile devices. Most mobile media players are in their nascent state and have limited functionality. Many do not enable rich multimedia interactivity with users. Designers are cognizant of these issues and are working to improve future versions of the software.

Communication and Social Software

Rich Site Summary (RSS) feeds, bulletins, updates, subscriptions, and shared content can be employed to keep users aware of dynamically changing instructional content between instructors, facilitators, and peers. Communication type software allows users to directly communicate in real-time. Popular communication software "messaging tools" include America Online Instant

Messenger (AIM), Google Talk, Jabber, Mabber, MSN Instant Messenger (Windows Live Messenger), and Yahoo Instant Messenger. Social utilities with instant messaging capabilities and chat functions such as Facebook and Myspace include other options that keep users connected to social networks. Staying connected to peers in a m-learning environment is an important factor of the social aspect of learning. Mobile compliant versions of these popular social utilities are available.

Mobile Web Browsers

The functionality of mobile web browsers is similar to that of full-fledged web browsers for desktop systems. Mobile web browsers however, are highly specialized for handheld devices and are limited to core navigation and display functionality. The more popular mobile web browsers include Android (Google), Internet Explore for Mobile (Microsoft), Minimo (Mozilla), Opera Mini, Safari (Apple), Skyfire, and Thunderhawk (Bitstream). Ensuring fast and error-free downloads are two future hurdles developers are addressing. Increasing relevant mobile web browsing functionality while providing user-friendly and clear navigation on small screens is also a concern.

Human Engagement as a Measurement of Usability

Defining Human Engagement

Traditional human engagement definitions are qualitative and subjective in nature. Measurement of human engagement was more commonly relegated to psychometric scaling techniques such as Likert Scaling. Other types of qualitative assessments of engagement, specifically engagement in learning, include works by the National Center for School Engagement (NCSE) and the Institute for Research and Reform in Education (IRRE). Human engagement is perceived as an important factor with researchers noticing a high positive correlation between increased levels of learning and achievement and increased levels of human engagement (Klem & Connell, 2004). It is believed that such discoveries in educational systems research are applicable to m-learning environments. The qualitative aspect of human engagement may facilitate undue bias during analysis; quantification would help mitigate this bias.

Modern methodological frameworks concerning human engagement are being developed that include quantified physiological response measures of arousal. These include the observation of cardiopulmonary, ocular, cardiac, and brain activities. Besides physiological response arousal measurements, approaches are being developed that combine qualitative psychological data and quantitative physiological data into a common framework (Butler, 2009).

Measuring Human Engagement

Psychological survey and questionnaire instruments with Likert Scaling methods are the most common approach to identifying human emotional, behavioral, and attitudinal states. Topics such as motivation, cognition, and perception may also be identified by intentional user response. The distinction of the psychological and social aspects of m-learning suggests that learning can be both an individual experience and a group experience. What is learned and what the learned content means is also significantly shaped by social norms. The future of psychological research, specifically cognition, will include attempts to integrate concepts such as motivation and perception collectively in cognitive frameworks.

Physiological response and arousal is a quantitative approach to measuring the degree of psychological response without the conscious intention of the student. One advantage to this approach is that bias, such as human error, may be mitigated. Table 1 shows common examples of physiological response measures.

Table 1. Physiological response measurements

Physiological Response Measurement	Unit of Measure	Observed Anatomical Entity
Electroencephalography *(EEG)*	Frequency (Hz) Amplitude	Brain/Electrical
Functional Magnetic Resonance Imaging *(fMRI)*	mL O_2/min Local SpO_2	Brain/Body
Functional Near Infrared Resonance Imaging *(fNIR)*	mL O_2/min SpO_2	Brain/Body/Blood
Blood Volume Pulse	Beats Per Minute (BPM) Pulse Waveform	Cardiac Activity
Blood Tissue Oxygen Saturation & Pulse Oximetry (Near Infrared Spectrophotometrics)	Beats Per Minute (BPM) SpO_2	Cardiac Activity Cardiopulmonary
Auscultation Technique	Frequency (Hz) Beats Per Minute (BPM)	Cardiac Activity
Arterial Pulse	Beats Per Minute (BPM)	Cardiac Activity
Arterial Blood Pressure	Millimeters of mercury (mmHg)	Cardiac Activity
Electrocardiogram *(ECG/EKG)*	Beats Per Minute (BPM) Amplitude Frequency (Hz)	Cardiac/Electrical Activity
Respiratory Sinus Arrhythmia *(RSA)*	Variation in Heart Rate during Respiration	Cardiopulmonary
Tidal Volume Estimation: Inductive Plethysmography	Functional Residual Capacity (FRC) Milliliters mL	Cardiopulmonary
Strain Gauge Analysis	Dimensional change in Length	Pulmonary
Carbon Dioxide (CO_2) Concentration Analysis	Parts Per Million (PPM)	Cardiopulmonary
Eye Movement Tracking	Saccade Dwell Time	Eye/Movement
Galvanic Skin Response *(GSR)*	Microsiemens Microhoms	Skin/Gross Body Electrical Conductivity-Resistance
Temperature	Fahrenheit Celsius	Gross Body or Local

The reason for the validity of physiological response and arousal as indicators of emotional state and engagement is that there is an identified connection between affective state (emotional state) and the sympathetic nervous system. Hence, emotion and engagement may trigger varying physiological responses in different people. According to McQuiggan, Lee and Lester (2006), "Because physiological responses are directly triggered by changes in affective state, biofeedback data such as heart rate and galvanic skin response can be used to infer affective state changes" (p. 60). However, it is still unclear which details and aspects of the emotional state are reflected by which particular physiological responses, although there have been advances in the temporal and spatial resolution of physiological response measurement devices. Regarding the current state of research concerning human physiological response and arousal, augmenting quantitative physiological response data with qualitative psychological data appears to be an appropriate approach. Overcoming research obstacles encountered in physiological response and arousal methodologies will help ensure increased human engagement in m-learning applications.

Challenges of Mobile Learning

Usability most often refers to the interaction between human beings and objects, but in a broader sense, it should include factors outside the immediate interaction that influence the quality of the human interactive experience. For example, psychological and social factors may influence the perceived usability and utility of a device. There are a number of challenges that may hamper the progression and incorporation of mobile devices in educational environments. First, the cognitive and physical ergonomic influence on m-learning may affect the overall performance and satisfaction levels of students. Jones (2009) conducted a study to examine the physical, physiological, and perceived subjective workload issues participants experienced while using a mobile device to read course material. The study resulted in mobile device users experiencing more physical discomfort such as eye fatigue, shoulder and neck pain, and lower back pain. Reasons for this stemmed from physical features of the device, sitting posture, and the duration of reading the material from the small screen. In addition, participants that used a mobile device to read course material experienced higher levels of perceived mental workload. Reasons for this included limitations of the mobile device preventing note taking and scrolling, and an overall preference for desktop computers for reading material. Despite participants experiencing higher levels of subjective mental workload and physical discomfort, 58 participants (of 84) indicated they would still use a mobile device for educational purposes. It is because of this desire from potential students that researchers need to

continue to evaluate the ergonomic issues that may negatively affect the performance and satisfaction levels of incorporating mobile devices in educational environments in hopes of making them conducive for learning.

Kukulska-Hulme (2002) conducted a study that examined students enrolled in an online course using a PDA for reading and note-taking to identify usability issues. Usability issues identified included:

1. Difficulty reading the text on the screen
2. Difficulty scanning the text when the font was enlarged
3. Difficulty with data entry
4. Eye ache and visual disturbance when looking at the screen

The usability issues identified key factors that need to be improved to make mobile devices effective, efficient, and satisfying to students using it for educational purposes. Evaluating ergonomic and usability issues caused when using mobile devices can aid in determining the appropriate method (i.e. content, representation of material, physical attributes of mobile devices, software applications, and physical environment to be used) of incorporating the devices in educational environments.

Despite its role in the progression and future of distance education, the lack of social cues when using mobile devices and computer mediated communication remains a concern. Crosby et al. (2001) conducted a study that analyzed physiological data feedback of distance education students. Since professors cannot view students' nonverbal cues (frustration and confusion over a topic), researchers looked for a way to communicate those emotions to instructors. Physiological data such as pulse, galvanic skin response, and general somatic activity were correlated with the emotion of the student through an Emotion Mouse. The study demonstrated that the physiological information provided insight about changes in users' emotional and subjective states while engaged in cognitive tasks. With m-learning further opening the lines of communication between professors and other classmates, understanding if it helps facilitate some of the challenges of distance education, such as learners having a sense of connectedness, needs to be explored. Fozdar and Kumar (2007) surveyed students to determine their perceptions on the effectiveness and preferences of m-learning. The survey results indicated m-learning could be used to improve student retention and enhance teaching and learning because students and teachers have access to each other and course material anytime, anywhere. An advantage of incorporating mobile devices in educational environments is that it provides students an opportunity to have a sense of community and be connected to their classmates and professors. However, the feeling of being "too connected" becomes a concern of students. Distinguishing the appropriate frequency of contact is important to keep students engaged however, not overwhelmed. Invading privacy or intruding on students' downtime becomes a concern, thus emphasizing the need to find the balance.

Future M-Learning Frameworks

Mobile Learning and Open Learning Environments

Engaging new users of m-learning technology in developing countries has become a research point of interest. Future research of mobile learning environments implementation is needed to minimize factors that may impede successful m-learning deployments. Mobile technology use has the potential to spur mass education initiatives in developing countries. One reason is that the penetration of Internet enabled mobile device usage significantly out paces Internet enabled desktop usage in many of these countries (Borau, Ullrich, & Kroop, 2009). Though the mobile technology aspect shows a positive trend, pedagogy and instructional delivery methods are not as modern.

The pedagogy in many developed countries is still teacher-centered however, there is a concerted move toward more interactive student-centered instruction; a goal of new projects such as Responsive Open Learning Environments (ROLE) (Borau et al., 2009). The ROLE consortium consists of more than a dozen respected research groups and companies. Major funding for this initiative is provided by the European Commission. This study of responsive open learning environments is composed of a systems view of learning environments, which also addresses m-learning technology.

Reduced Learning Cycles (Just in Time [JIT]) Knowledge

The rate at which knowledge may be obtained under suboptimal learning conditions (emergency and high stress situations) can be enhanced by on demand JIT customized student-centered and contextual instruction. Mobile devices have the potential to become a mirroring, metacognitive or guidance tool (Romero & Wareham 2009). Integrated intelligent agents that are conducive for learning environments and instruction material are possible because of the computing capability of modern and future mobile devices. An implemented JIT based learning model has the potential to bolster levels of user engagement in m-learning environments.

eReader Improvements

Display improvements and enhancements that include more than 16 gray levels of contrast, is a major challenge for eInk display (EPD) technology. eReader display technology is suitable for text only documents (Kao et al., 2009), though figures and images may be displayed with lower fidelity than the original non-converted images. This limitation may reduce reader satisfaction and may impede increased levels of m-learning student engagement. New research encompasses techniques that increase image quality post image file conversion and the potential for true color image reproduction.

Environmental Wireless Sensor Use

Enhancing the user experience and usability of m-learning device technology by using environmental sensors as proposed by Chang, Wang, and Liu (2009), provides a glimpse into the future of mobile learning. For example, gestural inputs would enable individual and group learning. Another type of sensor, Global Positioning System (GPS), could be used to enhance instruction. GPS sensors together with mobile devices could enhance student engagement by augmenting reality and placing the student partly in an immersive virtual learning environment. Other sensors such as thermal and force related sensors provide an unbounded platform for sensor use in m-learning environments. These various wireless sensors could become subcomponents of a larger connected wireless sensor network to enhance the user's experience. This wireless network could have vast scalability potential and could accommodate diverse instructional content.

User Perception in M-Learning

User perception is sometimes overlooked in mobile device design. Interactive Response Systems (IRS) are composed of handheld wireless devices that allow a user to answer questions posed by an instructor as instruction is administered. Implementation of this type of technology may have some challenges. Liu (2009) asserts that perceived usefulness of a newly introduced technology is the most important factor or challenge affecting successful technology use among students. This implies that instructors and administrators should focus on the usefulness of the proposed implemented technology and the benefit afforded by its usage. By doing so, satisfaction and human engagement increases and peer referral and recruit-

ment activities become more frequent. Concerted efforts to assure users of the positive aspects of newly implemented m-leaning technology should be addressed by system designers.

Wireless Networking Technology

Wireless networking technology provides a means of communication between instruction and student, allowing students to remain physically unbound to a fixed computing workstation. Standards such as IEEE 802.11 allow for the seamless interoperability among network designers and equipment manufacturers. Future challenges include ensuring higher bandwidth capacity, minimizing interference with other devices, and ensuring improved operational conditions.

CONCLUSION

Distance education and m-learning are rapidly growing and gaining support as viable educational modalities mainly because of the ubiquitous nature of wireless mobile devices. M-learning is further enabled by advances in wireless network technology including WiFi, 3G, and WiMAX networks. The application of wireless mobile devices in m-learning environments can be generalized into one overarching objective: to enable students to have access to class material without being subjected to a physical classroom or in front of a computer at a set point in time; which is also the main benefit of m-learning.

A key component of a m-learning environment is the software that mobile devices require. Traditionally, companies developed proprietary software and assumed all developmental costs and revenues. Open source software development allows free access for developers and users to software code for operating systems and third-party applications software. The benefits are lower shared developmental costs, increased creative solutions, and increased positive user experiences. Improved designs in specialized mobile learning software, mobile media player software, communication & social software, and mobile web browsers are expected from initiatives moving toward open source software. It is the goal that well-designed software will result in an engaged user.

Human engagement can be accessed from a psychosocial, physiological, or integrated approach. Questionnaires and survey instruments can provide useful qualitative data indicating a person's affective (emotional) state while physiological response measurements can provide insightful quantitative data. As with many devices human beings use, ergonomic complications and challenges may exist. Physical fatigue and eyestrain are common issues encountered when using handheld devices. Mental and cognitive issues can also arise.

With the advent of new mobile technology there are justified concerns and challenges. Being able to move users from an older paradigm of classroom-based instruction to m-learning and open educational environments is part scientific and part art. Future research will help in addressing issues such as social aspects, physical device design, wireless networking technology, and user psychological state. Although challenges do exist in m-learning and web-based instructional environments, they are not insurmountable. Researchers, scientists, and dedicated technologists are continuing to address concerns. If the trends remain true, m-learning and web-based instruction are poised to become the primary modes of instructional delivery in the near future.

REFERENCES

Borau, K., Ullrich, C., & Kroop, S. (2009). Mobile learning with Open Learning Environments at Shanghai Jiao Tong University, China. *Learning Technology publication of IEEE Computer Society Technical Committee on Learning Technology (TCLT)* . *Learning Technology Newsletter, 11*(1-2), 7–9.

Brown, T. (2005). Toward a model for m-learning in Africa. *International Journal on E-Learning, 4*(3), 299–315.

Butler, C. (2009). *The development of a human-centric fuzzy mathematical measure of human engagement in interactive multimedia systems and applications.* Unpublished doctoral dissertation, University of Central Florida.

Chang, B., Wang, H., & Lin, Y. (2009). Enhancement of mobile learning using wireless sensor network. *Learning Technology publication of IEEE Computer Society Technical Committee on Learning Technology (TCLT)* . *Learning Technology Newsletter, 11*(1-2), 22–25.

Corbeil, J., & Valdes-Corbeil, M. (2007). *Are You Ready for Mobile Learning? Frequent use of mobile devices does not mean that students or instructors are ready for mobile learning and teaching.* Retrieved July 17, 2009, from http://www.educause.edu/EDUCAUSE+Quarterly/EDUCAUSEQuarterlyMagazineVolum/AreYouReadyforMobileLearning/157455

Crosby, M., Auernheimer, B., Aschwanden, C., & Ikehara, C. (2001). Physiological data feedback for application in distance education. In *Proceedings of the 2001 Workshop on Perceptive User Interfaces.* Orlando, Florida.

Fozdar, B., & Kumar, L. (2007). Mobile learning and student retention. *The International Review of Research in Open and Distance Learning, 8*(2), from http://www.irrodl.org/index.php/irrodl

Guertin, L., Bodek, M., Zappe, S., & Kim, H. (2007). Questioning the student use of and desire for lecture podcasts. *Journal of Online Learning and Teaching, 3*(2), 133–141.

Informa Telecoms & Media. (2008). 3G Americas: Number of mobile devices passes 4 billion mark. Retrieved June 15, 2009 from http://www.fiercewireless.com/story/3g-americas-number-mobile-devices-passes-4-billion-mark/2008-12-23?utm_medium=rss&utm_source=rss&cmp-id=OTC-RSS-FW0

Jones, R. (2009). *Physical ergonomic and mental workload factors of mobile learning affecting performance of adult distance learners: Student perspective.* Doctoral dissertation, University of Central Florida, 2009.

Kao, W., Ye, J., Chu, M., & Su, C. (2009). Image quality improvement for electrophoretic displays by combining contrast enhancement and halftoning techniques. *IEEE Transactions on Consumer Electronics, 55*(1), 15–19. doi:10.1109/TCE.2009.4814408

Klem, A., & Connell, J. (2004). Relationships matter: Linking teacher support to student engagement and achievement. *The Journal of School Health, 74*(7), 262–273. doi:10.1111/j.1746-1561.2004.tb08283.x

Kukulska-Hulme, A. (2002). Cognitive ergonomic and affective aspects of PDA use for learning. In

Liu, Y. (2009). Exploring Students' Perceptions toward Using Interactive Response System. *Learning Technology publication of IEEE Computer Society Technical Committee on Learning Technology (TCLT). Learning Technology Newsletter, 11*(1-2), 29–32.

McQuiggan, S., Lee, S., & Lester, J. (2006). Predicting User Physiological Response for Interactive Environments: An Inductive Approach. In *Proceedings of the Second Conference on Artificial Intelligence and Interactive Entertainment.* Marina del Rey, California.

Proceedings of 2002 European Workshop on Mobile and Contextual Learning. University of Birmingham.

Romero, M., & Wareham, J. (2009). Just-in-time mobile learning model based on context awareness information. *Learning Technology publication of IEEE Computer Society Technical Committee on Learning Technology (TCLT). Learning Technology Newsletter, 11*(1-2), 4–6.

Thornton, P., & Houser, C. (2005). Using mobile phones in English education in Japan. *Journal of Computer Assisted Learning, 21*, 217–228. doi:10.1111/j.1365-2729.2005.00129.x

ADDITIONAL READING

AndroidU. R. L.http://www.android.com/

Apple iPod URL: http://www.apple.com/ipod-classic/

Apple iTunes U URL: http://www.apple.com/education/teachers-professors/mobile-learning.html

Apple SafariU. R. L.http://www.apple.com/iphone/iphone-3g-s/safari.html

BeBook URL. http://mybebook.com/

Brighthand.com URL: http://www.brighthand.com/best_pdas/default.asp

Creative ZenU. R. L.http://us.creative.com/

Google AndroidU. R. L.http://code.google.com/android/

iRex iLiad URL: http://www.irextechnologies. com/products/iliad

Java Mobile & Embedded Community URL: http://community.java.net/mobileandembedded/

Learning LightU. R. L.http://www.e-learningcentre.co.uk/eclipse/vendors/pdatools.htm

LiMo Foundation URL. http://www.limofoundation.org/

Linux Foundation URL. http://www.linuxfoundation.org/en/Mobile_Linux

Microsoft ZuneU. R. L.http://social.zune.net/

NintendoD. S. URL: http://www.nintendo.com/ds

Open Handset Alliance (OHS) URL. http://www. openhandsetalliance.com/

Open Source in Mobile (OSiM) URL. http://event. osimworld.com/

Opera MobileU. R. L.http://www.opera.com/ mobile/

SamsungY. P-P2 URL: http://pages.samsung. com/us/p2/

Skyfire Web BrowserU. R. L.http://www.skyfire. com/

SonyP. S. P. URL: http://www.us.playstation. com/PSP

Sun Microsystems (Java) URL. http://www.sun. com/software/opensource/java/

Sun Microsystems (Open Source) URL. http:// www.sunsource.net/

Symbian Foundation URL. http://www.symbian. org/

Chapter 7
Designing Animated Simulations and Web-Based Assessments to Improve Electrical Engineering Education

Douglas L. Holton
Utah State University, USA

Amit Verma
Texas A&M–Kingsville, USA

ABSTRACT

Over the past decade, our research group has uncovered more evidence about the difficulties undergraduate students have understanding electrical circuit behavior. This led to the development of an AC/DC Concept Inventory instrument to assess student understanding of these concepts, and various software tools have been developed to address the identified difficulties students have when learning about electrical circuits. In this chapter two software tools in particular are discussed, a web-based dynamic assessment environment (Inductor) and an animated circuit simulation (Nodicity). Students showed gains over time when using Inductor, and students using the simulation showed significant improvements on half of the questions in the AC/DC Concept Inventory. The chapter concludes by discussing current and future work focused on creating a more complete, well-rounded circuits learning environment suitable for supplementing traditional circuits instruction. This in-progress work includes the use of a contrasting cases strategy that presents pairs of simulated circuit problems, as well as the design of an online learning community in which teachers and students can share their work.

INTRODUCTION

Students often have specific difficulties understanding basic electricity concepts (e.g., Duit, et al., 1984; Caillot, 1991). One of the primary difficulties students have in learning about and understanding circuit behavior is the *current consumption model*, where current is viewed as a substance that is "consumed" by a device, such as a light bulb or resistor (Reiner et al., 2000). Students may conceive of a battery as a constant

DOI: 10.4018/978-1-61520-659-9.ch007

current source rather than a source of invariant voltage (Engelhart & Beichner, 2004). Students may also fail to differentiate between current and voltage, and power and energy (McDermott & van Zhee, 1984). Previous research has primarily been concerned with simple direct current (DC) circuit problems, and this may inadvertently guide one towards instructional decisions that reinforce misconceptions and difficulties students have when learning in other contexts. As part of an Office of Naval Research (ONR) funded project at Vanderbilt University, we extended research of student understanding of electric circuits into the domain of alternating current (AC) circuits. We were motivated by questions such as, to what extent do students exhibit the same misconceptions that they exhibit for DC circuits? How do students interpret time-varying phenomena?

Student Interviews

In interviews with students working on electrical circuit problems, we found that students had much greater difficulty understanding time-varying phenomena in circuits. We also found that students focused on manipulating formulas and performing numerical calculations during problem solving, and not applying the underlying principles or *invariants*, such as Kirchoff's or Ohm's laws, that govern circuit behavior. Analyzing common student difficulties that we identified, and by studying expert problem solving behavior, we developed a web-based tool (Inductor) for assessing and guiding students' learning of DC and AC circuits. Using Inductor we explored an additional research question: What are the effects of automated, invariants-based feedback on self-assessment and learning of electric circuit behavior? We found that by using this feedback students improved their problem solving performance in

Table 1. List of misconceptions

Misconceptions Related to AC Circuits
1. **Spatial AC misconception.** The sinusoidal AC voltage and current waveforms are not a representation of variation of these variables at a point in time. Rather they depict a variation of their magnitudes along the length of the wire in which the current is flowing. For example, students said that a string of identical light bulbs in series when connected to an AC source would light up in sequence, and some of the light bulbs may be on when others are off. At the same instant of time, the brightness of the bulbs would vary depending on their position in the circuit.
2. **Negative part of AC cycle is just a mathematical artifact.** No current flowing in circuit or power delivered during negative part of AC cycle. For example, a number of students said that a light bulb only lights up during the positive part of the sinusoidal cycle. Others said that there could be "no such thing as negative current. That is just a mathematical artifact. If current reverses, the electrons would reverse direction too. They would then run into each other, stopping flow, which implies there could be no current."
3. **Alternate form of this misconception.** The negative current "cancels" out the positive current. So bulb will never light up when you connect to true AC source.
4. **Empty pipe misconception.** During AC cycle electrons stop, turn around, and go the other way. In some cases when you have very long wires, they may never reach the light bulb connected to the end of the wire. Students thought that you would need two fuses to provide protection in an AC circuit, where you could do with one in a DC circuit.
5. **Incorrectly importing DC models to explain AC.** • Students often surmised that the alternating current going through a resistor was constant in time. • Students often hypothesized that a capacitor behaved the same in AC and DC circuits.
6. **Difficulties understanding circuit behavior when AC and DC signals are combined.** Students had difficulty "separating" or recognizing the AC and DC components of a signal in problems in which the midpoint of a sinusoidal voltage was not zero.
7. **More generally, difficulty thinking of circuit behavior when multiple waveforms, frequencies are combined.** Even advanced students stated that the number of channels you can got from cable TV was a function of the number of wires in the cable, or the thickness of the cable.

a short time, and were able to better explain their understanding of electric circuits.

Our protocol analysis of interviews with students solving circuit problems brought to light a number of difficulties students exhibit in both DC and AC circuit domains (Schwartz, et al., 2000; Biswas, et al., 2001). The misconceptions appeared to fall into three general categories: (i) those specific to particular AC or DC concepts (such as believing an AC voltage varies in space along the wire rather than in time), (ii) general difficulties (such as a failure to differentiate concepts, or incorrect simplifying assumptions when multiple invariants have to be applied to analyze circuit behavior), and (iii) lack of basic circuit knowledge, such as when to apply particular invariant properties and laws of circuit behavior, and in analyzing the behavior of dynamic elements, such as capacitors. We created a list of misconceptions related to understanding AC circuits (Table 1).

AC/DC Concept Inventory

The catalog of student difficulties had performing circuit analysis formed the basis for the development of a set of multiple-choice questions to assess student understanding with larger groups of students: the AC/DC Concept Inventory (Holton, Verma, & Biswas, 2008). The questions asked for qualitative (not quantitative) answers, and unlike traditional multiple choice tests in which only the correct answers matter, these questions have foil responses that are specifically linked to particular misconceptions our group and others have identified. The correct answers to our test questions matter as well, because they are written to specifically target core invariant principles of circuit behavior that experts use (see Table 2 below). We can analyze both correct responses and incorrect responses for information about students' understanding of invariants, their misconceptions and other learning difficulties.

We administered a paper and pencil version of the multiple-choice test to twenty 2nd year electri-

cal engineering students. We found that students had the most difficulty with the invariant principles underlying dynamic elements, such as capacitors (45% correct vs. 62% correct on questions not involving capacitors). Students appeared to have a better understanding of other invariant principles, such as Ohm's law, and applied them more correctly in circuit problems (63%). An analysis of incorrect answers revealed a significant number of misconceptions and difficulties (see Figure 1). Eight of twenty student answers indicated they possessed a current consumption (or "empty pipe") model of current, in which current flows from the positive side of a voltage source (DC or AC) sequentially and is consumed by the components. Five students revealed an "electron flow" model similar to the current consumption model except that flow starts from the negative terminal. Three students revealed a lack of knowledge about the relationship between power (light bulb wattage) and resistance, and six students tended to ignore the role of a capacitor in a circuit altogether. Students had the most difficulty with AC capacitor circuits (or filter circuits). The concepts of power (via bulbs) and the behavior of capacitors would thus later become a focus of the fourth phase of research utilizing an animated circuit simulation.

The context-dependent nature of students' knowledge of circuit behavior suggests that the difficulties students have in understanding electrical circuits are directly linked to instruction. Härtel (1982) believed that many learning difficulties can be traced to the fact that instruction is done in a piecemeal fashion, and students are never taught how to analyze a circuit as a system with interdependent components and constraints on behavior. This plus our own observations led us to develop an invariants-based framework that we believe experts apply in problem solving tasks, and we turned to a dynamic assessment approach (Campione and Brown, 1985, 1987; Bransford, et al, 1987; Magnusson, Templin, & Boyle, 1997) that focuses on how to prepare students to learn through instruction.

Table 2. AC & DC circuit invariants list

Invariant	Description
a. Ohm's Law	For resistors, capacitors, and inductors the current through the component is directly proportional to the voltage across the component. The ratio of voltage drop to current is the impedance of the component. For a resistor, the impedance is the resistance value, R. For capacitors (and inductors) the impedance is a function of the capacitance (or inductance) and the frequency of voltage and current.
b. Impedance of a Capacitor	The impedance of a capacitor is inversely related to the capacitance value and the frequency of the source. (Specifically the impedance of a capacitor is given by the expression: $X_C = 1/(2*pi*f*C)$, where f is the frequency, and C is the capacitance).
c. Charge held by a Capacitor	The charge held by a capacitor is directly proportional to the value of capacitance, C, and the voltage drop across it. ($Q = C*V$). Another way to express this relation is $I = C * dV/dt$, i.e., the current through a capacitor is related to the rate of change of the voltage across the capacitor.
d. Impedance of an Inductor	The impedance of an inductor is directly related to the inductance value and the frequency of the source. (Specifically the impedance of an inductor is given by the expression: $X_L = 2*pi*f*L$, where f is the frequency, and L the impedance.)
e. Inductor and Flux	The flux held by an inductor is directly proportional to the value of inductance, L, and the current through it. Another way to express this relation is $V = L * dI/dt$, i.e., the voltage drop across an inductor is related to the rate of change of current through the inductor.
f. Power	To determine the power dissipated by a resistor one has to know at least two of the three quantities for the resistor: its resistance, the voltage drop across the resistance, and the current through it. (Mathematically the power consumed $= V*I = V^2/R = I^2*R$)
g. Kirchoff's Laws of Conservation	Kirchoff's Voltage Law (KVL): Consider a closed loop consisting of one or more components. KVL states that the voltage drops across all elements in the loop at any instant of time must sum to zero. This relation holds universally for any set of components, and is independent of the frequency of the voltage and current. Kirchoff's Current Law (KCL): KCL states that the sum of the magnitudes of currents flowing into a point where a number of components are connected together must equal 0. (Current flowing away from the point is given a negative value). This relation holds universally at any point in time, and is not dependent on the frequency of the voltage and current.
h. Effective resistance	(a) Resistances in Series: The effective resistance of a set of resistances connected in series is the sum of the individual resistances. So in a series combination, the effective resistance always increases. (b) Resistances in Parallel: The effective resistance of a set of resistances connected in parallel is given by the formula: $1/R_{effective} = 1/R1 + 1/R2 + ...$ In a parallel combination, the effective resistance is always smaller.

Figure 1. Range of circuit misconceptions seen on misconceptions test

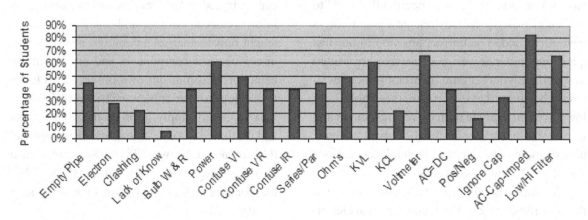

INDUCTOR: WEB-BASED DYNAMIC ASSESSMENT TOOL

Inductor was designed to be an online assessment tool in which students answer multiple-choice questions, select the invariant principle best applies to the circuit problem, and finally write an explanation for their answer. What makes Inductor a *dynamic assessment* environment is that students were provided opportunities to learn from outside resources while taking the test. Inductor not only provided instruction for remediating misconceptions, but it taught the invariants technique for circuit problem solving. After choosing the invariant principle involved in a problem and then selecting an answer, a student who is incorrect on either the invariant or the answer receives immediate feedback in the form of expert hints and explanations emphasizing the invariant properties of the circuit in the problem, and links to outside resources such as circuit diagrams and tutorials. Students could look up resources, then revise their answers or choice of invariant principles involved, and finally view a video of an expert explanation for the solution to the circuit problem.

In a related study, Leonard, Dufresne, and Mestre (1996) had physics students describe the principles involved in physics problems and write a justification for their answer. The instructors also discussed problem-solving strategies during their lectures, much like the invariant-based explanations and techniques for problem solving that we present through Inductor. They found that the students who were taught problem solving strategies generated more correct answers to problems, were less-dependent on surface features of problems for selecting the principles that governed problem solving, and better recalled the major principles covered in the course months later. The effort those instructors put into carefully reviewing and grading all the students' writings during the course provided valuable feedback and learning opportunities for the students, but also undoubtedly represented a significant investment of time and effort on the part of the instructors. The Inductor tool made a trade-off by providing automated feedback in the form of hints, expert explanations, and learning resources to students. Our focus was on self-assessment and providing Inductor as a supplementary resource to classroom instruction.

Pilot Study with Inductor

We ran a test study of the Inductor tool using our DC and AC test questions with a small group of first year electrical engineering students (N=6). All students completed two 14 item multiple choice tests using the online Inductor tool. The items in both tests were matched so that they both had the same level of difficulty. We wanted to see if performance improved from the first test to the second test, and also collect evidence for students' improving their explanations of circuit behavior.

Overall Results from Pilot Study

Overall, the participating students scored an average 61% correct answers on the first test, and 82% correct on the second test, an improvement of 21%. Five of the six participants showed an improvement from the first to second test. As in earlier research our group conducted, we found that students, at least initially, had the most difficulty with problems dealing with capacitors and other dynamic components. In this study, however, by the second test students were performing well in all categories, showing the largest improvement with DC capacitor circuit problems.

Student Explanations

Students initially revealed misconceptions ("higher resistance means more power is absorbed," "internal resistance rises," "the internal resistance is lower for low frequencies," "John's battery will be required to work harder to push current through the larger resistor"), errors ("after long time, all voltage will be across capacitor," "since they are in series the voltage should be the same across all of the bulbs"), and admitted to making some guesses. The responses of students who received invariants prompts and instruction revealed they more often attempted to revise and correct their misconceptions:

Example 1

- 1st explanation: "Since frequency and current are related linearly, and increase in frequency will increase current."
- 2nd explanation: "Actually, current and frequency have no relation so current does not change with frequency."

Example 2

- 1st explanation: "More current flows through the bulbs when they are in series. this makes a brighter light."
- 2nd explanation: "More voltage through the bulbs in parallel will make for brighter bulbs."
- 3rd explanation: "Power is what determines the light intensity. More power makes for more intensity."

The students also articulated some of the invariant principles they learned:

- "There is only one path for the current to flow through and both bulbs lie on this path."
- "With low frequencies most of the voltage is across the capacitor since its impedance is high when frequency is low."
- "because the lower resistance in Peter's circuit will result in more power consumption. lower resistance = more power"
- "since it is in series connection, the same resistances and current will produce the same power for all the bulbs. p=i^2R"

- "when you go from parallel to series, the intensity will decrease since the voltage is split across both bulb 1 and bulb 2."
- "because the lowere resistance will need more power." "the lower resistance will result in more power being dissapated because p=v^2/R"

It was clear that after receiving feedback on invariants, most students attempted to revise and correct their misconceptions.

Invariant Selection Analysis

For each circuit problem, our experts selected and agreed upon the invariants that were most relevant and helpful for solving the problem. They also identified which invariants were clearly irrelevant. The impedance of an inductor, for example, is irrelevant for a circuit with only capacitors. The remaining invariants were placed in a third category. These invariants were technically involved in the circuit's behavior, but were not necessarily useful for solving the problem asked about the circuit. Ohm's law, for example, is involved in many of the circuits used in our tests, but is not always an important one for answering a particular question about the circuit.

We needed a way to quantify how well students selected invariants for a particular problem. In this case a simple percent correct measure is insufficient for characterizing students' use of invariants, and would reward students who select more invariants regardless of their importance to the problem. To control for such response biases, we utilized a nonparametric discrimination measure known as Yule's Q. Nelson (1984) contrasted simple percentage correct measures, d' measures and Yule's Q, and advocated Yule's Q over d' on the basis that it was thought to make weaker assumptions about the data and required fewer observations. For our purposes this measure rewards the selection of invariants our experts agreed were appropriate for a problem, while controlling for the selection of clearly irrelevant invariants.

The Yule's Q measure was constructed first by calculating the percentage of invariants a student chose out of those invariants experts chose as relevant (h, or hit rate) as well as the percentage of invariants a student chose out of those invariants experts deemed clearly irrelevant (f, or false alarm rate). Invariants that are technically correct but less relevant to a problem were ignored in this computation. The Yule's Q score was then calculated by the formula: (h-f)/(h-2fh+f). A Q of one implies perfect discrimination of the relevant from irrelevant invariants, and zero implies chance performance.

Results of Invariant Selection Analysis

The average discrimination of invariants (as calculated by Yule's Q) when a correct answer to a circuit problem was chosen was 0.53. The average discrimination when an incorrect answer was chosen was 0.39. Thus students were more likely to select those invariants that experts deemed relevant on questions they answered correctly.

We found, however, that students' selection of relevant invariants declined from the beginning of the tests to the end. The graphs below reveal this pattern across each of the categories of questions and across both classes. Different explanations may be provided for this pattern of results. One is that the questions grow more difficult from the beginning to the end of a test. Another explanation is a fatigue or indifference factor. Students may have been concentrating only on getting the correct answer to questions, and gradually paid less attention to the invariant selection.

Student Survey Responses

After completing both tests, students answered questions on a follow-up survey. The students responded that they liked and used the outside resources and the hints we provided after answering

incorrectly. One student even suggested, "I was very impressed with the information provided to learn from mistakes. I think that that information should be provided regardless of whether the answer was right or wrong so that if the answer was just a guess I could solidify my understanding."

Students also mentioned they thought the test questions helped reinforce concepts they had learned and better apply what they had learned. When we asked those participants who received instruction on invariants what they believed invariants are and how they are used, the students primarily thought of invariants as a method for solving a problem. When asked what is an invariant, one student responded: "A circuit invariant is a certain method of circuit analysis needed to solve the problems presented.(ex: ohms law, power, etc)." When asked why it is useful to analyze a circuit by considering the invariants, that student also responded: "It gives the student an idea of where to start the problem." Other students also responded: "It lets you know how to solve the circuit and how to solve like circuits in the future," and "It allows one to find and use the necessary method of solution quicker."

Problems with Inductor

We identified certain problems from our tests of the Inductor environment as well. The student participation rate in our pilot studies was low, which we believe was partly due to the fact that use of the tool was not connected to their current class work. We found the outside resources used were not sufficient for addressing many of the difficulties students had in applying invariants to solve problems. Ultimately, Inductor was still primarily a "test", and not an engaging, motivating environment for learning about circuit behavior. We believe that the inclusion of more open-ended challenge problems that include diagnosis and design questions will motivate the students to think deeper and begin to see the importance of understanding how the bridges between invariants help to better structure problem solving tasks. First, however, our focus was on creating a better interactive resources for qualitatively understanding the dynamic behavior of electrical circuits–an animated circuit simulation.

THE DESIGN OF AN ANIMATED CIRCUIT SIMULATION

As mentioned above, the outside resources used in Inductor to provide feedback to students while they worked on circuit problems were not sufficient. Students still were not given a sense of how particular circuits behave in real-time. This motivated the creation of an animated circuit simulation. An interactive simulation environment may allow students to develop a more "voltage-centered" model of circuit behavior that experts use to understand the flow of current (Frederickson and White, 2000). Also, allowing students to experiment with a simulation will allow them to develop multiple context-dependent interpretations of the invariant laws that govern circuit behavior. Furthermore, from a design perspective, students can explore the role of circuit components by adding and removing them from the circuit, or by changing their values in a circuit. By showing an animation of current flow through a circuit in real-time, students can see Ohm's law (voltage equals current times resistance) in action and see the effects of more advanced components like capacitors and inductors on current flow.

This simulation can model DC and AC analog circuits, including components such as capacitors and inductors and transistors. There are two major design features however that distinguish it from existing simulations. One is that current flow is visualized as a single moving chain of dashes to help students understand the behavior of the various circuits they learn in introductory classes. From interviews we found that students knew the mathematical formulas related to circuits with capacitors, for example, but could not answer basic qualitative questions such as what happens in a circuit with a capacitor, light bulb and DC voltage source. Our circuit simulation shows them this behavior. Second, our circuit simulation uses a technique known as *enactive modeling* that allows for manipulating variables in real-time and immediately seeing their effects. Thus for example students can 'wiggle' the voltage and see its effects in a circuit with a capacitor or inductor, thereby helping students induce invariant principles related to the impedance of these components with respect to frequency changes. The enactive modeling strategy is described in more detail below.

Enactive Modeling

Scientists have routinely employed causal and mechanical models to help reason about events and communicate their understanding to other scientists (Salmon, 1998; de Regt & Dieks, 2002; Gooding, 1992), even if they later eschew these models for purely quantitative/mathematical descriptions. James Clerk Maxwell for example used a mechanical-fluid analogy for electro-magnetic fields that may have helped him deduce the quantitative relationships now known and taught as Maxwell's laws (Nersessian, 2002).

Researchers in the learning sciences and cognitive science are beginning to uncover more about the underlying basis for people's natural and informal reasoning about both physical and social events, and why people show a preference for causal and mechanical models. There appears to be a connection between our informal reasoning and the embodied nature of our thoughts and actions. We are beginning to pay attention to the role of one's body and intentional actions ("embodied cognition", "enactive learning") in order to better characterize the contextual constraints involved in natural reasoning about events. This applies to computer simulated events as well.

Roth and Lawless (2001) note for example that students' "gestures are an important means in the construction of perception and communication as students interact over and about a computer software environment." They suggest that learning environments that do not support students' use of body and gesture can limit what and how students learn.

Educational researchers have found evidence linking students' kinesthetic behavior to their understanding of dynamic systems. Clement (1994) and Reiner (2000) have found that both students and experts may sometimes "describe a system action in terms of a human action" and use gestures that depict changes happening in a system. They have interpreted these "self-projections" as evidence that a person is mentally enacting or simulating aspects of a system. Monaghan and Clement (1999) observed students performing hand motions and visualizations while using a relative motion simulation. Other researchers have referred to these kinds of self-projections as anthropomorphic reasoning (Zohar & Ginossar, 1998) or anthropomorphic epistemology (Sayeki, 1989).

Sometimes these self-projections may even underlie some of the misconceptions students have in science, but they can also be used positively, as a starting point for instruction. Susan Goldin-Meadow and others have found that teachers using gestures or attending to student gestures can make math instruction more effective (Goldin-Meadow, 1999). Physics education researchers have also found that kinesthetic real-time participation is a key component responsible for the success of microcomputer-based labs (MBL) in fostering understanding of physics concepts and graph interpretation skills (Beichner, 1990; Mokros & Tinker, 1987). In MBL activities, students use computers with sensors attached (distance, force, temperature, etc.) to explore the changes that occur in physical phenomena.

Enactive Modeling Hypothesis

Our hypothesis was that these kinesthetic activities are helping foster - yet also constrain - how students "intentionalize" the phenomena about which they are learning. They are connecting their natural, embodied experience of phenomena to the constraints and rules operating in scientific representations of the phenomena, and conversely, abstract scientific concepts are converted into embodied metaphors which students can use. This "exemplifies what we call symbolizing: a creation of a space in which the absent is made present and ready at hand" (Nemirovsky & Monk, 2000). More generally speaking, Roth and Lawless (2002), like Piaget, have argued that gestures can serve as a bridge between our everyday experiences in the physical world and the abstract scientific thinking that is a goal of science instruction.

If anthropomorphic reasoning, gestures, and self-projection help signify students' understandings and misunderstandings of complex systems, then it is possible that students may benefit by instructional interventions that facilitate and constrain their enactive participation with a complex system. For this research, I explored a new learner-centered simulation design strategy that may be uniquely suited to helping students understand complex changes happening in physical systems–*enactive modeling*.

A simple example of this enactive modeling strategy has been applied in physics education. Students have difficulties understanding how Newton's third law operates in static situations. Given a situation in which a book lies atop a table, students may recall that gravity pulls the books down, but they neglect the equal and opposite upward force that the table exerts on the book. Various strategies have been used to help students recognize this "passive" force, but an example of an inactive modeling (or enactive participation) strategy is for students to lie down on their backs and hold books up on their hands (Freudenthal,

1993). In a sense the students are enacting the role of the table and can sense that they have to push up harder if more books are added.

With more complex and simulated physical systems, however, determining how to facilitate such participation is more difficult. Most computer-based simulations are *symbolic* simulations, encapsulated representations of an external physical system such as an electric circuit. In contrast, *experiential* simulations are simulations in which the user or learner is a functional element, or agent, in the situation or system being modeled. An example of this is Model U.N., in which students from various schools take on roles of different countries in pretend meetings of the United Nations. The question I pursued though is can students learn by participating in simulations of complex *physical* systems as well as social systems. Does enacting the physical and temporal constraints that operate within a system help one understand the behavior of the system as a whole?

Enactive Modeling Interface for a Circuit Simulation

We applied the enactive modeling strategy to the design of an animated simulation of electrical circuit behavior. As mentioned before, electricity is one of the most difficult subjects for students to understand, and there are a great deal of misconceptions about circuit behavior. Particularly difficult are behaviors that change over time, as in alternating current (AC) circuits. In an AC circuit, the voltage changes very quickly, often switching from positive to negative values. Further complicating matters, some circuit elements such as capacitors and inductors respond differently based on the rate of change in voltage or current. A capacitor in an alternating current (AC) electrical circuit exhibits qualities similar to a resistor, a virtual impedance, but unlike a resistor, the impedance of a capacitor varies inversely with the frequency of the AC voltage source. This law of circuit behavior has consistently proved to be

one of the most difficult concepts for undergraduate students to learn, and is also very difficult to represent visually or explain verbally to students. Most students may only memorize a formula or a shortcut and never understand how or why a capacitor exhibits this impedance, or how this impedance characteristic is useful for designing or troubleshooting circuits (such as radio tuners), or how it is related to other invariant constraints on circuit behavior such as Kirchoff's laws or Ohm's law.

Imagine there is an interface to a circuit simulation that allows you to directly vary the voltage applied to a circuit, and the circuit responds in real-time. You could alternate the voltage from positive to negative just like an AC voltage source does, with zero voltage being the middle, or resting point. Imagine also that you could both feel the resistance of the circuit (Ohm's law), and see resistance in the form of reduced current flow. A circuit with high resistance would resist your applying voltage, and a circuit with low resistance would be easy to apply voltage. The flow of charge (current) through the circuit would also be visually depicted as an animation, to redundantly specify current and resistance.

A force-feedback steering joystick or steering wheel (commonly available as interfaces for computer games and simulations) may help students embody and understand the constraints of an AC voltage source. One may move the joystick or wheel to the right to increase the voltage positively, and to the left for negative voltages. Instead of setting the frequency of an AC voltage source by entering a numeric value or moving a slider control, students enact a change in frequency by changing how fast they move the steering wheel from side to side. The force feedback component allows one to make the joystick or steering wheel harder to move if there is a higher circuit resistance (or impedance), and easier for lower resistance. In the aforementioned AC capacitor circuit, the law governing how a capacitor's impedance varies inversely with frequency can be experienced directly, by sensing that the steering wheel is easier to move the faster you turn it and current flows faster, and harder to move and current flows slower when turned slowly or held at one position (as in a DC circuit). An inductor circuit component has the opposite relationship with frequency as a capacitor, and when combined with a capacitor may form a tuning circuit, in which there is one particular resonant frequency with the lowest resistance, or where the steering wheel turns the easiest.

Two sets of electricity misconceptions identified in earlier research also helped lead to the choice of a joystick or steering wheel for the input device to use with the circuit simulation. One set of misconceptions relates to AC circuits. Students may believe AC voltage varies spatially along a wire rather than temporally. Also, they may not understand what happens to current when voltage is in the negative part of a sinusoidal cycle. Another set of misconceptions concerns the relationship between resistance and current. For example, despite having two resistors instead of one, a parallel circuit can have lower total resistance than a circuit with only one of those resistors in series, and for many students this is counter-intuitive. Students may not even distinguish between voltage and current, and a constant voltage source versus constant current.

Pilot Study with Nodicity

A pilot study tested the use of this animated circuit simulation on individual undergraduate students, and assessed its effects of students' intuitive conceptions about circuit behavior. The study involved a mixed methods experimental design employing both quantitative and qualitative methodologies and analyses. The quantitative component consisted of a 20 item multiple choice circuit quiz derived from other quizzes. Students took the test as a pretest and posttest. The pretest helped show what preconceptions individual students had about electrical circuit behavior coming into the study.

After a individual tutorial session with the circuit simulation led by the first author, students took the same test again as a post-test. This revealed how students' conceptions may have changed as a result of instruction. As will be discussed in more detail below, some students used the simulation with a joystick interface to control voltage in real-time, while a second group of students used the simulation with only an on-screen graphical slider control to change voltage. The pretest and posttest measures allow one to quantitatively compare and contrast these two groups of students, to see what effects the joystick had. Furthermore, the students were videotaped during the session while using the computer simulation, capturing the computer screen and interface devices, the student, and the first author (acting as a tutor or guide) for later qualitative analysis. This allows for exploration of any potential links between particular events and actions by the student and myself to specific learning outcomes as measured by the test. With a qualitative analysis of the video, one may see if the students use gestures while using the simulation, and if there connections between what the students said or did and their developing understanding of electrical circuit behavior.

Participants consisted of 40 undergraduate electrical engineering majors recruited from introductory courses. Students were asked to volunteer to try out the animated circuit simulation during a tutoring session outside of class, and were paid approximately $15 for participation.

The entire tutoring session with each student lasted not more than an hour. This is a very short time for an instructional intervention, however, there were two factors influencing this time decision. One is practical. Few students have volunteered in previous studies even when their only commitment was to take a couple of short quizzes online on their own time. Secondly, the real-time reactive control feature this simulation has that allows for controlling voltage over time is most similar in spirit to microcomputer-based labs (MBL), and research on MBL has shown surprising learning gains in very short periods of time. In MBL, students for example might move a car back and forth, while a computer graphed its motion in real-time via a sonic distance sensor. Heather Brasell (1987) has shown marked improvement in younger students' graphing skills after just 40 minutes of instruction, and Linn et al. (1987) show that this improvement asymptotes rather quickly at about the 70% level even after a year of experience with MBL. Abbott et al. (2000) actually have examined the effects of one 2-hour active learning laboratory in electrical circuits (students worked with real bulbs and circuits). They did find some significant learning gains using a pretest and posttest as well.

Results from Pilot Study with Simulation

This study showed significant conceptual change gains, despite the short intervention. Overall, the gains were weak, with only a 12% increase in test scores, however, eight of the twenty questions on the misconceptions test showed more significant and larger gains (highlighted in bright red in Figure 2 and bright green in Figure 3). These eight questions did not have any structural characteristics in common. They included questions involving DC, DC capacitors, AC, and AC capacitor circuits. Some of the questions were very similar to the circuits explored in the simulation, and some were not (far transfer questions). What is the connection between the seemingly unrelated eight questions on which students showed significant gains? The connection appears to be that some questions on the test may force one to imagine the behavior of the circuit over time. A post-hoc analysis identified ten questions on the test that involve considerations of circuit behavior over time. Eight of these questions are the same eight questions identified before which showed significant gains. Figure 4 shows performance on these

ten time related questions versus the ten which were not time related. As the graph illustrates, students did worse on time related questions on the pretest than non-time related questions (53% vs. 68% correct, respectively). By the posttest, students answered on par in both categories (73% vs. 71% correct). Thus students showed a gain on time-related questions, yet not on non-time related questions. An ANCOVA analysis using the pretest as a covariant showed a difference between the two categories ($F(1,77)=9.81$, $p=.0025$).

Limitations of the Circuit Simulation

Students using our animated circuit simulation did not show improvement on *all* our concept inventory test questions in our pilot. In particular they showed no improvement on the non-temporal test questions. We believe part of the difficulty students have with these questions lies in the conflation of different variables (such as voltage vs. current) or a lack of distinction between components (capacitor vs. inductor) and circuit configurations (series vs. parallel). We have previously referred to this student difficulty as *undifferentiated concepts*. Students may either conflate or mix-up different concept pairs, such as voltage and current, voltage and power, AC and DC, series and parallel, capacitors and inductors, and low pass and high pass filter circuits. We have often observed that students have to resort to memorization to remember these distinctions, which means they may still not understand the inherent underlying reasons for these distinctions.

What then is required to get students to attend to these important distinctions and possibly improve on *all* the questions in our AC/DC Concept Inventory, including the non-temporal test questions? We describe an instructional strategy that we are in the process of pursuing known as *contrasting cases*.

Figure 2. Scores on individual test questions

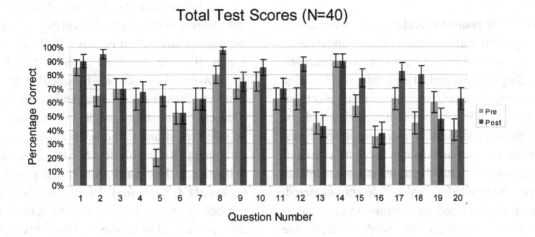

Figure 3. Chi square analysis of individual test questions

	Test Question:	1	2	3	4	5	6	7	8	9	10
Pre	Correct	34	26	28	25	8	21	25	32	28	30
	Incorrect	6	14	12	15	32	19	15	8	12	10
Post	Correct	36	38	28	27	26	21	25	39	30	34
	Incorrect	4	2	12	13	14	19	15	1	10	6
Chi Square Result:		.2918	.0000	1.0000	.4996	.0000	1.0000	1.0000	.0000	.4652	.0765

	Test Question:	11	12	13	14	15	16	17	18	19	20
Pre	Correct	25	25	18	36	23	14	25	18	24	16
	Incorrect	15	15	22	4	17	26	15	22	16	24
Post	Correct	28	35	17	36	31	15	33	32	19	25
	Incorrect	12	5	23	4	9	25	7	8	21	15
Chi Square Result:		.3006	.0000	.7491	1.0000	.0025	.7440	.0009	.0000	.1134	.0033

Figure 4. Time vs. non-time related questions

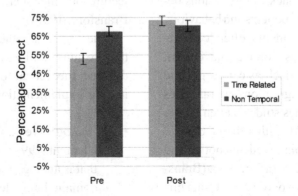

FUTURE DIRECTIONS

Our ultimate goal is to develop a complete learning environment that facilitates student understanding of electrical circuit behavior and principles, including all the invariant principles we identified as important for understanding analog electrical circuits. Holton (in press) has reviewed how, with the right supporting instructional strategies, simulations such as ours can serve as a foundation for a well-rounded learning environment that addresses knowledge-centered, learner-centered, assessment-centered, and community centered issues as delineated in the framework provided in the book *How People Learn* (Bransford et al., 1999).

However, as mentioned above, students only showed improvement on some of our test questions when using our animated circuit simulation. Furthermore, students in real classrooms still need a reason (raison d'etre) to use our simulation, as well as our dynamic assessment tool. Two additional strategies will be being pursued to address these issues and hopefully create a more well-rounded environment for learning about electrical circuit behavior. One is the contrasting cases strategy, and another is the creation of an online challenge-based community for learning about electrical engineering.

The Contrasting Cases Strategy

We believe difficulties on the non-temporal questions in our AC/DC Concept Inventory primarily revolve around students' lack of distinctions between different circuit components and behaviors. As mentioned above, students often conflate various variables such as voltage and current and various components such as capacitors and inductors and circuit configurations such as series versus parallel. Sometimes students even ignore the fact that a circuit is AC rather than DC.

One strategy for helping students notice and make new distinctions is contrasting cases (Bransford, Franks, Vye, & Sherwood, 1989). Using this strategy, one shows learners two related cases at a time, rather than one at a time. This approach has theoretical roots in ecological psychology and perception and action research (Gibson, 1979). Essentially one can direct attention to a feature by showing two cases side by side that differ by that very feature on which attention needs to be focused. Contrasting cases has been shown to be an effective strategy in other domain areas such as algebra (Rittle-Johnson & Star, 2006; Derry, Wilsman, & Hackbarth, 2007) and psychology (Schwartz & Bransford, 1998). However, it has not been tested in the domain of engineering or science education before. Related research has also found giving students contrasting cases activities or simulation activities is more effective when employed *before* a class lecture rather than afterward (Brant, Hooper, & Sugrue, 1991; Schwartz & Bransford, 1998). That supports the notion that a supplementary online resource for learning and investigating circuit behavior may be effectively integrated into a circuits class, especially when students explore the simulation before hearing a lecture about the related concepts and formulas. Even though students may or may not fully understand the model underlying the simulation they use, they may form questions or strategies for learning about the domain after exploring the circuit. So even when a simulation is not completely understandable to a student, it may engender a preparedness for future learning, and the students may become more curious or attentive when presented with a subsequent lecture or other learning activity (Schwartz & Bransford, 1998).

There are also other instructional techniques that are quite similar to contrasting cases but have different theoretical foundations and differences in their implementation. These include:

- *Analogies* (e.g., Gentner & Gentner, 1983; Clement, 1993)–show a system or scenario that is analogically isomorphic to the target domain being learned. This strategy has roots in cognitive information processing models and mental models research.

- *Variation strategy* (e.g., Marton & Pang, 2006; Pang & Marton, 2005)–show students variations in a scenario along one dimension while keeping other aspects invariant. This approach has roots in phenomenology.
- *Multiple representations* (e.g., Kozma et al., 1996; Schnotz & Bannert, 2003)–provide multiple linked representations of phenomena, such as graphs and pictorial and textual representations. This strategy has roots in multimedia learning theory and human-computer interaction (HCI) research.
- *Embodied metaphors and gestures* (e.g., Glenberg et al., 2004; Núñez, Edwards, & Matos, 1999; Goldin-Meadow, 2005)– Essentially use one's own body and actions as an analogy to or enactment of the target domain. The aforementioned enactive modeling strategy falls in this category, as well. These approaches have roots in embodied cognition (Gibbs, 2006).

This plethora of independently-derived instructional approaches raises numerous research questions, however. For not only do each of these techniques have distinct theoretical roots, they also have specific differences in how they have been applied to instruction. Moreover only the analogies technique has been sufficiently researched in domain areas involving misconceptions that are resistant to traditional instruction, including electrical circuits (e.g., Gutwill, Fredericksen, & White, 1999). The limitations of the analogy strategy have been well identified (e.g., limitations of using the water analogy to electrical circuits, Gentner & Gentner, 1983).

We will be conducting future meta-analyses and pilot studies that explore how presenting circuit simulation problems in pairs rather than one at a time may better facilitate student understand-ing and noticing of critical features in electrical circuit problems.

Online Challenge-Based Community

One other larger challenge for our research is addressing student learning of electricity concepts at the K-12 level. Currently most high schools only spend a week or two on electrical circuits over the course of four years of instruction. Curricula exists for more in depth learning about digital and analog electrical circuits (such as Project Lead the Way), yet it also may have limitations. One is that the curricula is very expensive and not openly available to the public. Only schools in richer school districts may adopt specialized technology education courses on electrical circuits. Also, this leaves out not only poorer schools, but the ever increasing number of online virtual high schools. More and more high school students in the United States and elsewhere are beginning to take more, if not all, of their courses online. Existing technology education curricula are designed for face to face courses only, and use expensive and dangerous equipment that could not be utilized by students on their own from home.

We have proposed the creation of an online/ blended learning community for students learning science, technology, engineering and math (STEM) concepts known as BOOSTEM. Students can use animated simulations such as our own, and also collaborate and compete with one another on engineering and design projects and problems. Similar online portals have been developed for science education, including the ITSI project at the Concord Consortium and the EU-funded SCY project (Science Created by You).

CONCLUSION

This chapter has discussed initial tests of two software tools for facilitating student understanding of electrical circuit behavior: the Inductor dynamic assessment tool and the Nodicity animated circuit simulation software. The design of both tools and the impetus for much our work is improving student performance on a qualitative multiple-choice questionnaire–the AC/DC Concept Inventory. Students using the simulation showed improvement on approximately half the questions, those involving temporal considerations of circuit behavior, and our group is beginning to research strategies such as contrasting cases to improve performance on the other, non-temporal questions. The ultimate goal is creating a complete, well-rounded environment for understanding how to analyze, design, and troubleshoot electrical circuits. Hopefully, the data and lessons learned that are presented in this chapter will help others attempting to design better instruction in engineering and science.

REFERENCES

Beichner, R. J. (1990). The effect of simultaneous motion presentation and graph generation in a kinematics lab. *Journal of Research in Science Teaching, 27*(8), 803–815. doi:10.1002/tea.3660270809

Bransford, J. B., Brown, A. L., & Cocking, R. R. (1999). *How People Learn*. Washington, D.C.: National Academy Press.

Bransford, J. B., Franks, J. J., Vye, N., & Sherwood, R. (1989). New approaches to instruction: Because wisdom can't be told. In Vosniadou, S., & Ortony, A. (Eds.), *Similarity and analogical reasoning* (pp. 470–497). Cambridge: Cambridge University Press. doi:10.1017/CBO9780511529863.022

Brant, G., Hooper, E., & Sugrue, B. (1991). Which comes first: The simulation or the lecture? *Journal of Educational Computing Research, 7*(4), 469–481.

Caillot, M. (Ed.). (1991). *Learning Electricity and Electronics with Advanced Educational Technology*. New York: Springer-Verlag.

Campione, J. C., & Brown, A. L. (1985). Dynamic assessment: One approach and some initial data. Tech Rep. No. 361, Univ. of Illinois at Urbana-Champaign, Champaign, IL.

Campione, J. C., & Brown, A. L. (1987). Linking dynamic assessment with school achievement. In Lidz, C. S. (Ed.), *Dynamic Assessment: An Interactional Approach to Evaluating Learning Potential* (pp. 479–495). New York: Guilford Press.

Clement, J. J. (1993). Using bridging analogies and anchoring intuitions to deal with students' preconceptions in physics. *Journal of Research in Science Teaching, 30*(10), 1241–1257. doi:10.1002/tea.3660301007

Clement, J. J. (1994). Use of physical intuition and imagistic simulation in expert problem solving. In Tirosh, D. (Ed.), *Implicit and Explicit Knowledge*. Norwood, NJ: Ablex Publishing Corp.

De Regt, H. W., & Dieks, D. (2002). A contextual approach to scientific understanding. Retrieved from March 15th, 2003 from http://philsci-archive.pitt.edu/documents/disk0/00/00/05/53/index.html

Derry, S. J., Wilsman, M. J., & Hackbarth, A. J. (2007). Using contrasting case activities to deepen teacher understanding of algebraic thinking and teaching. *Mathematical Thinking and Learning, 9*(3), 305–329.

Duit, R., Jung, W., & von Rhoneck, C. (Eds.). (1984). *Aspects of Understanding Electricity*. Kiel, Germany: Verlag, Schmidt, & Klaunig.

Engelhart, P. V., & Beichner, R. J. (2004). Students' understanding of direct current resistive electrical circuits. *American Journal of Physics, 72*(1), 98–115. doi:10.1119/1.1614813

Freudenthal, H. (1993). Thoughts on teaching mechanics: Didactical phenomenology of the concept of force. *Educational Studies in Mathematics, 25*(1/2), 71–88. doi:10.1007/BF01274103

Gentner, D., & Gentner, D. R. (1983). Flowing waters or teeming crowds: Mental models of electricity . In Gentner, D., & Stevens, A. L. (Eds.), *Mental Models*. Erlbaum.

Gibbs, R. W. (2006). *Embodiment and Cognitive Science*. Cambridge University Press.

Gibson, J. J. (1979). *The Ecological Approach to Visual Perception*. Boston: Houghton Mifflin.

Glenberg, A. M., Gutierrez, T., Levin, J. R., Japuntich, S., & Kaschak, M. P. (2004). Activity and Imagined Activity Can Enhance Young Children's Reading Comprehension. *Journal of Educational Psychology, 96*(3), 424–436. doi:10.1037/0022-0663.96.3.424

Goldin-Meadow, S. (1999). The role of gesture in communication and thinking. *Trends in Cognitive Sciences, 3*(11), 419–429. doi:10.1016/S1364-6613(99)01397-2

Goldin-Meadow, S. (2005). *Hearing Gesture: How Our Hands Help Us Think*. Belknap Press.

Gooding, D. (1992). Putting agency back into experiment . In Pickering, A. (Ed.), *Science as Practice and Culture*. University of Chicago Press.

Gutwill, J. P., Fredericksen, J. R., & White, B. Y. (1999). Making their own connections: Students' understanding of multiple models in basic electricity. *Cognition and Instruction, 17*(3), 249–282. doi:10.1207/S1532690XCI1703_2

Härtel, H. (1982). The electric circuit as a system: A new approach. *European Journal of Science Education, 4*(1), 45–55.

Holton, D.L. (in press). How people learn with computer simulations. *Handbook of Research on Human Performance and Instructional Technology*. IGI Global.

Holton, D. L., Verma, A., & Biswas, G. (2008). Assessing student difficulties in understanding the behavior of AC and DC circuits. In *Proceedings of the 2008 ASEE Annual Conference*. Pittsburgh, PA.

Kozma, R., Russell, J., Jones, T., Marx, N., & Davis, J. (1996). The use of multiple, linked representations to facilitate science understanding . In Vosniadou, S., Glaser, R., De Corte, E., & Mandl, H. (Eds.), *International Perspectives on the Psychological Foundations of Technology-Based Learning Environments* (pp. 41–60). Hillsdale, NJ: Erlbaum.

Leonard, W. J., Dufresne, R. J., & Mestre, J. P. (1996). Using qualitative problem-solving strategies to highlight the role of conceptual knowledge in solving problems. *American Journal of Physics, 64*(12), 1495–1503. doi:10.1119/1.18409

Magnusson, S. J., Templin, M., & Boyle, R. A. (1997). Dynamic science assessment: A new approach for investigating conceptual change. *Journal of the Learning Sciences, 6*, 91–142. doi:10.1207/s15327809jls0601_5

Marton, F., & Pang, M. F. (2006). On some necessary conditions of learning. *Journal of the Learning Sciences, 15*(2), 192–220. doi:10.1207/s15327809jls1502_2

McDermott, L. C., & van Zee, E. H. (1984). Identifying and addressing student difficulties with electric circuits . In Duit, R., Jung, W., & von Rhoneck, C. (Eds.), *Aspects of Understanding Electricity*. Kiel, Germany: Verlag, Schmidt, & Klaunig.

Mokros, J. R., & Tinker, R. F. (1987). The impact of microcomputer-based labs on children's ability to interpret graphs. *Journal of Research in Science Teaching, 24*(4), 369–383. doi:10.1002/tea.3660240408

Monaghan, J. M., & Clement, J. (2000). Algorithms, visualization, and mental models: High school students' interactions with a relative motion simulation. *Journal of Science Education and Technology, 9*(4), 311–325. doi:10.1023/A:1009480425377

Nelson, T. O. (1984). A comparison of current measures of accuracy of feeling-of knowing predictions. *Psychological Bulletin, 95*, 109–133. doi:10.1037/0033-2909.95.1.109

Nemirovsky, R., & Monk, S. (2000). "If you look at it the other way…": An exploration into the nature of symbolizing . In Cobb, P. (Eds.), *Symbolizing and communicating in mathematics classrooms: Perspectives on Discourse, Tools, and Instrumental Design* (pp. 177–221). New Jersey: Erlbaum.

Nersessian, N. J. (2002). Maxwell and "the Method of Physical Analogy": Model-based reasoning, generic abstraction, and conceptual change. In D. Malament (Ed.), Essays in the History and Philosophy of Science and Mathematics (pp. 129-166). Lasalle, Il: Open Court.

Núñez, R. E., Edwards, L. D., & Matos, J. F. (1999). Embodied cognition as grounding for situatedness and context in mathematics education. *Educational Studies in Mathematics, 39*(1-3), 45–65. doi:10.1023/A:1003759711966

Pang, M. F., & Marton, F. (2005). Learning theory as teaching resource: Enhancing students' understanding of economic concepts. *Instructional Science, 33*(2), 159–191. doi:10.1007/s11251-005-2811-0

Reiner, M. (2000). Thought experiments and embodied cognition . In Gilbert, J. K., & Boulter, C. J. (Eds.), *Developing Models in Science Education* (pp. 157–176). Netherlands: Kluwer Academic Publishers.

Reiner, M., Slotta, J. D., Chi, M. T. H., & Resnick, L. B. (2000). Naïve physics reasoning: A commitment to substance-based conceptions. *Cognition and Instruction, 18*(1), 1–34. doi:10.1207/S1532690XCI1801_01

Rittle-Johnson, B., & Star, J. (2007). Does comparing solution methods facilitate conceptual and procedural knowledge? An experimental study on learning to solve equations. *Journal of Educational Psychology, 99*(3), 561–574. doi:10.1037/0022-0663.99.3.561

Rittle-Johnson, B., & Star, J. (in press). Compared to what? The effects of different comparisons on conceptual knowledge and procedural flexibility for equation solving. *Journal of Educational Psychology*.

Roth, W. M., & Lawless, D. V. (2002). Scientific investigations, metaphorical gestures, and the emergence of abstract scientific concepts. *Learning and Instruction, 12*, 285–304. doi:10.1016/S0959-4752(01)00023-8

Salmon, W. C. (1998). *Causality and Explanation*. Oxford University Press. doi:10.1093/0195108647.001.0001

Schnotz, W., & Bannert, M. (2003). Construction and interference in learning from multiple representation. *Learning and Instruction, 13*(2), 141–156. doi:10.1016/S0959-4752(02)00017-8

Schwartz, D., Biswas, G., Bransford, J., Bhuva, B., Balac, T., & Brophy (2000). Computer tools that link assessment and instruction: Investigating what makes electricity hard to learn. In S. Lajoie (Ed.), *Computers as Cognitive Tools, Vol. II* (pp. 273-307). Mahwah, NJ: Lawrence Erlbaum Associates.

Schwartz, D., & Bransford, J. (1998). A time for telling. *Cognition and Instruction, 16*(4), 475–522. doi:10.1207/s1532690xci1604_4

Zohar, A., & Ginossar, S. (1998). Lifting the taboo regarding teleology and anthropomorphism in biology education: Heretical suggestions. *Science Education, 82*, 679–697. doi:10.1002/(SICI)1098-237X(199811)82:6<679::AID-SCE3>3.0.CO;2-E

Chapter 8
Use of Living Systems to Teach Basic Engineering Concepts

Kauser Jahan
Rowan University, USA

Jess W. Everett
Rowan University, USA

Gina Tang
Rowan University, USA

Stephanie Farrell
Rowan University, USA

Hong Zhang
Rowan University, USA

Angela Wenger
New Jersey Academy for Aquatic Sciences, USA

Majid Noori
Cumberland County College, USA

ABSTRACT

Engineering educators have typically used non-living systems or products to demonstrate engineering principles. Each traditional engineering discipline has its own products or processes that they use to demonstrate concepts and principles relevant to the discipline. In recent years engineering education has undergone major changes with a drive to incorporate sustainability and green engineering concepts into the curriculum. As such an innovative initiative has been undertaken to use a living system such as an aquarium to teach basic engineering principles. Activities and course content were developed for a freshman engineering class at Rowan University and the Cumberland County College and K-12 outreach for the New Jersey Academy for Aquatic Sciences. All developed materials are available on a dynamic website for rapid dissemination and adoption.

DOI: 10.4018/978-1-61520-659-9.ch008

INTRODUCTION

An aquarium is an exquisite combination of interacting systems which can be analyzed using multidisciplinary engineering principles. Children typically have personal aquariums for their pet fishes and visit some large aquarium as part of a school field trip or as part of their family outing. Movies such as Disney-Pixar's "*Finding Nemo*", *Epcot's Living Seas* also make tremendous impact on a young audience. While these activities apparently raise the knowledge base in terms of nature and the environment, children seldom make a connection to the engineering principles playing out in the maintenance of a natural, commercial or personal aquarium. Thus the idea of using an aquarium to promote engineering concepts for a wide audience is innovative and exciting. A creative initiative between the College of Engineering at Rowan University, Cumberland County College (CCC) and the New Jersey Academy of Aquatic Sciences (NJAAS) to enhance STEM (**S**cience, **T**echnology, **E**ngineering, **M**athematics) education at all levels has been undertaken by receiving support from the National Science Foundation. There is a growing realization among engineering faculty that a new vision for the education of engineers needs to evolve to keep this country at the forefront of technology. Science and engineering are essential for paving the way for America's future through *discovery, learning and innovation*[1]. A recent report[2] indicates that the United States lags behind the world in technological innovation because of its poor performance in teaching math and science. This eliminates many of the best and brightest schoolchildren from the ranks of future scientists and engineers. Many students who do undertake science and engineering studies in college are unprepared and drop out in frustration, while other potentially capable students never consider these subjects in the first place. In both cases, precious human and institutional resources are squandered. Enhanced engineering education in our K-12 classrooms can provide students at an earlier age with a more specific understanding of what a technical career entails.

The College of Engineering at Rowan University is always seeking innovative teaching methods to excite freshman engineering students about engineering design (Jahan, K., Hesketh, R. P., Schmalzel, J. L. & Marchese, A. J., 2001; Harvey, R., Johnson, F., Marchese, A. J., Newell, J. A., Ramachandran, R. P., & Sukumaran, B., 1999; Hesketh, R.P., Farrell, S., & Slater, C.S., 2003; Schmalzel, J. L., Marchese, A. J., Mariappan, J., & Mandayam, S., 1998; Hesketh, R. P., Jahan, K., & Marchese, A. J., 1997; Marchese, A. J., Newell, J., Ramachandran, R. P., Sukumaran, B., Schmalzel, J. L & Maraiappan, J. L., 1999; Jahan, K.., & Dusseau, R.A., 1998; Jahan, K., Marchese, A. J., Hesketh, R.P., Slater, C.S., Schmalzel, J.L., Chandrupatla, T.R., & Dusseau, R.A., 1998; Jahan, K., & Dusseau, R.A., 1998; Ramachandran R. P., Schmazel, J., & Mandayam, S., 1999; Marchese, A. J., Ramachandran, R. P., Hesketh, R., Schmalzel, J., & Newell, H. L., 2003; Farrell, S., Hesketh, R. P., Newell, J. A., & Slater, C. S., 2001). The aquarium project was selected to expose K-12 students/educators, freshman students in engineering at Rowan and CCC to basic science and engineering concepts. Students can easily be introduced to chemical, mechanical, electrical, civil and environmental principles such as mass and energy balances; fluid flow; work, energy, and efficiency; forces and levers; material strength and stresses; water quality and treatment; and electrical signal processing via this project. The aquarium theme also adds to the need for an understanding of biological systems, ecosystems, pollution and sustainable development. These are concepts that have been absent in typical traditional engineering courses.

PROJECT PARTNERS

NJAAS[3] and CCC[4] were selected as partners for enhancing the broader impacts of the project. Both CCC and NJAAS are located within 30 minutes of the Rowan University campus and both these locations are within the two New Jersey Federal Empowerment Zones (EZ). The EZ programs are designed to empower people and communities across the United States by inspiring communities to work together to develop a strategic plan designed to create jobs and opportunities in the nation's most impoverished urban and rural areas. Our three-way partnership is summarized in Figure 1.

PROJECT ACTIVITIES

Project activities include the development of hands-on activities, implementation of the activities in courses, workshops and in outreach activities, dissemination via a dynamic website and other traditional mechanisms.

DETAILED PROJECT PLAN

The proposed project is comprised of three major areas that introduce students to fundamental principles of science and engineering relevant to an aquarium. These areas include (1) Engineering Fundamentals, (2) Instrumentation and Control and (3) Ethics and Sustainability.

1. Engineering Fundamentals

This group of experiments allows students to be exposed to various common engineering principles that govern material properties, gas and heat transfer, hydraulics, light, and water treatment.

Material Testing

This module focuses on material properties such as stress, strain, hardness, density, refractive index, insulation, and even toxicity as some materials may have additives that are toxic to aquatic and marine life. Students specifically focus on acrylic (commonly used for aquariums) and compare it to glass.

Figure 1. Chart indicating partners and their respective roles

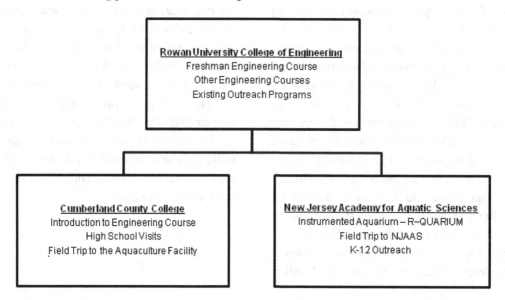

Gas Transfer

This experiment exposes students to the principles of gas transfer. Students will conduct a visual simple batch experiment with various commercial diffusers to transfer oxygen into the water and compare diffuser efficiencies.

Respiration

Students also investigate the differences in air and water as respiratory environments. The most significant difference is that water contains only $1/13$ as much O_2 as air does, or 1% to 21% (water to air) by volume. The energy that is required to move water into and out of a respiratory organ is much more than that used to move air because water is more dense and viscous.

Hydraulics

Water pressure in an aquarium is a major parameter in design considerations. Students conduct experiments to determine the pressure distribution and net forces on the aquarium walls using conventional hydraulics equipment. They also conduct experiments on head losses in pipes and porous media such as sand.

Light

Light intensity is of utmost importance in the operation of an aquarium. Students design a lighting system to simulate day and night, while the temperature is maintained within a certain range. Students also learn to use timers to conduct open loop control, and use thermostats as sensors for feed-back control. Simulink is used to simulate various lighting scenarios.

Water Treatment Processes

Students observe demonstrations of various water treatment processes such as Sand Filtration, Re-verse Osmosis, Ultrafiltration and Ion Exchange. Specifically they work on an experiment on determining the adsorption coefficient of activated carbon in removing organic pollutants.

2. Instrumentation and Control

These experiments focus on the use of remote sensing, real time data and images on the WWW. A 100 gallon marine aquarium that is maintained at the NJAAS distance learning classroom has been instrumented with a webcam and sensors to display real time images and water quality data through the project website. This instrumented aquarium is titled **R~QUARIUM.** A web-camera and a sensor system are used to instrument the marine aquarium. The sensor system allows real time values of water pH, Dissolved Oxygen, Conductivity and Temperature.

Data Acquisition and Remote Control

This module exposes students to data acquisition techniques. Students' prototype a relatively simple but representative example of a technologically complex feedback control system to monitor water quality parameters (e.g., aquarium temperature, pH, Conductivity, and Dissolved Oxygen). Students also identify all relevant electronic components (e.g., sensors, amplifier, A/D, and D/A, etc.) and their interrelations.

WWW and Networking

One important feature of the *R~Quarium* is on-line real-time data acquisition. This module focuses on the basic networking technology in support of this theme. Students are introduced to a seven-layer networking model, distributed systems as exemplified in the Internet, and database fundamentals with emphasis on their application to the aquarium. A website has also been dedicated for rapid information dissemination[5].

3. Environmental Ethics and Sustainability

These modules specifically focus on the impact of engineered activities on the environment. Students learn about the effect of pollutants on receiving waters, the benefits of alternate energy and watch videos on marine pollution and protection.

Toxicity of Pollutants to Aquatic Organisms

Typically engineering students are never exposed to experiments dealing with toxicity measurements of pollutants. Ironically most engineering facilities contribute to pollution of water via industrial, municipal or accidental discharges. Students use the HACH BOD Trak Equipment to determine the toxicity of synthetic wastewaters typically generated from Chemical, Electrical and Mechanical Manufacturers.

Alternate Energy

Students use solar panels to deliver energy to run a small aquarium pump and compare the energy costs to traditional delivery methods. Students also work on a simple experiment on making biodiesel from vegetable oil and methanol (Meyer, S.A., & Morgenstern, M.A., 2005).

Videos

In order to emphasize the role of environmental ethics and environmental justice issues, students watch videos and discuss specific case studies. Specifically students watch the impact of the Exxon Valdez Spill on the fishing community in Cordova, Alaska (CNN, 1999) and the impact of technology on driving species to extinction (*EmptyOceans Empty Nets,* 2003).

Each learning module has been developed such that it has broad appeal to different learning styles, is simple, cost effective and allows students to see applications of their math and science courses. Gender sensitive information is also provided such as contribution of male and female scientists to the specific topic. The contributions of Rosalind Franklin to the structure of activated carbon, the contribution of Stephanie Kwolek for the development of Kevlar and the contributions of Rachael Carson to environmental protection and regulations are specifically discussed in the water treatment, materials and ethics modules. All activities have been posted on the project website with laboratory handouts, K-12 lesson plans and video clips of the activity for rapid and easy adoption.

IMPLEMENTATION OF ACTIVITIES

The hands on activities that have been developed have been implemented in the Freshman Engineering course at Rowan University and CCC. Both NJAAS and CCC allow the Rowan University students to visit the state aquarium and the aquaculture centers as part of their engineering course. These field trips allow the students to connect concepts learnt in the classroom to the real world. Apart from the above experimental experiences, students are also exposed to traditional experiences such as engineering drawing using software (AutoCAD, PCAD and Solid Works), technical report writing and oral presentations.

We have also worked with NJAAS personnel to instrument an existing 100 gallon marine aquarium as seen in Figure 2 in their distance learning classroom to provide real time images and water quality data that will be posted on their website and titled the R~QUARIUM. This is a unique opportunity for NJAAS as currently none of the NJ State Aquarium exhibits have real time data and images online. The R~QUARIUM site will also generate excitement among students of all ages. Students will not only visualize activities of different living organisms in the aquarium, they will also see values of water quality parameters such as pH, dissolved oxygen, conductivity and temperature in the aquarium. Educators will be able to download data and help students plot and conduct statistical analyses of the data.

Hands on activities are also used in different workshops offered for teachers and K-12 students at Rowan University. Developed experiments can be used in a variety of engineering courses as shown in Table 1.

DISSEMINATION

The aquarium project has disseminated in a variety of ways. A dynamic website been established to disseminate our Aquarium Project.

The right hand corner of the website presents basic information for prospective educators and students of any age group. The photos section has some striking images of the Rowan engineering students working on data acquisition as presented below in Figures 3(a) and 3(b):

All laboratory activities, handouts, experimental methods, quizzes and exams are posted on this website. Navigation through the webpage has been made easy with titles to useful links. The webpage is presented in Figure 4.

The Sensors option provides information about the various types of water quality probes being used for collecting information on the water quality of the marine aquarium at NJAAS.

The LIVE DATA section displays real-time pH, temperature and Dissolved Oxygen data for the instrumented aquarium located at NJAAS. The WEBCAM provides live feed of the images of the marine aquarium. Educators and students

Figure 2. Instrumented marine aquarium at the NJAAS

Table 1. Courses incorporating proposed experiments

Civil	Electrical and Computer	Mechanical	Chemical
Water Treatment and Design	Networks 1	Applied Heat Transfer	Biochemical Engineering
Wastewater Treatment	Networks 2	Biomechanics	Bioprocess Engineering
Solid and Hazardous Waste Management	Sustainable Design	Thermodynamics	Reaction Engineering
		Mechanics of Materials	Thermodynamics
			Unit Operations

Figure 3. (a) and (b), data acquisition images

Figure 4. Webpage for the R-QUARIUM project

also have the opportunity to look up historic data under the ARCHIVES option. This option provides daily, monthly and annual water quality data in a tabular form or in the form of an EXCEL spreadsheet. Plots can also be obtained for the water quality parameters for a select time period. Educators can also show and teach simple statistics such as average, MIN, MAX, standard deviation and median values using this data. A sample plot is presented in Figure 5 for weekly temperature data. The plot options are available for all water quality parameters from the sensors in the instrumented aquarium.

The Meet the Fish option allows children to learn about various types of fish species in the marine aquarium while the Fish Tank Science option allows visitors to learn about aquarium care. The Virtual Fish Tank option allows children to play with an animated aquarium and learn about engineering and maintenance of an aquarium. Children can click on various options to learn about different fish and their needed water quality environment. Images for this option are provided in Figure 6.

The ACTIVITIES section provides experimental handouts for K-12 and college educators and students. Activities have been divided according to school standards. An example laboratory handout for the Gas Transfer is provided below.

Gas Transfer

Objective: This experiment will allow students to visualize how oxygen is transferred from the gas phase (air) to a liquid phase (water).

Applications to an Aquarium

Fish need a certain level of oxygen for survival. As such oxygen gas needs to be transferred to water efficiently.

Other Engineering Applications

Gas transfer is a vital unit operation in many environmental engineering processes. It involves either desorption or adsorption of gas. The transfer of oxygen to liquid systems is particularly important in oxidation of iron and manganese in water treatment and in the biological treatment of wastewaters.

Figure 5. Weekly sample plot of temperate from the ARCHIVE DATA section

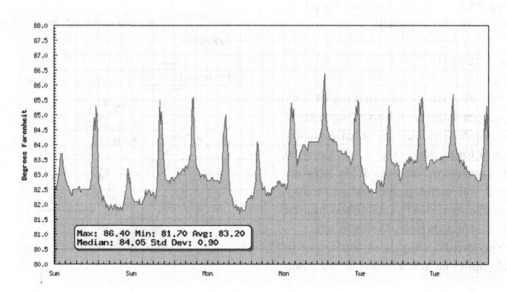

Figure 6. Virtual fish tank

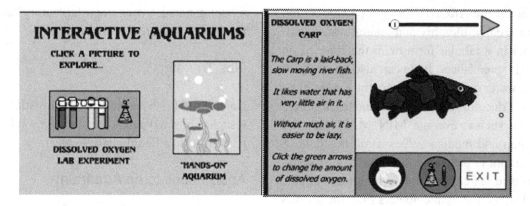

Video Clip

A video clip of the experiment is available on-line so that students/educators can visualize the experiment online.

Upper Level Courses

The following courses in engineering at the Junior/Senior years can use this experiment. Water and Wastewater Treatment and Design; Chemical Reaction Kinetics; Unit Operations

Lesson Plans for Teachers

Age appropriate lesson plans for K-12 teachers are also available on the website.

Famous Scientists

Students can also see the contribution of both men and women in this area. Robert Boyle made important contributions to experimental chemistry and is known for his ideal GAS LAW. Cornelia Clapp was a professor of zoology at Mt. Holyoke College and was active in the marine biology research at the then newly established Marine Biology Lab at Woods Hole, Massachusetts.

The experimental setup is shown above in Figure 7. Supplies include a beaker, an air pump

an air diffuser, a stopwatch, and a DO meter with probe (or a DO measurement colorimetric kit).

Experimental Setup and Data Analyses

The setup is simple and inexpensive. Cheap DO kits can be used instead of a DO probe. Aeration kinetics can be expressed as $\frac{dC}{dt} = K_L a(C^* - C)$ where C is the DO concentration at time t and C* is the saturated DO concentration at the experimental temperature (Davis, M.L., & Cornwell, D.A., 2001).

Figure 7. Experimental setup of the gas transfer experiment

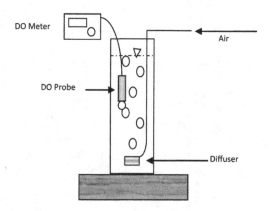

The above equation can be integrated as follows

$$\int_{Co}^{Ct} \frac{-dC}{C*-C} = K_L a \int_{o}^{t} -dt$$

to obtain $\ln(C* - C_t) = \ln(C* - C_o) - K_L at$

Students collect C_t versus Time (t) data and plot $\ln(C* - C_t)$ versus time to obtain a linear line. The slope of this line is equal to $K_L a$. The experiment can be run with various diffusers and the results can be compared to determine the best performance. The experiment can be instrumented by hooking up the DO meter to a computer for data acquisition. For K-12 educators they can have students plot the DO versus time profiles with their students. Students can also evaluate various air diffusers that can be purchased from common

pet stores. Raw data is available on the website for K-12 educators. Samples of data analyses (Davis, M.L., & Cornwell, D.A., 2001) and data plots are provided below for two diffusers Airmax and Wonder in Table 2 and Figure 8.

The slope of the lines provides the values of the gas transfer coefficient. The entire exercise teaches students many engineering skills. The experiment is very visual and simple to understand. The data collection is easy and the analysis exposes students to integral calculus, data plotting and trendline fitting using a spreadsheet such as EXCEL.

CONCLUSION

The aquarium project generates a lot of enthusiasm and excitement in our freshman class and K-12

Figure 8. Determining gas transfer coefficients by plotting data

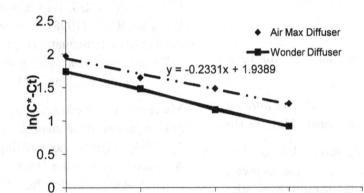

Table 2. Oxygen transfer data analyses

Time (min)	Air Max C_t mg/L	Wonder C_t mg/L	Air Max (C* -Ct) mg/L	Wonder (C* - Ct) mg/L	Air Max ln(C*-Ct)	Wonder ln(C*-Ct)
0	2	3.5	7.2	5.7	1.974081	1.740466
1	4	4.8	5.2	4.4	1.648659	1.481605
2	4.8	6	4.4	3.2	1.481605	1.163151
3	5.7	6.7	3.5	2.5	1.252763	0.916291

educators and students. The overall impact of the projects appears to be positive and exciting. The theme is also extremely useful for outreach activities that range from K-12 education through college students. The three-way partnership has also added great value to the project. Initial responses to the project from female students have also been positive. This project can be a national model for other institutions and aquariums for adoption.

ACKNOWLEDGMENT

This project is being funded by the National Science Foundation (DUE-0737277). Any opinions, findings and conclusions or recommendations expressed in this material are those of the author(s) and do not necessarily reflect the views of the National Science Foundation (NSF).

REFERENCES

CNN (1999). *Recovering from the Exxon Valdez oil spill in Alaska*.

Davis, M. L., & Cornwell, D. A. (2001). *Introduction to Environmental Engineering*. McGraw Hill.

Farrell, S., Hesketh, R. P., Newell, J. A., & Slater, C. S. (2001). Introducing freshmen to reverse process engineering and design through investigation of the brewing process. *I. J. E. E (Norwalk, Conn.)*, *17*(6), 2001.

Harvey, R., Johnson, F., Marchese, A. J., Newell, J. A., Ramachandran, R. P., & Sukumaran, B. (1999). Improving the Engineering and Writing Interface: An Assessment of a Team-Taught Integrated Course. *ASEE Annual Meeting*, St. Louis, MO.

Hesketh, R. P., Farrell, S., & Slater, C. S. (2003). An Inductive Approach to Teaching Courses in Engineering. 2003 ASEE Annual Conference, Session 2531, June 2003.

Hesketh, R. P., Jahan, K., & Marchese, A. J. (1997) Integrating Hands-on Education to Freshman Engineers at Rowan College. *1997 ASEE Zone 1 Spring Meeting*. West Point, NY, April, 1997.

Jahan, K., & Dusseau, R. A. (1998). Teaching Civil Engineering Measurements through Bridges. In *Proceedings of the 1998 Annual Conference of ASEE*, Seattle, Washington, June, 1998.

Jahan, K., & Dusseau, R. A. (1998). Water Treatment through Reverse Engineering. In *Proceedings of the Middle Atlantic Section Fall 1998 Regional Conference*, Washington D.C., November 6-7, 1998.

Jahan, K., Hesketh, R. P., Schmalzel, J. L., & Marchese, A. J. (2001). Design and Research Across the Curriculum: The Rowan Engineering Clinics. *International Conference on Engineering Education. August, 6 – 10, 2001 Oslo, Norway.*

Jahan, K., Marchese, A. J., Hesketh, R. P., Slater, C. S., Schmalzel, J. L., Chandrupatla, T. R., & Dusseau, R. A. (1998). Engineering Measurements and Instrumentation for a Freshman Class. In *Proceedings of the 1998 Annual Conference of ASEE*, Seattle, Washington, June, 1998.

Marchese, A. J., Newell, J., Ramachandran, R. P., Sukumaran, B., & Schmalzel, J. L & Maraiappan, J. L. (1999). The Sophomore Engineering Clinic: An Introduction to the Design Process through a Series of Open Ended Projects. In *Proc. Conf. Amer. Soc. Eng. Edu*, Charlotte, NC.

Marchese, A. J., Ramachandran, R. P., Hesketh, R., Schmalzel, J., & Newell, H. L. (2003). The competitive assessment laboratory: Introducing engineering design via consumer product benchmarking. *IEEE Transactions on Education*, *46*(1), 197–205. doi:10.1109/TE.2002.808216

Meyer, S. A., & Morgenstern, M. A. (2005). Small scale biodiesel production: A laboratory experience for General Chemistry and Environmental Science students. *Chemical Educator*, *10*, 1–3.

Empty Oceans Empty Nets (2003). Bullfrog Films.

Ramachandran, R. P., Schmazel, J., & Mandayam, S. (1999). Engineering Principles of an Electric Toothbrush. ASEE annual conference and Exhibition, Charlotte, North Carolina, Session 2253, June 20-23, 1999.

Schmalzel, J. L., Marchese, A. J., Mariappan, J., & Mandayam, S. (1998). The Engineering Clinic: A Four-Year Design Sequence. *2nd Annual Conference of National Collegiate Inventors and Innovators Alliance,* Washington, DC.

ENDNOTES

1.   http://www.nsf.gov
2.   See http://www.engineeringk12.org/Engineering_in_the_K-12_Classroom.pdf and http://www.nae.edu/nae/naehome.nsf
3.   See http://www.njaquarium.org/index2.html and http://www.njaquarium.org/Community/PISEC.html
4.   http://www.cccnj.edu/
5.   http://elvis.rowan.edu/rquarium/

Chapter 9
The Use of Applets in an Engineering Chemistry Course:
Advantages and New Ideas

B.M. Trigo
Polytechnic School of the University of São Paulo, Brazil

G.S. Olguin
Polytechnic School of the University of São Paulo, Brazil

P.H.L.S. Matai
Polytechnic School of the University of São Paulo, Brazil

ABSTRACT

This chapter deals with the use of Applets, which are examples of software applications, combined with a specific methodology of teaching, based on Paulo Freire's education concepts. According to his methods, co-creation between its participants is fundamental for the effectiveness of learning process. In that way, to promote a cooperative learning, the Applet should have interactive features. The Chemistry course of Polytechnic School of the University of São Paulo, in which students take in the first semester of the first year of the engineering course, was the case study. First, a research with the teachers of the Chemical Engineering Department was carried out, to identify the main problems and difficulties teachers and students face. Then, a topic was selected to be explored with the Applet, which was developed and applied to a small group of students. To identify the success of this experiment a questionnaire was created and the results are presented in this chapter. Some conclusions were drawn and the interactive features of the Applet received a positive feedback.

INTRODUCTION

Since the sprouting of the personal computers, around 1980, computer science has been gaining importance, becoming a supporting instrument for other daily activities of ours. With the introduction of computing in a certain area or activity, barriers and difficulties can be overcome. Consequently, new paradigms, possibilities and challenges are created.

In education it is not different, computing is more and more present, assisting the learning process in a variety of ways, creating new challenges

DOI: 10.4018/978-1-61520-659-9.ch009

that compel us to re-think the way education is performed, considering new delivery methods instead of only the traditional chalk-and-talk method. "While we may feel comfortable with traditional approaches, the new technologies provide us with the tools to challenge these positions, and open up the teaching/learning questions for some rethinking" (Roy & Lee, 1999).

With the sprouting of the Internet, it was created what we today understand as the web-based teaching, a method that brings innumerable benefits and challenges for the educators. There are many reasons and motivations to develop new processes of learning and teaching. One of the most important ones are:

- To provide an environment where the student can have access to the studying material and develop his learning in his own pace and in his own environment.
- To provide to the professors quality resources to improve the quality of teaching and learning.
- The increasing growth of the Internet will promote new possibilities. However, what educators must inquire themselves is how to use this new media so that its application doesn`t consist, only, of a substitution of the current media.

TEACHING METHODOLOGIES AND TECHNOLOGIES IN EDUCATION

To achieve those goals, we have to understand the problems we are dealing with. Following this thought, different ways of transmitting knowledge have been searched before. According to Freire, "the communication is the relation which becomes effectuated by the co-participation of the subjects in the act of knowing. It is considered that the educative process is a particular form of communication. In the social relations among the learning subjects, a dialectic synthesis happens, as moments of a communicative and educative process, in a given socio-cultural context, which comprehension requires considering its inter-subjective nature, or either, the active participation of the subjects of the process" (Aragão, 2004). For being a mean of communication, we cannot overlook the new communication technologies. Their development modifies, however, the education, when seen as a media. Or else, today, the formation is directly related to the way in which the professor uses the media. Paulo Freire's vision is clear in what concerns the active participation of the student in the learning process. Without this participation, the learning process becomes a reproduction of events, cases and facts, previously lived deeply by the student, what, according to Freire, can be considered a banking education. This term comes from the fact that the student becomes a "bank deposit", where the professor "only deposits" the knowledge. Therefore, it becomes of full responsibility of the professor "to nourish" the deposit and the student only needs to absorb the maximum of what is presented to him, for future reproduction.

We know that it is not possible to present all the knowledge to a student, especially nowadays, where the barriers of the knowledge are being broken in a higher and higher speed. But what happens when this "banking student" comes across a new and challenging situation? The answer is simple: he simply cannot face this difficulty, for this knowledge has never been deposited on him. Following this line of thought, an education where the student participates actively, treading his way, seems more reasonable. Thus, a new perspective of education would be an active, interactive, dynamic education to build a better learning in spite of the traditional pedagogical approach, based on the transmission and reproduction.

"The Digital style produces, obligatorily, not only the uses of new equipment for the production and apprehension of knowledge, but also new behaviors of learning, new rationalities, new percipient stimulus" (Aragão, 2004). Many educators possess an ingenuous, "magical" vision, towards the technologies and its potentialities, thinking that with the insertion of new technologies learning problems will finish. The advances on new technologies will be of little help without changes in the ways of education, of learning and communication. "We assume that the effectiveness of computer simulations is determined by two factors: the simulation design and the context in which simulations are used" (Zhou, Brouwer, Nocente & Martin, 2005). In order to change the context in which we live, we need to fully involve the student, by creating an environment of co-creation of the knowledge, where professors as much as students are active parts, treading together the way to be covered.

THE LABORATORY CLASSES IN CHEMISTRY TEACHING

The laboratory classes, in particular, can obtain great advantage when we examine the new technologies and methods of education. Even more if we come to analyze the difficulties that have been faced in these classes, since its introduction until the present. "Chemistry as we know it, did not exist before Lavoisier (1743-1794). The wealth of facts that had been accumulated by the practitioners of the chemical arts were gathered by Lavoisier and others and interpreted correctly to start what some have called the chemical revolution" (Ellio, Stewart & Lagowski, 2008). The practice of instrumental chemistry did not exist in an academic environment before this period, in spite of the fact that some remarkable chemists of the time used to give classes. Few students had access to the practice of chemistry, through these professors' private laboratories. At the beginning of the nineteenth century, the first chemistry laboratories started to appear, with Goettingen, J.N. Von Fuchs, Dobereiner, N.W. Fischer and Liebig`s. "Liebig`s laboratory-oriented teaching methods stand out among this group because his

methods seem be the antecedents of those found in the modern graduate-level chemical research program" (Ellio, Stewart & Lagowski, 2008).

"Laboratories are one of the characteristic features of education in the sciences at all levels" (Reid & Shah, 2007). Everybody knows of the importance of the laboratory classes, however a few manage to justify its presence. Could it be fundamental for the learning process? "Many research studies have been conducted to investigate the educational effectiveness of laboratory work in science education in facilitating the attainment of the cognitive, affective, and practical goals. These studies have been critically and extensively reviewed in the academic literature. From these reviews it is to clear that in general, although the science laboratory has been given distinctive rolls in science education, research has failed to show simple relationships between experiences in the laboratory and student learning" (Hofstein & Mamlok-Naaman, 2007). Moreover, we can criticize the existence of the laboratory classes for the high cost of its programs, as much in terms of material as in terms of the involved staff. "University student's reactions to the practical work are often negative and this may reflect student perception that there is lack of any clear purpose for the experiments: they go through the experiment without adequate stimulation" (Reid & Shah, 2007).

Liebig's style of chemical instruction can be described as "problem-based", because in this way the student developed his abilities and strategies of research better, in contrast to many laboratories where the students only observed the professors making the experiments. We notice that, to reach the goals which have been proposed by a laboratory class, it is necessary to use an adequate methodology, where the student feels motivated and perceives the importance of the experiments and their relations to the theory. But will an adequate methodology, by itself, be enough? Perhaps yes, but certainly with the introduction of computing in the academic world, some difficulties which have been found before can be overcome, improving the learning process as a whole. We are not trying to say that a virtual laboratory can replace the traditional laboratory classes. Instead, it could complement it, for many reasons, including financial, security, simulations that are microscopic, ease of information search over the internet, and others.

CASE STUDIES

"Essential Simulation is an engineering tool used by both students and practitioners to gain insight into a system's behavior" (Veith, Kobz & Koelling, 1998). Simulations can very be useful when we want to analyze the behavior of systems, without having to act in the systems themselves, either for their complexity, for their availability or for the cost of the system. Next an example of author Tamie L. Veith is presented, where the simulation brings innumerable benefits, because the analyzed system is very complex, expensive and of difficult access by students: "As part of an engineering problem-solving class, a student is given a simulation model of a manufacturing line that consists of several machines in sequence. The cost of holding partially completed items, revenue associated with completed items, and the relationship between production rates and per-unit production costs are embedded in the simulation model. The student specifies the production rates at the different machines and runs the simulation. The model shows items moving through the manufacturing line, the queues of items at the different machines and the customers picking up completed items at the end of the line. After execution, the simulation computes an average per-unit profit or loss for the manufacturing line. Through repeated interaction with the simulation the student identifies the production rates that maximizes the long-run profits" (Veith, Kobz & Koelling, 1998).

Applets are software applications, which are executed in the context of another program, in the majority of the times a web browser, to create an interface with the user. The Applet normally interacts with this host program through restricted privileges of security. The use of applets can be very interesting; therefore these simulations can present characteristics that make possible the active learning by the students. The use of these software can overcome many of the difficulties professors face. One example is the applet used in the University of São Paulo by Professor Eduardo Toledo dos Santos, for the Descriptive Geometry online teaching. One problem found in the teaching of Descriptive Geometry is that the resolution of exercises on the blackboard requires great skills and precision. The use of a simulation would help the professors with these kinds of issues. "Another problem refers to the fact that the individual learning speed varies greatly among the students of a same classroom, demanding that each step is explained several times, which can compromise the attention needed for the learning, especially in exercises of difficult space visualization" (Santos, 1999).

A CHEMISTRY APPLET APPLICATION

The Polytechnic School of the University of São Paulo offers 15 engineering courses, which all have a first year in common. In this common year, all basic courses are taught, including a chemistry course called General Technological Chemistry (QTG). The official contents of the discipline are: "Topics: the chemical composition in microscopic level and how the constituent units of materials for Engineering are arranged and interact with each other; concepts on chemical behavior of materials, that is, the reactions of degradation of metallic materials (electrochemical and corrosion); the mechanisms of action and the main uses of substances that act as surfactants; the use of fuels; aspects related to environmental chemistry. Objectives: develop critical awareness about the importance of environmental management in the exercise of Engineering" (General Technological Chemistry contents). Beyond the expositive lessons, the students have activities in laboratory classes, with the intention to create more proximity with the chemistry and the concepts which have been worked in classroom.

A previous interview with the QTG professors was made to list the difficulties of the students and the professors (Trigo, Olguin & Matai, 2008). The main one was the problem the students face to interrelate the concepts approached in the course. This issue comes from another one, the problem the professors have to create a better chronogram for the discipline. Because QTG is a basic course, the number of concepts taught is very high and the number of lessons is restricted, what causes a difficulty for the professors, once they cannot deepen the concepts and do not have enough time to relate them in the way that they would like. Thus, the students don't see how one subject is related to another one, which causes a student's loss of interest. The same issue affects the laboratory classes of QTG. Students feel that the laboratory classes have no relation with the theory, because they don't understand how the laboratory classes complement what is seen in the classroom.

Taking this into consideration, in the interview with the professors, a proposal came up to develop Applets covering the topics of the discipline and join them in a main "host" Applet, creating a virtual chemistry laboratory, which would be available for the students in the discipline's website. The intentions of this work would be to show, in an interactive way, how the topics interrelate with each other. The purpose of the virtual laboratory is not to substitute the traditional laboratory classes or the lessons, but to be a tool that complements what is taught in the discipline, improving the students learning.

To reach the expected effect, these Applets have to present some characteristics discussed above. High interactivity is certainly a main one. A clear and efficient interconnection among the topics covered in the Applets is another important one. To achieve these, a conceptual map was created, to be a guideline for the development process. Conceptual maps are graphical representations that aim to show how the concepts relate to each other. The concepts appear inside boxes and the relations are specified through linking phrases, which connects the boxes. From the conceptual map, the topic electrochemistry was chosen to start the development of the Applets. An electrochemistry Applet was developed, using methodologies that make the student participate actively with the simulation, to contribute to the QTG discipline. The Applet was used in a group test in the first semester of 2009, and a questionnaire was developed to try to create metrics to calculate the effectiveness of the Applet in the student. The results were collected and analyzed, showing some positive feedback from the students. QTG discipline is planning on using the Applet in the first semester of 2010.

METHODOLOGIES

After the elaboration of the electrochemistry Applet, an activity was conducted with a test group, composed of 30 students of the second year of engineering at University of São Paulo. These students participate in this activity in a voluntary basis, and all of the second year students of the Chemistry, Materials, Metallurgy and Mine courses were invited, for the proximity of these areas with the knowledge area of Chemistry.

The activity, accomplished in May of 2009, had the presentation of two different computational simulations for the students, called expositive simulation and interactive simulation. The expositive simulation, found in the tutorvista web site, is an electrolysis simulation that has animations and audio explanations. The student could control the simulation with a play/stop button. During the presentation, the student only watched the simulation, that's the reason why we called the expositive simulation. This simulation was chosen for this purpose because it has a good graphics quality, correct explanation of the theory involved, and it was a method of presentation close with the banking education described above, since it was no interactive with the student in the learning process. The interactive simulation is the electrochemistry Applet developed through the opinion of QTG professors at University of São Paulo. The student can interact with the simulation, changing the value of different variables. For that reason we called it the interactive simulation. In the application of this simulation, the student received an activity itinerary, where each step was an interaction of the simulation. In some cases, for example, the student had to verify the changes that occur when a variable, like temperature, changed from 100 degrees Celsius to 200 degrees.

The idea was to apply both simulations and compare the teaching methodologies, according to Freire, the banking education and the emancipatory education. The first simulation applied was the expositive and then the interactive. In the end, a questionnaire was applied to know the reactions of the students. The questionnaire was composed of multiple answer questions, and some of those questions could have more than one answer and an open field for commentaries.

The first two questions had the intention of identifying the group, to know them better.

- Question 1: In what kind of high school did you study?
  A. Public.
  B. Private.
- Question 2: Did you go to a pre-college school?
  A. No.
  B. Yes, for one year.

C. Yes, for two years.
D. Yes, for more than two years.

In Brazil there is a big difference in quality of teaching when you compare the public and the private school systems. With few resources, the public schools don't have the proper infrastructure, and in most cases, they don't have computers for the students do their researches, papers or to use computer simulations in the teaching process. Is Brazil there are also what we called pre-college schools. They are a preparatory school for the students that are going to take the college entry exams.

Questions 3 and 4 were made to check the previous experiences that the student had with computational simulations. In the following questions, the students are asked about the quality of the simulation, so it's necessary to know the previous experiences they had with those. A student that came from a private school, and were in contact with various simulations, will have a different opinion from the ones that never saw a computational simulation before.

- Question 3: Have you ever used a computational simulation related to your school contents?
  A. Never.
  B. Once only.
  C. More than once.
- Question 4: In what discipline?
  A. Physics.
  B. Chemistry.
  C. Math.
  D. Others.
  E. Never used a computational simulation before.

The other six questions deal with the application of the simulation. The question 5 makes a comparison between both simulations. Questions 6 and 7 evaluate what points of the simulations could be improved.

- Question 5: Regarding the application of the computational simulations that was made today, which one you thought was the most interesting?
  - A. The expositive simulation.
  - B. The interactive simulation.
- Question 6: What could be better in the expositive simulation?
  - A. User friendly.
  - B. Interactive.
  - C. Content.
  - D. Graphics.
- Question 7: What could be better in the interactive simulation?
  - A. User friendly.
  - B. Interactive.
  - C. Content.
  - D. Graphics.

The questions 8, 9 and 10 deal with the application and contents of the simulations in the context of the QTG discipline. The results of these questions are going to be the base of the future adaptations of the Applet. With that, we plan to use the Applet in the QTG discipline by 2010.

- Question 8: These simulations complement the contents taught in the QTG discipline?
  - A. No.
  - B. Little.
  - C. Regular.
  - D. A lot.
- Question 9: What is the best way to present these simulations for the students?
  - A. Leaving it in the QTG`s website.
  - B. Asking for a work assignment involving the simulation, done at home.
  - C. Presenting it in the class.
  - D. Doing an assignment involving the simulation in the class.
- Question 10: What subjects could be deepen in the interactive simulation?
  - A. Electricity.
  - B. The graphics of the Tafel`s equation.
  - C. The chemistry equations.
  - D. Application of the day-by-day of engineering.

DATA

With the data from questions 1 and 2 (Figure 1 and 2), it's possible to notice that most of the students studied at private institutions and made at least one year of pre-college schools, before joining The Polytechnic School of the University of São Paulo.

Figure 1. In what kind of high school did you study?

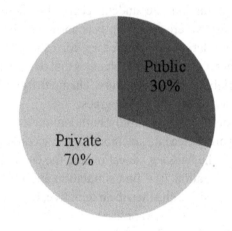

Figure 2. Did you go to a pre-college school?

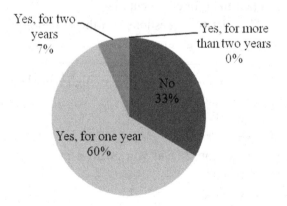

From the results of questions 3 and 4 (Figures 3 and 4), we see that 80% of the students have already seen an Applet before, in different areas of knowledge. Differently from the hypothesis raised, students from public and private schools have already seen computational simulation before. The percentage means that the group of students has, in its majority, previous experiences with simulations, and their opinion about it is more accurate and structured.

From the graphic 5 (Figure 5) it's possible to see that the interactive simulation had bigger acceptance. Analyzing the answers for questions 6 and 7 (Figures 6 and 7), we can conclude that interactive is the main differential between the simulations. From the open commentaries, we could realize that the critics referring the improvement on the contents of the interactive simulation were due to conceptual errors identified by the students. For example, one critic was that they were able to choose a gas for the electrode. In the simulation, there is this possibility, but no variables appear when you choose a gas. Still, for the students, the best choice is that a gas could not be chosen for an electrode.

Figure 3. Have you ever used a computational simulation related to your school contents?

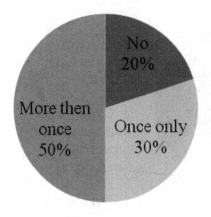

Figure 4. In what discipline?

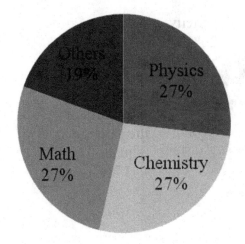

Figure 5. Regarding the application of the computational simulations that was made today, which one you thought was the most interesting?

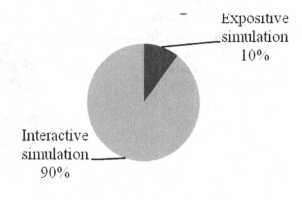

Figure 6. What could be better in the expositive simulation?

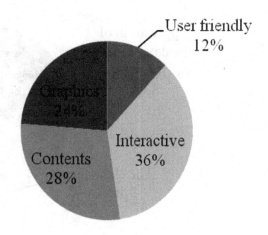

Figure 7. What could be better in the interactive simulation?

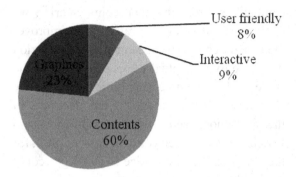

Figure 8. Simulations complement the contents taught in the QTG discipline?

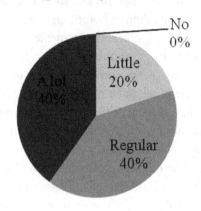

Figure 9. What is the best way to present these simulations for the students?

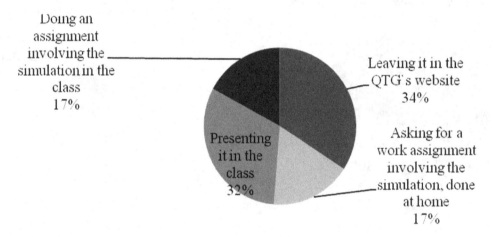

Figure 10. What subjects could be deepen in the interactive simulation?

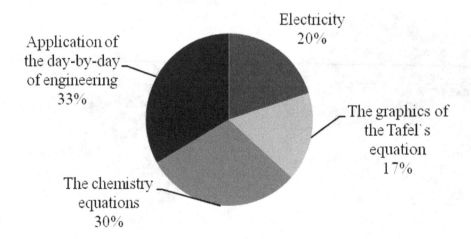

Finally, we can see from the graph 8 (Figure 8) that both simulations would help the discipline and, according to question 9 (Figure 9), their applications could be done in different ways, being a class assignment or a home assignment.

To give continuity to the work done so far, question 10 (Figure 10) provided us some indicatives of how the Applet could be expanded. We can see that the answers given by the students had a balance percentage.

CONCLUSION

New technologies can help the improvement of education, if used in a correct way. Taking this in consideration, different education methodologies were studied in order to develop a series of Applets for a discipline in the Polytechnic School of the University of São Paulo. The one chosen to be the guideline of these programs is in line with Paulo Freire's view of education, where the student has an active participation in the learning process. Without this participation, the learning process becomes a reproduction of previous events and facts lived by the student, which represents Paulos Freire's banking education. Co-participation is fundamental to have a better learning process, according to him.

In order to achieve this, interactive needs to be a strong feature of the Applets. The professors of the discipline QTG, together with the development team, created the idea of a virtual chemistry laboratory to help the students understand and have a better view of the whole process of the discipline and how the concepts approached in the course interrelate with each other. The first Applet was created and it was tested in a group of students. The students sent some positive feedback, which indicates that some of the objectives proposed by the Applets are being accomplished. One important aspect of the test group is that they had already been in contact with computational simulations before. That indicates that the group is critic and

can make a good judgment on the effectiveness of the simulations. Another important result was that the students felt that the Applet develop was better than the other simulation present to them, mainly because of the interactive aspect of the Applet.

With all that was discussed we feel that the use of simulations with interactive aspects is very important and can make the learning process very interesting for the students and the professors. There are several disciplines that can make use of computational simulations, remembering that, if used with the different methodology, one that tries to create a co-participation of everyone involved in the process, it can become a powerful tool to complement classes and laboratory classes already used.

REFERENCES

Aragão, C. R. D. (2004). A interatividade na prática pedagógica online: relato de uma experiência. *Revista da FAEEBA Educação e Contemporaneidade, 13*(.22), 341.

Ellio, M. J., Stewart, K. K., & Lagowski, J. J. (2008). The role of the laboratory in chemistry instruction. *Journal of Chemical Education, 85*(1), 145–149. doi:10.1021/ed085p145

General Technological Chemistry contents (2008). Retrieved from http://sistemas2.usp.br/jupiterweb

Hofstein, A., & Mamlok-Naaman, R. (2007). The laboratory in science education: the state of the art. *Chemistry Education Research and Practice, 8*(2), 105–107.

http://www.tutorvista.com/content/chemistry/chemistry-ii/electrolysis/electrolysis-animation.php

Reid, N., & Shah, I. (2007). The role of laboratory work in university chemistry. *Chemistry Education Research and Practice, 8*(2), 172–185.

Roy, G. G., & Lee, P. L. (1999). Interactive web-based teaching for computing in an engineering degree. *International Journal Engineering, 15*(5), 358–364.

Santos, E. T. (1999). *Um applet para o ensino de geometria descritiva na internet* (pp. 2519–2526). Anais XXVII Cobenge.

Trigo, B. M., Olguin, G. S., & Matai, P. H. L. S. (2008). *The Use of Applet in an Engineering Chemistry Course: a Case Study of the Polytechnic School of the University of São Paulo.* ICEE.

Veith, T. L., Kobz, J. E., & Koelling, C. P. (1998). World wide web-based simulation. *International Journal Engineering, 14*(5), 316–321.

Zhou, G. G., Brouwer, W., Nocente, N., & Martin, B. (2005). Enhancing conceptual learning through computer-based applets: the effectiveness and implications. *JI of Interactive Learning Research, 16*(1), 31–49.

Chapter 10
Competitive Design of Web–Based Courses in Engineering Education

Stelian Brad
Technical University of Cluj-Napoca, Romania

ABSTRACT

Developing engineering study programs of high quality, able to satisfy customized needs, with flexible paths of study, with easy and rapid access to the most appropriate educational facilities and lecturers is a critical and challenging issue for the future of engineering education. The latest developments in communication and information technologies facilitate the creation of reliable solutions in this respect. Provision of web-based courses in engineering education represents one of these solutions. However, the absence of physical interactions with the training facilities and the specificity of remote collaboration with lecturers rise up additional challenges in designing a high-quality web-based engineering course. In order to define superior solutions to the complex set of requirements expressed by several stakeholders (e.g. students, lecturers, educational institutions and companies), a comprehensive planning of quality and an innovative approach of potential conflicting problems are required during the design process of web-based engineering courses. In this context, the present chapter introduces a generic roadmap for optimizing the design process of web-based engineering courses when a multitude of requirements and constrains are brought into equation. Advanced tools of quality planning and innovation are considered to handle the complexity of this process. The application of this methodology demonstrates that no unique, best-of-the-world solution exists in developing a web-based engineering course; therefore customized approaches should be considered for each course category to maximize the impact of the web-based educational process.

DOI: 10.4018/978-1-61520-659-9.ch010

INTRODUCTION

Today's evolutions in science and technology lead to a rapid depreciation rate of knowledge in engineering. There are areas where this rate is less than one year; however, countless opinions consider an average depreciation rate of knowledge in engineering around three years. Producing companies operate in environments influenced by globalization, emphasising horizontal integration, innovation and customer satisfaction, while focusing on small number of business areas. In this very demanding economic environment, producing companies expect from engineers to excel from graduation to retirement. Therefore, continuous training of engineers is vital for ensuring business competitiveness from technological perspectives, this issue subjecting engineering education to tremendous pressures, either directly or indirectly.

The very wide areas in engineering study rise up many challenges on how to approach properly the educational process. Experience clearly shows there is no general pattern for success. Depending on the subject area, personalized models and means are required to maximize the impact of the educational process (Barros, Read & Verdejo, 2008; Brad, 2005; Ogot & Okudan, 2007; Popescu, Brad & Popescu, 2006). It should be also noticed that specific engineering theory needs to be reformulated and often interrelated with elements from other theories, with practical knowledge and with skills development before it can be applied in real-life problem solving (Brackin, 2002; Kolmos & Du, 2008; Yeo, 2008). For example, in engineering education, skills development includes many other aspects than technical or technological ones, like: team working, communication, project management, learning to learn, visioning, change management, leadership (Hutchings, Hadfield, Horvath & Lewarne, 2007; Kaminski, Ferreira & Theuer, 2004).

Dynamics of changes in the economic environment determines both undergraduate and postgraduate students in engineering to look for flexible, high quality and financially affordable paths of study, for easy and rapid access to the most appropriate educational facilities and to the most appropriate lecturers and trainers to satisfy specific needs. A good opportunity in front of such expectations stands in web-based education, which exploits the facilities provided by the latest developments in communication and information technologies to remotely access, either off-line and/or on-line, real and virtual labs, libraries, documentation, tutorials, seminars, courses, etc. (e.g. Bhatt, Tang, Lee & Knovi, 2009; Callaghan, Harkin, McGinnity & Maguire, 2008; Du, Li & Li, 2008, Ebner & Walder, 2008; Helander & Emami, 2008).

Provision of web-based courses is not a simple task (Finger, Gelman, Fay & Szczerban, 2005; Lau, Mak & Ma, 2006; Li & Wang, 2007). Beyond the immanent technological challenges, there are other issues that require meticulous treatment. In this respect, a web-based course should be viewed as an educational product that asks for high quality on four generic development axes: need-requirement axis, education provider axis, study program axis and teaching-learning process axis (Popescu, Brad & Popescu, 2006). The needs and related requirements are closely linked to the challenges that students encounter onto the workforce market, characterized by high fluidity and volatility. The education providers should permanently adapt their study programs to the latest technological developments and to the newest approaches in education. The teaching-learning process should be strongly directed towards students, offering them the possibility of customising the learning process and its outcomes. The four axes should be simultaneously tackled and the process should consider the dynamics of the need-requirements axis. Because the life-cycles of most engineering courses are very short, there is a major concern on designing and developing high-quality courses from the very first time; that is, a concern for effective and efficient engineering course design. Web-based engineering courses rise up supplementary requirements and must overpass additional constrains than classical, face-to-face courses, like: virtual collaborative experimentation, interactive remote approach, collaborative remote learning, off-line active learning, cross-institutional collaboration, remote test and assessment (e.g. Helander & Emami, 2008; Hutchings, Hadfield, Horvath & Lewarne, 2007; Jou, Chuang, Wu & Yang, 2008; Mackey & Ho, 2008; Rizzotti & Burkhart, 2006; Wang, Dannenhoffer, Davidson & Spector, 2005). Therefore, a careful and comprehensive planning is required to design a high-impact web-based engineering course (Bier & Cornesky, 2001; Brad, 2005; Brad, 2009; Koksal & Egitman, 1998; Ogot & Okudan, 2008).

These circumstances ask for competitive design of web-based courses, where competitive design should be viewed as the implementation of an optimal framework and the use of adequate concepts, methods and tools in all aspects that define the engineering study in order to define, design and develop, from the very beginning and in a timeframe required by the educational market dynamics, of a high-impact engineering course. In other words, quality must be "designed" within the course before delivering the course to the students (Brad, 2005).

This chapter is going to introduce a roadmap for quality planning and innovation of web-based engineering courses. The areas of intervention are firstly highlighted. Then, the methodology is revealed, together with the challenges in designing and instituting a high-impact web-based engineering course. The chapter ends with an exemplification of the approach.

BACKGROUND

Educational product, as any other complex product, requires a careful planning in the early stages of its development (Brier & Cornesky, 2001; Brad, 2005). Moreover, during the development phase, a lot of negative correlations between

various technical characteristics that define the performance of an educational product must be solved without compromises (e.g. time allocated to prepare the web-based course versus life-cycle of the web-based course). This necessitates innovative problem solving approaches (Altshuller, 2000; Brad, 2005). Basic elements about performance planning and innovative problem solving are further given.

Performance Planning

Performance planning means setting up performance expectations towards achieving some goals (Koksal & Egitman, 1998; Kumar & Labib, 2004; Popescu, Brad & Popescu, 2006; Suliman, 2006). For the case of a web-based engineering course, a result of the planning process is the formulation and structuring of various data that are afterwards used to support the conceptualization and synthesis process of the course. These data include, but are not limited to, the following:

- A well-structured set of technical characteristics that describes the web-based engineering course (they are also called performance metrics);
- Value weights and target values of the technical characteristics;
- Functions of the web-based engineering course and their value weights;
- Correlations between various technical characteristics;
- Relationships between requirements and technical characteristics, etc.

Technical characteristics are measurable, quantifiable indicators that altogether reveal the performance achieved by the web-based engineering course. From this perspective, technical characteristics can be planned (e.g. levels of performance over the life-cycle, target values, etc.).

Performance can be measured in absolute terms, by relating the performance of the given system to an ideal state, or in relative terms, by comparing the system with other competing systems (if they exist). Requirements offer information of what the system's stakeholders want from that system, whereas system's functions are entities that describe how the system operates. The process of planning the performance of a given system can be run for a single objective-function or concurrently, for several objective-functions (Brad, 2008).

When a single objective-function is considered, it deals with quality issues in most of the cases. Design for quality is actually the major concern in the field of engineering course development, too, as several research papers reveal (Bier & Cornesky, 2001; Brackin, 2002; Kaminski, Ferreira & Theuer, 2004; Koksal & Egitman, 1998; Kumar & Labib, 2004; Raharjo, Xie, Goh & Brombacher, 2007; Suliman, 2006; Yeo, 2008). In education, quality is about satisfying students' requirements with respect to a given course, as well as with respect to the whole educational system—in a broader sense (Koksal & Egitman, 1998; Brad, 2005; Popescu, Brad & Popescu, 2006). However, there is no relevant research that is reported in the literature highlighting results about quality planning of web-based engineering courses; the published researches being mainly focused on classical courses.

A web-based engineering course brings supplementary quality-related requirements than a classical one. Beyond requirements related to the course content and structure of the information, in the case of web-based courses implicit requirements become expressed requirements. For example, interactivity, collaboration, tangibility are major performance characteristics in designing a web-based engineering course (Baros, Read & Verdejo, 2008; Du, Li & Li, 2008; Finger, Gelman, Fay & Szczerban, 2005; Hamada, 2008; Helander & Emami, 2008; Jou, Chuang, Wu & Yang, 2008; Nedic & Machotka, 2006; Rojko, Hercoq & Jezemik, 2008).

To plan quality in an educational product, tools like quality function deployment (QFD) and analytic hierarchy process (AHP) are often used (Bier & Cornesky, 2001; Brad, 2005; Kaminski, Ferreira & Theuer, 2004; Kumar & Labib, 2004; Ogot & Okudan, 2007; Popescu, Brad & Popescu, 2006; Raharjo, Xie, Goh & Brombacher, 2007; Suliman, 2006). Basic information about these tools is further introduced.

About Quality Function Deployment

QFD is a structured planning and communication methodology consisting of a reunion of methods and means linked by special algorithms that together provide a robust way by which a multi-functional team identifies and transfers the needs and expectations of the stakeholders through each stage of system development and implementation (Brad, 2008; Brad, 2009; Kumar & Labib, 2004).

A QFD-project usually starts with the definition of stakeholders' requirements about the system under consideration. From this perspective, it is important to understand that requirements belong to various categories, as follows: (a) high-level requirements, which express the vision about system development along some vectors of competitiveness; (b) functional requirements, which express what actually the system has to do (fundamental functionalities); (c) performance requirements, which express the level of performance the system has to achieve; (d) resource-related requirements, which express the amount of resources (human, financial, material, informational) the stakeholders agree to allocate for system development; (e) design constrains, which express design ideas that have to be incorporated into the system; (f) conditional constrains, meaning additional performance requirements in relation to resources, being generated by the functional requirements; in this category, conformity requirements are also included. From innovation point of view, performance requirements are by far the most challenging ones. Requirements are further ranked to reveal their relative importance in the equation of quality. Tools like AHP are very often used to rank requirements (Brad, 2008; Raharjo, Xie, Goh & Brombacher, 2007; Saaty & Vargas, 2001).

The key performance characteristics of the system are defined and linked to the stakeholders' requirements by means of a relationship matrix. This matrix uses a set of coefficients to establish the level of relationship between each requirement and each key performance characteristic. For deciding upon the key performance characteristics from a comprehensive, initial set of performance characteristics, an evaluation process should be conducted. A key performance characteristic fully satisfies the related stakeholders' requirements only when it attains its target value (Brad, 2008). A key performance characteristic can be easily identified from the relationship matrix in the sense that at least a strong or very strong relationship level of the respective performance characteristic with a certain requirement exists.

Setting target values is of very great importance. Traditional approaches heavily rely on experience and intuition. This leads most probably to feasible solutions rather than optimal ones. The target values will drive all subsequent development activities. In this respect, the approach for defining the right target value of each performance characteristic is of major concern. When establishing target values, several aspects should be considered (e.g. correlations between performance characteristics, relationships between stakeholders' requirements and performance characteristics, current performances of systems in the same category with the new system, available budget, etc.).

A very critical issue in planning performance is related to the evolution of stakeholders' requirements over the life-cycle of the new system (Brad, 2008). In a complex, non-linear evolving environment, stakeholders could change their requirements both in terms of contents, ranks, target values and minimum acceptable values (Liang, 2009; Lindroos, Malmivuo & Nousianinen, 2007;

Salihbegovic & Tanovic, 2008; Takago, Matsuishi, Goto & Sakamoto, 2007). What today is of lower importance tomorrow could become of very high importance. Issues that today are missing in the list of stakeholders' requirements could arise in the future. Thus, input data in the planning process are more complex from the perspective of system's life-cycle.

Actually, several lists of stakeholders' requirements, including their related ranks, should be elaborated even before starting the conceptualization of the new system. Each list should consider an interval of time within the life-cycle of the new system. Each list is actually related to a certain release of the system. These aspects come up from the business strategy associated to the new system. When a new system is going to be developed, people should establish some business goals in relation to the respective system, including the estimated time of its life-cycle and the number of releases over the estimated life-cycle. If for the first release of the system, stakeholders' requirements and their ranks can be determined with classical tools (e.g. deep market research based on interviews, questionnaires, focus groups, conjoint analyses, etc.), for medium and long term perspectives non-conventional approaches must be considered in defining the list of stakeholders' requirements. They should include multiple scenarios, probabilities of occurrence, internal and external influence factors, etc.

However, the development of each release of the system must consider a clear defined list of requirements, ranks, target values and minimum acceptable values during performance planning process. The solution proposed for the first release meets a lot of challenges because it should smoothly and cost-effectively translates to the next releases, too.

Correlations between performance characteristics are also analyzed. Critical for system design are the negative correlations occurring between performance characteristics, meaning that the attempt to improve a certain performance characteristic could harm the improvement of the performance characteristic with which it is negative correlated. To solve the conflict without compromises, innovative problem solving tools should be considered (Altshuller, 2000; Brad, 2005; Brad, 2008).

Knowing the ranks of stakeholders' requirements and the relationship coefficients between stakeholders' requirements and performance characteristics, the value weight of each performance characteristic can be determined (Bier & Cornesky, 2001; Brad, 2008; Brad, 2009). Value weight reveals the maximum impact that each performance characteristic brings in satisfying the set of stakeholders' requirements.

Performance characteristics could be further deployed into functions, further into technical modules and then into components of the system. Having established a cost-objective and/or a time-objective for developing the overall system, as well as the value weights of each module and component, value could be effectively engineered within the system during the design process (Brad, 2008). For example, having determined the relative value weight of module X of 20% and the cost-objective to realize the system is set to 10 000$, then the upper cost limit to develop module X is 2 000$. This limit could generate design constraints that might require innovative approaches in system design (technical, organizational, etc.).

About Analytic Hierarchy Process

AHP is a highly developed mathematical system for priority setting (Raharjo, Xie, Goh & Brombacher, 2007; Saaty & Vargas, 2001). The main operational steps of the AHP method are:

- Define the problem;
- Structure a hierarchy representing the problem;
- Perform pair-wise comparison judgments on the requirements with respect to the goal of the system.

When two requirements are compared, the Saaty's eigenvector method is mostly considered (Saaty & Vargas, 2001). In AHP, each pair-wise comparison represents an estimate of the ratio of the priorities of the compared elements. Estimates of the priorities (or weights) are calculated for each pair-wise comparison matrix for each level of the hierarchy. To synthesize the results over all levels, the priorities at each level are weighted by the priority of the higher-level criterion with respect to which the comparison was made. AHP/Saaty method is a powerful tool for consistent ranking of requirements; therefore it is very often used in the framework of system planning.

Innovative Problem Solving

Innovating solutions is the process of bringing creative ideas into practice. According to countless opinions there is no unique way to solve a particular problem. However, the process of solving the problem innovatively could follow a systematic approach. The innovative problem solving process comprises five stages: problem formulation, problem analysis, solution generation, selection and implementation.

The problem formulation is about asking appropriate questions to "visualize" the problem as correct and clear as possible. In order to formulate the problem, data about context, initial situation and goal should be collected. This means to see the background, factors and events leading to problem occurrence, to see what is not satisfactory today in the current system and to "draw the map" of the future desirable system.

Problem solving is about "translating" the system from the initial situation to the desired situation. By understanding the context properly and by analyzing the difference between the current situation and the desired situation, the problem could be better formulated. The problem formulation must reflect the context, the initial system and the future system (the vision). A good formulation of the problem means: not too broad and not too narrow, not driven by solution, not driven by assumptions.

Problem analysis is focused on causes leading to the problem. In this respect, relevant data about problem should be gathered and structured. Various sides of the problem should be explored, as well as viewed from different perspectives. This very often could lead to problem reformulation.

The process of generating solutions deals with all about finding various ways to remove the causes that generate the harmful effects and/ or transforming the disturbing factors into useful factors. Each potential solution must be clear formulated both in terms of concept and steps on how to solve the problem.

The selection process of the most appropriate solutions from the generated set requires the use of ranked criteria. These criteria should consider benefits (e.g. cost-effectiveness over the life-cycle, technical strengths, etc.) and disadvantages (e.g. risks, costs, technical weaknesses, etc.).

The implementation stage requires careful project planning and organization to put the solution into practice. Specific project management tools are considered in this respect.

The most challenging phase of the innovative problem solving process is the solution generation phase because it requires finding mature ways to solve the problem without compromises. There are more than one hundred tools and algorithms used in practice for supporting the creative process. Sometimes it is quite difficult to select the best tool for a given problem. However, literature published in the last decade mostly promotes a limited set of inventive problem solving tools. In the top of the list is the theory of inventive problem solving method (TRIZ) and its derivates: ARIZ, ASIT, USIT, CSDT, etc. (Altshuller, 2000; Brad, 2005; Brad, 2008). Basics about TRIZ method are further highlighted.

About Theory of Inventive Problem Solving

TRIZ is a method that helps in solving difficult problems through identification and elimination of conflicts that are presented within systems. TRIZ is a systematic approach for generating creative solutions using the present day scientific progress (Altshuller, 2000). TRIZ uses 39 standard technical characteristics (also called standard parameters) that may cause conflicts within a system. Examples of elements from this set of standard technical characteristics are: capacity, complexity of control, accuracy of measurement, waste of substance, stability, reliability, harmful factors acting on object, brightness, area of moving object, etc. These characteristics are very generically formulated in order to comprehend any particular problem; therefore, their meaning should not be interpreted in a strict sense. To apply the TRIZ procedure, any particular problem should be translated into a generic problem, where the conflicting specific technical characteristics find equivalents within the set of TRIZ standard technical characteristics. According to countless opinions, this is one of the most difficult tasks in TRIZ implementation, because the applicant(s) must associate a particular, tangible problem to a very generic one. For example, the amount of information delivered in a course could be associated to the TRIZ standard parameter "amount of substance"; and the level of interest of students about the course could be associated to the TRIZ standard parameter "temperature", etc.

A conflicting problem is about the conflict occurring between two parameters x and y when attempting to improve the performances of both parameters; for example, by trying to improve the performance of parameter x, a harmful effect will be generated upon parameter y lowering its performance. Innovative problem solving means to improve the performance of parameter x without affecting the performance of parameter y, even if the two parameters are negative correlated. For example, between the volume of information in a course and the time required to deliver this information is a negative correlation as long as the goal is to increase the volume of information and to reduce the course duration. Innovative problem solving means to deliver more information in less time. The targets related to parameters "amount of information" and "time" will define the level of challenges in innovative problem solving.

The standard parameters are used by a so-called "contradiction matrix" to reveal the paths of innovation for the given problem. The contradiction matrix consists of 39 columns and 39 rows, corresponding to the number of TRIZ standard parameters. At the intersection between a row and a column of the contradiction matrix a set of numerical values is revealed. The size of this set is between 0 and 4 elements. The numerical values associated to these elements represent the indexes of the so-called TRIZ inventive principles. There are 40 inventive principles in TRIZ. According to the contradiction matrix, when a conflicting problem occurs in the analyzed system between two parameters, a limited set of TRIZ inventive principles (between 0 and 4) could be associated to the respective conflicting problem. This set recommends the most appropriate paths (directions of intervention) where innovative solutions should be identified. If the set has 0 elements, the meaning is the conflicting problem hardly can be solved without affecting the current structure of the system. For example, considering the above-mentioned problem "amount of information" versus "time", the related TRIZ parameters are "amount of substance" (index 26) and "speed" (index 9). For these standard parameters, the contradiction matrix reveals the inventive principles "change the volume and/or density and/or the degree of flexibility and/or physical state" (index 35), "replace solid parts with fluid parts" (index 29), "after an element of the system has completed its function it should be rejected or

modified" (index 34) and "replace stationary fields with moving fields" (index 28). Innovative solutions suggested by these vectors could be: replace the standard way of information delivery (e.g. text and drawings) with a non-conventional way (e.g. movies, sound and animations/simulations), use of virtual reality, use of games, use of didactical prototypes, consideration of site visits, etc. For more complex problems, the matrix of contradiction and the TRIZ inventive principles are integrated into systematic algorithms of inventive problem solving (e.g. ARIZ, Su-Field, etc.). Reader is encouraged to consult TRIZ literature in order to find out more about methods, tools and methodologies of creativity and innovative problem solving.

AREAS OF INTERVENTION

Designing a web-based course in engineering is a complex task. Special attention should be given to many aspects like course structure, course subject, communication technology, teaching process management, costs, effort and time to prepare the course, dynamics of information, dynamics of the educational models, technology to provide the information, etc. (Barros, Read, & Verdejo, 2008; Du, Li & Li, 2008; Ebner & Walder, 2008; Helander & Emami, 2008; Li & Wang, 2007; Saygin & Kahraman, 2004; Sessink, van der Schaaf, Beeftink, Hartog & Tramper, 2007; Toral, Barrero & Martinez-Torres, 2007; Vargas, Sanchez, Duro, Dormido, Farias, Dormido, Esquembre, Salzmann & Gillet, 2008; Xuelian, 2008). All these areas of intervention are important for setting up a competitive web-based course in engineering. From this perspective, a web-based engineering course is revealed as a complex adaptable system (Brad, 2008). A web-based engineering course, seen as a "living" complex adaptable system, should be able to internalize information, to modify its parameters and behavior over its life-cycle, to compress information gathered from the interaction with the external environment and to "learn" from this interaction. Thus, a web-based engineering course should be evolutionary; and from this point of view the search is not for the "best of the world" solution (because it does not exist) but for mature solutions able to balance various dimensions characterizing a web-based engineering course. Elementary aspects about these dimensions are further brought into discussion.

The first dimension is referring to the course structure. The web-based course structure usually has some additional elements than a classical course. On one hand, this is because a web-based course is provided remotely, via electronic means. On the other hand, the limitations in terms of interaction between student and teacher require additional sections in the course structure, which could contain in some cases explanations on how to use the course material, as well as additional materials in electronic format that support the core unit of the course.

Content is another dimension of analysis for a web-based course. In principle, the content should be similar as in the case of a face-to-face course. However, a web-based course should contain supplementary questions and answers and/or supplementary case studies and exercises. Also, the information itself should be user-centric organized such as the information to be delivered in a more intuitive way–because of the remote, limited and less natural interaction of student with teacher and with other students. Sometimes, a web-based course could include supplementary explanations to compensate the lack of student interaction with the teacher. The volume of information should be carefully dimensioned, too, in order to avoid overloading the student's study time.

For storing the information of a web-based course, electronic formats are mainly used. Nowadays, a quite large variety of electronic formats for storing information is met. This ranges from usual "static" electronic files where information is stored in text, graphics and picture formats to more animated formats like web pages, electronic

books, movies, simulations, interactive games, remote test benches, virtual instrumentation, etc. Focal points in this area are also about easy access of students to information using web technologies and about handling with minimum effort the course files on computers.

Communication technology is another key issue in the case of web-based engineering courses. It varies from already well-known, usual communication means like e-mails, web conferences and chats with web cameras as part of the system (e.g. Yahoo Messenger, Skype), to complex web-based video conference systems, web-based smart boards and remote desktop control server applications for interacting in the teaching/learning environment.

Management of the teaching process rises up specific challenges in the framework of web-based courses in engineering, too. A lot of web-based software platforms are available onto the market for handling various aspects of the teaching process like course scheduling, course content administration, student profiles, student attendance and results administration, forums, role management, access rights and so on. More and more popular are the open-source learning management systems (LMS) or virtual learning environments (VLE) like, for example, Moodle.

Besides the above mentioned characteristics, any initiative of setting up and running web-based courses requires a careful planning from financial, as well as economic points of view. This involves calculation of financial indicators for the investment in a web-based course like return on investment and internal rate of return, to which economic indicators like economic net present value can be added. The cost-benefit analysis should represent a major element for deciding the form in which a web-based course in engineering has to be designed and delivered.

Moreover, the dynamics and evolution of the educational environment should not be excluded from the list of relevant dimensions to which a web-based course is planned, designed and implemented. Dynamics of the educational environment is reflected in the interaction be-tween the life-cycles of four sub-systems that describe the educational product–market needs and requirements, teaching-learning process, study program and educational provider. The four life-cycles must be well-balanced in order to create favourable premises for the provision of successful web-based courses in engineering (Popescu, Brad & Popescu, 2006). Evolution of the educational environment is reflected in the evolutions of the four sub-systems mentioned before as a consequence of various attractors. Attractors have to be seen as aggregated effects of some external influence factors acting with a relevant intensity upon these sub-systems, thus generating bifurcation points in their evolutions. For example, technological progress in the area of remote communication tools will influence the evolution of the teaching-learning process; as well, various transformations in the social and economic environments will influence the evolution of the educational provider models, etc.

Even if several new tools and paradigms have been developed and tested in the attempt of adapt-ing the training in engineering to nowadays' needs, there are still many unknowns on how to develop a web-based course as a whole, not only limited to the teaching and learning methods. Almost any topic in engineering has particular characteristics, which implicitly requires a deep customization of a web-based course. Thus, a collection of complex problems must be crossed over to set up mature web-based courses in engineering. In conclusion, a web-based course in engineering must be "designed for customization" as any other kind of tailor-made product or service in order to achieve high quality levels (Brad, 2008). But such approaches require radical innovation and use of adequate planning tools during the conceptualiza-tion process of the web-based course. The next section of this chapter is going to propose such methodology.

METHODOLOGY

The methodology for systematic planning and innovation of a competitive web-based course in engineering integrates in a novel way QFD-based planning matrices with TRIZ inventive vectors in order to display and rank priorities of intervention and to formulate innovative solutions to various challenges. The methodology consists of several steps that are further revealed.

Step 1

The methodology starts with definition of business objectives in relation to the web-based course. Here, they are denoted with BO_1, BO_2, ..., BO_m, where m is the number of business objectives. In order to lead the planning process under various constrains (e.g. time, budget, technology, etc.), business objectives have to be ranked. The AHP method is proposed for performing this task. At the end of this process the following set of ranked business objectives is obtained: $BO(R(t))$ $= \{BO_1(\underline{R}_1(t)), BO_2(\underline{R}_2(t)), ..., BO_m(\underline{R}_m(t))\}$, where $\underline{R}_i(t)$, $i = 1, ..., m$, is the estimated degree of im-portance of the business objective BO_i, $i = 1, ..., m$, at the moment in time t when the course is delivered. Working with estimates instead of static values underlines the inconsistency of informa-tion between the moment of web-based course planning, conceptualization and development and the moment when the course is commercialized; thus showing the dynamic and evolutionary nature of the educational program. In other words, the importance given now to the business objectives might be different after N months (time t) when the course is delivered. Thus, because of the evolving educational environment, ranks at moment t can be only estimated, the accuracy of the estimates depending on the level of understanding the evolu-tion and interaction of the influential factors of the educational environment. This philosophy is also transferred to the next steps of the methodology.

Step 2

In order to meet the business objectives, a set of target-functions is further formulated $TF = \{TF_1, TF_2, ..., TF_n\}$, where n is the number of target-functions. To see the consistency of the target-functions with the business objectives, a relationship matrix is worked out. It relates the business objectives (located on the rows of the matrix) with the target-functions (located on the columns of the matrix) by means of the so-called relationship coefficients (located at the intersections between the rows and the columns of the matrix). The following categories of relationships could exist: 0–no relationship; 1–weak/possible relationship; 3–medium relationship; 9–strong relationship; 27–very strong/critical relationship (Brad, 2008). Along each column, at least a strong relationship must exist such as the corresponding target-function to be relevant. The estimated value weight $\underline{W}_j(t), j = 1, ..., n$, at the moment t, of each target-function is calculated with the formula:

$$\underline{W}_j(t) = \sum_{k=1}^{m} \underline{R}_k(t) \cdot a_{kj}, \quad j = \overline{1, n}, \tag{1}$$

where $a_{kj}, k = 1, ..., m; j = 1, ..., n$ are the relationship coefficients between the business objectives and the target-functions. It is simple to deduce that designing for life-cycle of a web-based course requires consideration of various value weights of the target-functions at various life-time moments.

Step 3

For each target-function, requirements are formulated and ranked. For ranking, tools like AHP could be also used. Further, measurable performance characteristics are defined and related to requirements. There are denoted with $RE_h[TF_j]$, $h = 1, ..., g[TF_j]$ the requirements related to the target-function $TF_j, j = 1, ..., n$, and with $\underline{U}_h(t)$ $[TF_j], h = 1, ..., g[TF_j]$ the estimated degrees of importance of requirements $RE_h[TF_j], h = 1, ..., g[TF_j]$ at the moment t. There are denoted with $PC_f[TF_j], f = 1, ..., d[TF_j]$ the performance characteristics related to the target-function $TF_j, j = 1, ..., n$. The estimated value weights of these performance characteristics, at the moment t, are denoted with $\underline{V}_f(t)[TF_j], f = 1, ..., d[TF_j]$ and calculated with the formula:

$$\underline{V}_f(t)[TF_j] = \sum_{l=1}^{g[TF_j]} \underline{U}_l(t)[TF_j] \cdot b_{lf}[TF_j], \quad f = \overline{1, d[TF_j]}, \quad j = \overline{1, n}$$

$$(2)$$

where $b_{lf}[TF_j], f = 1, ..., d[TF_j]$ are the relationship coefficients of the planning matrix corresponding to the target-function $TF_j, j = 1, ..., n$. The value weights $\underline{V}_f(t)[TF_j], f = 1, ..., d[TF_j]$ give a perspective of the most relevant performance characteristics in relation to the web-based course with respect to the target-function $TF_j, j = 1, ..., n$, at the moment t.

Step 4

The correlations between performance characteristics to the level of each target-function are identified. The negative correlations are extracted. A negative correlation between two performance characteristics occurs when attempting to improve one of the performance characteristics a harmful effect is generated upon the other one. Where negative correlations occur, innovation is required to define mature solutions of the web-based course. This kind of innovation is called in this paper "performance characteristic-oriented innovation." TRIZ method is applied at this step for each negative correlation in order to define vectors of innovation of the web-based course.

Step 5

For each performance characteristic a target is established. Achieving in a very short time high levels of quality of the web-based course represents a big challenge for any educational provider. This automatically requires innovation, which is here called "target-oriented innovation." For setting up appropriate lines of innovation of the web-based course, TRIZ method is applied at this step of the methodology. This step is related to each performance characteristic of each target-function.

Step 6

Development of a web-based course could meet some supplementary barriers coming up from financial, organizational and human resources constrains. This category of barriers necessitates further innovative solutions in setting up a competitive web-based course. Innovation in this area is called here "system-oriented innovation." TRIZ method could be also used at this point for tackling properly the innovative problem solving process. This step is related to each target-function.

Step 7

The set of vectors of innovation identified at steps 4, 5 and 6 are exploited for formulating local solutions of the web-based course. Each local solution should move towards satisfying as much as possible the requirements of the corresponding target-function, towards a so-called "local qualitative optimality." Local variants of the web-based course effectively show the "multiple" facets of the same system. They reveal the differences in relevance of the modules of the same system when it is placed in different contexts. This step actually brings to live one of the most important properties of complex systems: in a complex world there is no single best solution–otherwise the problem would not be considered complex. Thus, a set of n local solutions of the web-based course are elaborated at the end of this step. The set of local solutions is denoted $S = \{S_1, S_2, ..., S_n\}$.

Step 8

Local solutions have to be aggregated into a final solution, which, in principle, should comprise to the maximum possible extend the strengths of all local solutions. In order to generate the final solution of the web-based course as an aggregated result of the local solutions, a specific algorithm is required. An aggregation algorithm in five stages is further proposed.

The aggregation algorithm starts with the identification of the correlation types and correlation levels between the target-functions TF_1, $TF_2, ..., TF_n$. The correlation coefficients between the target-functions are denoted here with $C_{jk}, j = 1, ..., n, j \neq k$. They could have negative values, positive values or null values, depending on the characteristics of the target-functions. For defining the level of correlation, the following values are used in practice: 0–no correlation; 1–weak positive correlation; 2–medium positive correlation; 3–strong positive correlation; -1–weak negative correlation; -2–medium negative correlation; -3–strong negative correlation.

Further, the aggregation algorithm considers information from step 2 of the methodology. The target-function with the highest value weight in the set will be taken as the starting point in the aggregation algorithm. It is symbolized with *PTF* the target-function having the highest value weight from the n target-functions TF_1, TF_2, ..., TF_n. If $\underline{W}_{max}(t)$ denotes the maximum value weight at the moment t, the following relationship comes up:

$$\underline{W}_{max}(t) = \max\{\underline{W}_1(t), \underline{W}_2(t), ..., \underline{W}_n(t)\}.$$

(3)

In the next stage, the other target-functions are grouped relative to *PTF*. *PTF* is correlated with the other $n-1$ target-functions in various ways:

positive, negative or not correlated, as well as at various strengths. The type of correlation between two target-functions is determined by the correlations between their constitutive requirements. In this respect, the rest of the $n-1$ target-functions can be sorted on three categories: the group of those target-functions that are positive correlated with *PTF*, the group of those target-functions that are not correlated with *PTF* and the group of those target-functions that are negative correlated with *PTF*.

Further, the algorithm asks to order the target-functions inside each of the three groups. Considering that the *PTF* is the k-th target-function in the set *TF* from step 2, for the target-functions that are positive or negative correlated with *PTF* an index will be calculated; index denoted with $\underline{H}_j(t)$, $j = 1, ..., n$, $j \neq k$, as the product between the value weight $\underline{W}_j(t)$, $j = 1, ..., n$, $j \neq k$ and the correlation coefficient C_{jk}, $j = 1, ..., n$, $j \neq k$. This formula is shown below:

$$\underline{H}_j(t) = \underline{W}_j(t) \cdot C_{jk}, \ j = \overline{1, n}, \ j \neq k, \ C_{jk} \neq 0 .$$

(4)

In the group of target-functions that are positive correlated with *PTF*, the target-functions will be ordered starting with the one having the highest H and ending with the one having the lowest H. The same rule is kept for the group of target-functions that are negative correlated with *PTF*. It is highlighted the fact that $C_{jk} < 0$ in the group of negative correlated target-functions, so the one with the highest H will have the lowest magnitude in absolute value, too. The target-functions that are not correlated with *PTF* will be ordered starting with the one having the highest value weight W and ending with the one having the lowest value weight W.

In the last stage, the aggregated solution is generated following an iterative approach. The aggregated solution will result as a "compromise & combination" of the set of n local solutions. In this respect, the following rule is applied:

a. The solution corresponding to the *PTF* will be taken and analyzed together with the solution corresponding to the first target-function in the group of the target-functions that are positive correlated with *PTF*. Because the two target-functions are positive correlated, the best ideas from the local solutions will be combined, resulting an improved hybrid solution.

b. The hybrid solution from (a) will be then analyzed against the local solution corresponding to the second target-function in the group of the target-functions that are positive correlated with *PTF*. The new variant will result as a combination of the best ideas from the hybrid solution generated at phase (a) and from the current local variant.

c. The process will go on in the manner above described until all target-functions from the group of target-functions that are positive correlated with *PTF* are consumed. After that, the group of no correlated target-functions is taken into account and the process is continued until all of these target-functions are consumed. At the end, the group of target-functions that are negative correlated with *PTF* will be taken into account. Because at this phase potential conflicts could occur, they have to be solved without compromises, if possible. In this respect, it is firstly required to identify pairs of conflicting problems between the compared variants. Afterwards, innovative solutions have to be formulated. Methods like TRIZ could offer a real support in this respect. At the end of this process, the complete overall solution will be defined.

Step 9

The result is reviewed with respect to the most relevant performance characteristics of each target-functions and with respect to a cost-objective. Relevance of the performance characteristics for each target-function TF_j, $j = 1, ..., n$, is given by their value weights $\underline{V}_f(t)[TF_j]$, $f = 1, ..., d[TF_j]$. If failures are identified both in terms of cost and technical performances, improvements are further formulated. To perform this task properly, various innovative problem solving tools could be also considered (e.g. TRIZ).

CASE STUDY

In order to exemplify the methodology, a case study is elaborated. The application example is referring to a web-based course about engineering design of industrial robots, which has to be delivered to a branch X of the university Y, placed in other locality. The course includes lectures and project work (e.g. engineering design of a robotic module). To keep the case study relative simple, as well as in a reasonable paper volume, only a limited set of target-functions and requirements describing the target-functions will be further considered.

Thus, in this application example, the following business objectives have been considered to frame the analysis and design space of the web-based course: BO_1–to minimize possible frustrations of the students located in the facility X because of less face-to-face contact with the lecturers [37%]; BO_2–to be relative easy accepted and adopted by the lecturers with less experience in using information and communication technologies and tools [12%]; BO_3–the quality of the teaching process not being affected because of the remote performance [33%]; BO_4–the operation and maintenance costs to be acceptable [9%]; BO_5–the return on investment to be attractive [9%]. Values between brackets reveal the relative ranks of the business objectives, as they have been determined with the AHP method.

In relation to the business objectives, the following target-functions will be taken into account: design for easy and fast scalability $[TF_1]$, design for cost affordability $[TF_2]$, design for fast development and implementation $[TF_3]$, and design for reliability $[TF_4]$. Figure 1 reveals the deployment process of the business objectives into target-functions. The relationship matrix, the correlation matrix, as well as the calculated value weights of the target-functions are put into evidence. According to the results in figure 1, for this case study a special attention should be given to design for reliability and design for cost affordability.

For each target-function the set of requirements and related performance characteristics are defined. The deployment process of requirements into performance characteristics is revealed for each target-function in Figures 2, 3, 4 and 5.

Figure 1. Deployment of business objectives into target-functions

		TF_1	TF_2	TF_3	TF_4
	TF_4				
	TF_3				+1
	TF_2				-2
	TF_1		-1	+2	
	Target-functions	TF_1	TF_2	TF_3	TF_4
Business objectives	Degree of importance	Relationship matrix (bottom) and Correlation matrix (up)			
BO_1	3.7	3		1	27
BO_2	1.2	3		27	9
BO_3	3.3	9		9	27
BO_4	0.9	9	27	1	9
BO_5	0.9	9	27	27	
	Value weight [%]	14.9	11.9	22.3	50.9

Figure 2. House of scalability

	$PC_1[TF_1]$	$PC_2[TF_1]$	$PC_3[TF_1]$	$PC_4[TF_1]$	$PC_5[TF_1]$	$PC_6[TF_1]$	$PC_7[TF_1]$	$PC_8[TF_1]$
$PC_8[TF_1]$								
$PC_7[TF_1]$								+1
$PC_6[TF_1]$								
$PC_5[TF_1]$								+1
$PC_4[TF_1]$					+1		+1	+1
$PC_3[TF_1]$								
$PC_2[TF_1]$			+1	+1	+1			
$PC_1[TF_1]$								
Metric	$PC_1[TF_1]$	$PC_2[TF_1]$	$PC_3[TF_1]$	$PC_4[TF_1]$	$PC_5[TF_1]$	$PC_6[TF_1]$	$PC_7[TF_1]$	$PC_8[TF_1]$
Optimization trend	min	max	min	min	min	max	min	min
Requirement $U \setminus D$	1	2	3	4	2	2	2	2
$RE_1[TF_1]$ 3.2	9	27	1	27	27		1	1
$RE_2[TF_1]$ 3.2	27	3	27	3	9		3	1
$RE_3[TF_1]$ 1.8		1		9	9	9	3	9
$RE_4[TF_1]$ 1.8		3		3	9	27	9	27
Value weight	115.2	103.2	89.6	117.6	147.6	64.8	34.4	71.2
Value weight [%]	15.5	13.9	12.1	15.8	19.8	8.7	4.6	9.6
Global weight [%]	2.3	2.1	1.8	2.4	2.9	1.3	0.7	1.4
Most relevant	x	x		x	x			
Target	< 60 s	All	< 600 s	< 1 h	→ 0	> 50	< 600 s	→ 0

Figure 3. House of affordability

	$PC_1[TF_2]$	$PC_2[TF_2]$	$PC_3[TF_2]$	$PC_4[TF_2]$	$PC_5[TF_2]$
$PC_5[TF_2]$					
$PC_4[TF_2]$					+2
$PC_3[TF_2]$					
$PC_2[TF_2]$			+2	+1	+2
$PC_1[TF_2]$		+1	+1	+2	+1
Metric	$PC_1[TF_2]$	$PC_2[TF_2]$	$PC_3[TF_2]$	$PC_4[TF_2]$	$PC_5[TF_2]$
Optimization trend	max	max	max	min	min
Requirement $U \setminus D$	5	4	3	3	2
$RE_1[TF_2]$ 4.0	9	27	27	27	27
$RE_2[TF_2]$ 3.0	27	27	9	27	
$RE_3[TF_2]$ 3.0	27	27	3	9	1
Value weight	198	270	144	216	111
Value weight [%]	21.1	28.8	15.3	23.0	11.8
Global weight [%]	2.5	3.4	1.9	2.7	1.4
Most relevant	x	x		x	
Target	> 250%	> 0	> 1	< 1.5 years	< 5%

In the case of target-function TF_1 (design for easy and fast scalability), the following requirements are considered: $RE_1[TF_1]$–Good functional scalability (system enhancement by adding new functionalities at minimal effort: fast and easy adding new files in any format, fast and easy adding or removing communication tools, fast and easy adding or removing new multi-media applications, etc.); $RE_2[TF_1]$–Good load scalability (fast and easy expanding or contracting its resource pool, fast and easy adding new files to the course, fast modification of files' content and structure, etc.); $RE_3[TF_1]$–Good geographic scalability (maintenance of performance, usefulness and usability regardless of expansion and concentration to a more distributed geographic

Figure 4. House of development

		$PC_1[TF_3]$	$PC_2[TF_3]$	$PC_3[TF_3]$	$PC_4[TF_3]$	$PC_5[TF_3]$
$PC_5[TF_3]$						
$PC_4[TF_3]$						
$PC_3[TF_3]$					+1	+1
$PC_2[TF_3]$						
$PC_1[TF_3]$			+1			
Metric		$PC_1[TF_3]$	$PC_2[TF_3]$	$PC_3[TF_3]$	$PC_4[TF_3]$	$PC_5[TF_3]$
Optimization trend		min	min	min	min	min
Requirement	$U \setminus D$	3	5	2	3	2
$RE_1[TF_3]$	2.0	27	27	9	9	9
$RE_2[TF_3]$	3.5	9	1	27	1	1
$RE_3[TF_3]$	1.5		9	9	27	1
$RE_4[TF_3]$	3.0	3	9	9	3	27
Value weight		94.5	98.0	153.0	71.0	104.0
Value weight [%]		18.2	18.8	29.4	13.6	20.0
Global weight [%]		4.1	4.2	6.5	3.0	4.5
Most relevant		x	x	x		x
Target		< 20% of the time required to elaborate a classical course	< 20% of the costs required to elaborate a classical course	< 1 h	< 2 h	< 300 s

Figure 5. House of reliability

		$PC_1[TF_4]$	$PC_2[TF_4]$	$PC_3[TF_4]$	$PC_4[TF_4]$	$PC_5[TF_4]$	$PC_6[TF_4]$	$PC_7[TF_4]$	$PC_8[TF_4]$
$PC_8[TF_4]$									
$PC_7[TF_4]$									+1
$PC_6[TF_4]$								+1	+1
$PC_5[TF_4]$									+2
$PC_4[TF_4]$						+2			
$PC_3[TF_4]$					-2	+1			
$PC_2[TF_4]$				-1	-2				
$PC_1[TF_4]$			-2	+1	+1				
Metric		$PC_1[TF_4]$	$PC_2[TF_4]$	$PC_3[TF_4]$	$PC_4[TF_4]$	$PC_5[TF_4]$	$PC_6[TF_4]$	$PC_7[TF_4]$	$PC_8[TF_4]$
Optimization trend		max	min	min	max	max	min	min	max
Requirement	$U \setminus D$	5	5	4	3	5	4	3	4
$RE_1[TF_4]$	2.5	27	27	27	27				3
$RE_2[TF_4]$	2.5	3	27	27	27	27			9
$RE_3[TF_4]$	1.0		1				3	9	
$RE_4[TF_4]$	1.5						27	27	
$RE_5[TF_4]$	1.5	27	27	9	9				3
$RE_6[TF_4]$	1.0		3			9			27
Value weight		115.5	179.5	148.5	148.5	76.5	43.5	49.5	61.5
Value weight [%]		14.0	21.8	18.0	18.0	9.3	5.3	6.1	7.5
Global weight [%]		7.1	11.1	9.2	9.2	4.7	2.7	3.1	3.8
Most relevant		x	x	x	x				
Target		> 12 h continuous work	< 6 Mb/s	< 40 ms	> 1024 × 768	> 90%	< 150 s	< 600 s	> 90%

area); $RE_4[TF_1]$–Good administrative scalability (capacity to increase easy and fast the number of terminals for sharing a single distributed system). The degrees of importance for these requirements are determined with the AHP method and have the following values: $U[TF_1] = \{32\%, 32\%, 18\%, 18\%\}$. The related performance characteristics are: $PC_1[TF_1]$–Time required adding new files; $PC_2[TF_1]$–Number of file formats allowed; $PC_3[TF_1]$–Time required modifying information; $PC_4[TF_1]$–Time required integrating new tools; $PC_5[TF_1]$–Cost required integrating new tools; $PC_6[TF_1]$–Number of remote terminals; $PC_7[TF_1]$–Time required integrating/removing new terminals; $PC_8[TF_1]$–Cost required integrating/removing new terminals.

For the target-function TF_2 (design for cost affordability), the related requirements are as follow: $RE_1[TF_2]$–Affordable life-cycle costs (operating costs, maintenance costs, replacement costs, scalability costs, disposal costs, support costs, administrative costs, etc.); $RE_2[TF_2]$ –Affordable level of the initial investment (system's price, installation and testing costs, training costs, etc.); $RE_3[TF_2]$–Attractive financial indicators. The degrees of importance for these requirements are determined with the AHP method and have the following values: $U[TF_2] = \{40\%, 30\%, 30\%\}$. The main performance characteristics related to the target-function TF_2 are the followings: $PC_1[TF_2]$–Return on investment; $PC_2[TF_2]$–Net present value; $PC_3[TF_2]$–Benefit/cost ratio; $PC_4[TF_2]$–Payback period; $PC_5[TF_2]$–Percentage of the annual operational costs from the initial investment.

In this case study, in relation with the target-function TF_3 (design for fast development and implementation) the key requirements for web-based course planning are the followings: $RE_1[TF_3]$–Easy and fast adaptation of a classical course to a web-based course; $RE_2[TF_3]$–Short time required to train students and lecturers in using the system; $RE_3[TF_3]$–Short time for setting up the system; $RE_4[TF_3]$–Short time for initiating work-

ing sessions (lectures, project work, consultancy, etc.). The AHP method is also applied to calculate the degrees of importance for TF_3-related requirements. They have the following values: $U[TF_3] = \{20\%, 35\%, 15\%, 30\%\}$. The performance characteristics associated to the target-function TF_3 are the followings: $PC_1[TF_3]$–Time required to adapt a classical course to a web-based course; $PC_2[TF_3]$–Costs required to adapt a classical course to a web-based course; $PC_3[TF_3]$–Time required to train students/lecturers to use the system; $PC_4[TF_3]$–Time required to set up the system (hardware and software); $PC_5[TF_3]$–Time required to initiate the working session.

For the target-function TF_4 (design for reliability) the selected requirements for system's planning are: $RE_1[TF_4]$–High quality of signal transmission (signal continuity, sound quality, image quality, etc.); $RE_2[TF_4]$–High capacity of interaction student-teacher (speaking, seeing, writing, drawing, cooperation); $RE_3[TF_4]$–Easy testing of the system; $RE_4[TF_4]$–Easy error tracking and fast problem fixing; $RE_5[TF_4]$–Low internet resources required; $RE_6[TF_4]$–Good course management (session record, file management, student work management, etc.). In the case of the target-function TF_4 the degrees of importance of the related requirements are: $U[TF_4] = \{25\%, 25\%, 10\%, 15\%, 15\%, 10\%\}$. The main performance characteristics taken into account for the target-function TF_4 are the followings: $PC_1[TF_4]$–Mean time between failures; $PC_2[TF_4]$–Average bandwidth required; $PC_3[TF_4]$–Delay; $PC_4[TF_4]$–Resolution; $PC_5[TF_4]$ –Level of interaction relative to a face-to-face meeting; $PC_6[TF_4]$–Time required to identify the problem; $PC_7[TF_4]$–Time required to fix the most complex problem and restart the system; $PC_8[TF_4]$–Monitoring level of students with respect to a face-to-face meeting.

Figure 2 illustrates the deployment of TF_1-related requirements into TF_1-related performance characteristics. The set of matrices from figure 2 is called "house of scalability." According to the results in Figure 2, the most relevant performance characteristics for web-course optimization in relation to TF_1 are: time required integrating new tools, cost required integrating new tools, time required adding new files and number of file formats allowed.

With D, in Figure 2, as well as in Figures 3, 4 and 5 is denoted the degree of difficulty to solve performance characteristics. The rank of D is given on a five-point Likart scale (from 1 "less difficulty" to 5 "very high difficulty"). Difficulty is seen from organizational and financial points of view. With respect to D, for the target-function TF_1, innovation should be mainly directed towards the performance characteristic $PC_4[TF_1]$ (time required integrating new tools). According to the results in Figure 2, no "performance characteristic-oriented innovation" is specially required because only positive correlations exist between performance characteristics (see step 4 of the methodology). With respect to the current state of technological progress, "target-oriented innovation" (see step 5 of the methodology) should be mainly directed towards $PC_5[TF_1]$ (cost required integrating new tools).

Thus, with respect to $PC_4[TF_1]$, the mini-problem which necessitates innovation could be formulated as: (a) to reduce the time for integrating new tools by involving low costs; and (b) to reduce the time for integrating new tools by involving low efforts. In TRIZ-language, this problem shows like: time [9] versus amount of substance [26]; and time [9] versus energy spent by moving objects [19] (Altshuller, 2000). Numbers in brackets show the position of the generic parameters in the "TRIZ table of generic parameters causing conflicts" (Altshuller, 2000). For the first set of generic conflicts, the following generic directions of intervention are recommended by TRIZ: use the resonance frequency [18], replace a continuous action with an impulse (and if it necessary use pauses between impulses to introduce additional actions) [19], replace "solid" parts with "fluidic" parts [29], enrich the "atmosphere" with supplementary components [38]. Numbers

in brackets show the position of the directions of intervention in the "TRIZ table of inventive vectors" (Altshuller, 2000). For the second set of generic conflicts, the following generic directions of intervention are recommended by TRIZ: compensate a certain "weight" with forces coming from the external environment [8], make the system more dynamic (with components able to change their position, with higher mobility and interchangeability, with capacities of automatic adjustments) [15], change the degree of flexibility or density or volume [35], enrich the "atmosphere" with supplementary components [38].

In accordance with the above recommended directions of innovation, a possible solution of the web-based course to meet the major scalability requirements is the following: materials can be prepared in any format (e.g. .ppt, .pdf, .doc, .xls, .jpg, .cdr, .avi, etc.)–principle 18; students can download files from a repository system (e.g. Moodle)–principle 19; students can open any file on their computers (alternatively, the file can be displayed on a screen by means of a multimedia projector)–principle 35; multiple modes of communication are ensured (e.g. off-line via e-mails; on-line via Skype, Yahoo messenger, etc.)–principle 29; web-cameras are used for visual contact–principle 38; alternatively, for a better interaction with the classroom, a video-conference system replaces the communication on web messenger systems and web-cameras–in this respect, two video cameras are used: one camera broadcasts person(s) A (e.g. the teacher) and displays the image that the other camera broadcasts; the other camera broadcasts person(s) B (e.g. the students) while displaying image received from the place where person(s) A is(are) located–principle 38; teacher can handle remotely each camera–principle 8; both teacher and student have to have direct internet connections and real IP addresses (a broadband internet connection is recommended)–principle 15; teacher works on his/her tablet PC, which has to have a direct internet access–principle 38; teacher uses a remote desktop control server application (e.g. a free license, like VNC) in order to display remotely any application that runs on his/her computer–principle 35; if the PC is a tablet PC, teacher can use a virtual whiteboard to interact even more with students (e.g. CorelGrafigo, Windows Journal, etc.)–principle 38; moreover, teacher can run a system application for remote monitoring of computers in the network–principle 35. These solutions allow a comprehensive and multimodal provision of the course content (drawings, films, text, simulations, dynamic annotations, on-line hand written and sketches, on-line running of any application on computers, etc.).

In order to meet the target with respect to the performance characteristic $PC_5[TF_1]$, the following TRIZ mini-problems are revealed: (a) amount of substance [26] versus harmful side effects [31]; (b) amount of substance [26] versus loss of information [24]; and (c) amount of substance [26] versus convenience of use [33]. The generic directions of interventions recommended by TRIZ are: group (a): transition from a homogeneous structure to a heterogeneous structure of the system or of the external environment [3], change the degree of flexibility or density or volume [35], define a composite structure of the system [40], introduce a neutral "substance" or "additive" [39]; group (b): use a mediator to do the action [24], replace "mechanical" components with "simple or complex fields" [28], change the degree of flexibility or density or volume [35]; group (c): change the degree of flexibility or density or volume [35], replace "solid" parts with "fluidic" parts [29], make use of "waste material" or make the system to serve itself [25], carry out in advance some actions or put the components of the system such as to come fast into action when necessary [10].

In addition to the technical elements defined a paragraph before, the following ones should be considered to meet the major scalability requirements: build the web-course system having both web-cameras and videoconference units–principle 3; provide also materials with detailed information and explanations on how to use the course, considering supplementary case studies and exercises, thus facilitating off-line learning–principle 28; provide records of the course, and simulations, etc.–principle 28; use local assistants to support teacher for any small inconvenient caused by the remote interaction–principle 24; record the session for later use–principle 25; combine remote sessions with one or two face-to-face sessions–principle 40; train the users in advance on how to use the system (e.g. demos, intuitive guiding manuals, face-to-face practical sessions)–principle 10.

An analysis of these technical elements against the most relevant scalability performance characteristics highlighted in Figure 2 shows the web-based course system meets the intended targets (e.g. any file format is supported, any software application running on the teacher's desktop is seen on students' screens, time required to use any new software application is almost instantaneous, etc.).

Figure 3 illustrates deployment of TF_2-related requirements into TF_2-related performance characteristics. The set of matrices from Figure 3 is called "house of affordability." According to the results in Figure 3, the most relevant performance characteristics for web-course optimization in relation to TF_2 are: return on investment, net present value and payback period. In terms of "system-oriented innovation" (factor D in Figure 3), innovations from organizational and financial points of view should be mainly directed towards high return on investment and net present value. In other words, this requires relative low initial investments, as well as continuous and significant savings, relevant incomes even from early phases of course launching, supplemented by low operating and administrative costs. As in the case of design for scalability, no innovations generated by conflicts between performance characteristics are necessary. In terms of targets, innovations are mainly on achieving a high rate of the return on investment.

In this case, the following conflicting problems are formulated: (a) durability of the moving object [15] versus amount of substance [26]; and (b) brightness [18] versus amount of substance [26]. With respect to these conflicting problems, the following generic directions of innovation are suggested by TRIZ: group (a): transition from a homogeneous structure to a heterogeneous structure of the system or of the external environment [3], change the degree of flexibility or density or volume [35], carry out in advance some actions or put the components of the system such as to come fast into action when necessary [10], define a composite structure of the system [40]; group (b): increase the segmentation of the system [1], replace a continuous action with an impulse (and if it necessary use pauses between impulses to introduce additional actions) [19].

In order to meet cost-affordability requirements (see results in figure 3), the web-based course could comprise the following main technical elements: use open source repository systems (e.g. Moodle), where any file format can be uploaded/downloaded–principle 35; multiple free communication modes are ensured (e.g. off-line via e-mails; on-line via Skype, Yahoo messenger) combined with usual web-cameras for visual contact–principle 40; teacher uses a free license remote desktop control server application (e.g. VNC) in order to display remotely any application that runs on his/her computer–principle 35; file formats are limited to those which are immediate accessible to the teacher, with no supplementary costs for realization–principle 1; train the users in advance on how to use the system by means of face-to-face practical sessions and/or a simple guiding procedure–principle 10; use local assistants from the audience–principle 3; let students know in advance how to set up the system for the next session–principle 10; use asynchronous means to learn (e.g. links to specialized web-sites, electronic books, etc.)–principle 19; compensate the lack of good interactivity with more detailed written materials and run the course for larger groups of students at a time–principle 1. This solution involves low investment. Combined with qualified teachers and high quality written materials, this solution could lead to attractive values of the return on investment.

Figure 4 illustrates deployment of TF_3-related requirements into TF_3-related performance characteristics. The set of matrices from Figure 4 is called "house of development." According to the results in figure 4, the most relevant performance characteristics for web-course optimization in relation to TF_3 are: time required to adapt a classical course to a web-based course, costs required to adapt a classical course to a web-based course, time required to train students/lecturers to use the system and time required to initiate the working session. There are no conflicts between the performance characteristics, thus no innovations are required in this dimension. However, organizational innovations should be considered in relation to factor D, mainly for reducing costs necessary to adapt a classical course to a web-based course (see Figure 4), as well as to meet the target (less than 20% from the development costs of a classical course).

In the case of target-function TF_3, the following generic conflicting problems are formulated with respect to organizational challenges: (a) amount of substance [26] versus convenience of use [33]; (b) amount of substance [26] versus capacity [39]. With respect to meeting key targets, the following TRIZ-type challenges are defined: (c) "tension (pressure)" [11] versus volume of non-moving object [8]; (d) "tension (pressure)" [11] versus harmful side effects [31]. TRIZ recommends the following paths of intervention for these problems: group (a): change the degree of flexibility or density or volume [35], replace "solid" parts with "fluidic" parts [29], make use of "waste material" or make the system to serve itself [25], carry out in advance some actions or put the components of the system such as to come fast into action when necessary [10]; group (b): inversion (do the things vice versa than usual)

[13], replace "solid" parts with "fluidic" parts [29], transition from a homogeneous structure to a heterogeneous structure of the system or of the external environment [3], replace an expensive system with several inexpensive systems, comprising some properties [27]; group (c): change the degree of flexibility or density or volume [35], use a mediator to do the action [24]; group (d): extract or separate some disturbing parts from the system [2], apply homogeneity principles [33], replace an expensive system with several inexpensive systems, comprising some properties [27], use the resonance frequency [18].

The solution related to the target-function "design for fast development and implementation" shall follow several directions of intervention that are common with the target-function "design for scalability." In this respect, materials can be prepared in any format (e.g. .ppt, .pdf, .doc, .xls, .jpg, .cdr, .avi, etc.) and a repository system is used for file management (e.g. Moodle) to upload and download files on computers, too–see principle 35. Multiple communication modes are implemented (e.g. e-mail; Skype, Yahoo messenger, web-cameras)–see principle 18. For a better interaction with the classroom, a videoconference system can be taken into account. Teacher uses a remote desktop control server application (e.g. VNC) for handling remotely any application that runs on his/her computer. If a tablet PC is part of the system, teacher can use a virtual whiteboard to interact even more with students (e.g. CorelGrafigo, Windows Journal, etc.). Moreover, teacher can run specialized applications for monitoring computers in the network. In addition, students can work in close connection with the teacher for structuring the classical course to a web-based course. In this respect, each student can receive a small homework with the purpose of adapting a sub-chapter of the classical course into a web-based one. This means they will reformulate the content according to their understanding, will add questions and will indicate where more explanations and exercises are required. Thus, teacher can easier and faster

intervene to develop the web-based course. This action is in accordance with principle 13. Principle 27 encourages the use of virtual modules to train students (e.g. for the course of industrial robot design that is treated in this case study, students can use robot off-line programming and modelling systems like RobotStudio, which have the capability to be installed on a server and accessed via internet from any location). The use of local assistants for supporting teacher in classes is also indicated in this case (principle 24). According to principle 2, it is encouraged to replace the classical pedagogical methods with problem-based and project-based learning methods. Cooperative learning strategies are highlighted by principle 33. The solution meets the targets established for fast development and implementation of the web-based course (review the targets in Figure 4).

In Figure 5, deployment of TF_4-related requirements into TF_4-related performance characteristics is revealed. The set of matrices from Figure 5 is called "house of reliability." Major considerations in web-course design should be given to improve the following metrics: mean time between failures, average bandwidth required, signal delay and resolution. With respect to this target-function, several innovations are required to overpass the conflicts existing between various metrics (see Figure 5). Thus, innovative solutions are necessary to avoid signal loss at low bandwidth, to avoid signal delays at low bandwidth, to ensure high resolution at low bandwidth and to avoid delays if the resolution has to be kept high. From organizational point of view, innovation is required for keeping a high level of interaction between students and teachers. From targets point of view, the main focus is on student-teacher interaction and quality of signal transmission (see Figure 5).

In the case of TF_4, the vectors of innovation are directed towards three major areas: (a) to approach the conflicts between performance characteristics, (b) to approach targets, and (c) to approach organizational challenges. For the first category of problems, TRIZ reflects the following generic conflicts: (a_1) loss of information [24] versus speed (time) [9]; (a_2) stability of object [13] versus speed (time) [9]; (a_3) brightness [18] versus speed (time) [9]; (a_4) brightness [18] versus harmful side effects [31]. For the second category of problems, the following challenges have to be tackled: (b_1) accuracy of process [29] versus power [21]; (b_2) accuracy of process [29] versus volume of moving object [7]; (b_3) accuracy of process [29] versus amount of substance [26]; (b_4) capacity [39] versus power [21]; (b_5) reliability [27] versus speed (time) [9]. For the third category of problems, the following TRIZ-related generic conflicting parameters have to be approached: (c_1) convenience of use [33] versus complexity of system [36]; (c_2) convenience of use [33] versus "shape" (type of course) [12].

In relation to this, the following generic inventive vectors are revealed by TRIZ matrix of contradictions: group (a_1): use simple and inexpensive "copies" of the system [26], use "additives" to see processes that are difficult to see or change the "colour" or the level of transparency of the system [32]; group (a_2): apply homogeneity principles [33], make the system more dynamic (with components able to change their position, with higher mobility and interchangeability, with capacities of automatic adjustments) [15], replace "mechanical" components with "simple or complex fields" [28], use the resonance frequency [18]; group (a_3): carry out in advance some actions or put the components of the system such as to come fast into action when necessary [10], inversion (do the things vice versa than usual) [13], replace a continuous action with an impulse (and if it necessary use pauses between impulses to introduce additional actions) [19]; group (a_4): change the degree of flexibility or density or volume [35], replace a continuous action with an impulse (and if it necessary use pauses between impulses to introduce additional actions) [19], use "additives" to see processes that are difficult to see or change the "colour" or the level of transparency of the system [32], introduce a neutral "substance" or

"additive" [39]; group (b_1): use "additives" to see processes that are difficult to see or change the "colour" or the level of transparency of the system [32], extract or separate some disturbing parts from the system [2]; group (b_2): use "additives" to see processes that are difficult to see or change the "colour" or the level of transparency of the system [32], replace "mechanical" components with "simple or complex fields" [28], extract or separate some disturbing parts from the system [2]; group (b_3): use "additives" to see processes that are difficult to see or change the "colour" or the level of transparency of the system [32], use "flexible membranes" [30]; group (b_4): change the degree of flexibility or density or volume [35], make objects of the system to operate continuously, without pauses [20], carry out in advance some actions or put the components of the system such as to come fast into action when necessary [10]; group (b_5): perform harmful operations at very high speeds [21], change the degree of flexibility or density or volume [35], compensate for the relatively low reliability of the system by taking countermeasures in advance [11], replace "mechanical" components with "simple or complex fields" [28]; group (c_1): use "additives" to see processes that are difficult to see or change the "colour" or the level of transparency of the system [32], use simple and inexpensive "copies" of the system [26], apply the equipotentiality principle [12], use multi-level assembly of the system / move the problem into several "dimensions" [17]; group (c_2): make the system more dynamic (with components able to change their position, with higher mobility and interchangeability, with capacities of automatic adjustments) [15], reject and regenerate parts of the system [34], replace "solid" parts with "fluidic" parts [29], replace "mechanical" components with "simple or complex fields" [28].

In the case of the last target-function, a large set of vectors of innovation are proposed to support the conceptualization process of the web-based course system. Many of them are already met in relation to the previous target-functions (e.g. principles 33, 35, 29, 15, 28, 18, 10, 19, 39, 2, 11). Solutions proposed in those cases are naturally included here, too. Thus, materials can be prepared in any format and students can download files from a repository system (e.g. Moodle). Students can open any file on their computers (alternatively, the file can be displayed on a screen by means of a multi-media projector) and multiple modes of communication are provided (e.g. e-mails, Skype, Yahoo messenger, etc.). Web-cameras and/or videoconference systems are integrated. Teacher uses a tablet PC and a remote desktop control server application (e.g. VNC) in order to handle remotely any application that runs on his/her tablet PC. The tablet PC, together with specific software applications (e.g. CorelGrafigo, Windows Journal, etc.), acts as a virtual interactive whiteboard. A specific application for remote computer monitoring is integrated to "see" and "record" any action of each student on his/her terminal. Thus, a comprehensive and multimodal provision of the course content is ensured. Materials with detailed information and explanations on how to use the course material, with supplementary case studies and exercises are also provided, thus facilitating off-line learning. Local assistants are included in the scheme. Sessions are recorded for further use. Remote sessions are combined with one or two face-to-face meetings. All users are trained in advance for handling the system (e.g. demos, intuitive guiding manuals and face-to-face practical meetings). A group of students could also cooperate with the teacher before course delivery to "translate" the classical course to a web-based course (students are granted for this effort). Problem-based and project-based learning methods should be the core pedagogical approaches for course delivery. Cooperative learning strategies are also included. To these components of the web-based course system, some additional elements should be considered. They are suggested by the inventive principles 26, 32, 30, 20, 21, 26, 12, 17 and 34. Thus, films and simulations that are free on in-

ternet could be included in various places of the course to improve information content–principle 26. In the same spirit, courses can be recorded in advance and uploaded on the repository system. Complex designs (as it is the case of industrial robots) and complex hardware technical systems can be replaced with virtual systems (today, there are simulators for robots, PLCs, controllers, etc.). Development of partnerships with local companies for facilitating access of the students to various hardware solutions should be taken into account–principle 32. According to the same principle, students can be pre-grouped to work together in activity centres (when this thing is possible). According to principle 30, it is encouraged that some students to mentor other students. Principle 20 suggests a continuous evaluation of students, not only at the end phase of the course and encourages project-based and exercise-based course delivery. Principle 21 focuses on quality of transmission. In this respect, videoconference systems with facilities of fast resetting should be the option. In addition, the consideration of redundant solutions for communication is required (web-cameras, internet messaging systems), thus acting as a "plan B." Moreover, courses have to be recorded in advance, as a "plan C" for the cases of longer period internet connection breakdown–principle 26. Principle 12 clearly asks to consider multiple technologies of course delivery that can be easy and fast adapted to meet varying student needs. Principle 17 recommends the use of experiential learning rather than theoretical learning; therefore simulation tools which replicate accurately the real systems are necessary (e.g. in the case of industrial robot design, students can use web-based modelling and simulation tools like RobotStudio, which replicate real systems with very high accuracy). Projects can be remotely analyzed by the teacher, too, using the virtual whiteboard and the tablet PC, as well as a remote desktop control server application and a videoconference system. Principle 34 suggests the use of practical activities in companies. This is possible even for web-based courses. Also, this principle encourages for periodical change of the instructional activities in order to keep student interest for the course. This can be done by means of exercises.

The solutions proposed above converge to the main targets expressed by the function "design for reliability." Specifically, they are able to ensure a good interaction between student and teacher, and from technical point of view (e.g. signal quality) are feasible, too.

The next step of the methodology requires aggregation of the local solutions. According to the proposed algorithm, *PTF* is in this case study associated to TF_4. The aggregation flow for this case study starts by combining the local solution related to TF_4 with the local solution related to TF_3. The hybrid result, here named HF_1, is further combined with the local solution of TF_1. The new hybrid, named HF_2, is further combined with the local solution of TF_2.

From aggregation point of view, the first two steps raise up no problems, because solutions are either positive (TF_4 with TF_3) or not correlated (HF_1 with TF_1). The challenge occurs when aggregation is required between HF_2 and the local solution related to TF_2. Thus, HF_1 will incorporate all common components of the local solutions related to TF_4 and TF_3, to which all individual components of the local solutions related to TF_4 and TF_3 are added (please review the local solutions for TF_4 and TF_3). Moreover, HF_2 will include all components of HF_1, to which the individual components of the local solution related to TF_1 are added (please review the local solution for TF_1).

Analyzing the main characteristics of the local solution related to TF_2 against the hybrid solution HF_2, the following major conflicting problems occur: (a) need of videoconference system (HF_2) versus avoidance of expensive equipment (TF_2); (b) need of simulation tools which accurately replicate real systems (HF_2) versus avoidance of expensive software (TF_2). Both cases are subjected to the same TRIZ generic problem: capacity (TRIZ parameter 39) versus harmful side effects (TRIZ parameter 31); and convenience of use (TRIZ parameter 33) versus amount of substance (TRIZ parameter 26). The inventive principles proposed by TRIZ in connection to these problems are: apply the equipotentiality principle [12], use the "resonance frequency" [18], convert harm into benefit [22], change the concentration and flexibility [35], and introduce a "neutral substance" or an "additive" [39].

Principle 12 highlights the importance of analyzing conflicts in the context of each course. In this respect, for the course of industrial robot design, principle 18 suggests selecting software solutions for robot simulation from providers that have educational licenses (e.g. ABB is the producer of RobotStudio; the company provides free educational licenses for 50 users; similar educational programs have other companies like Fanuc, Siemens, etc.). Principle 39 suggests renting videoconference systems, if attractive offers exist. However, an alternative is to buy the videoconference system using a long term payment scheme (e.g. quarterly rates, over several years time horizon)–principle 35. Principle 22 suggests exploiting various opportunities for attracting funds to buy a high quality technology for web-based courses. In this respect, this principle asks applying for grants, sponsorships from several available programs (e.g. in Europe: European Social Fund, Leonardo da Vinci Program, Erasmus Program, etc.).

Comprising all recommendations, the final solution of the web-based course in industrial robot design comprises five units. They are further detailed.

1. **Course structure unit:** The course contains lectures, seminars, project work, practical activities and individual work–all these components are delivered remotely; however, at least two face-to-face meetings between lecturer and students should be considered (this action is feasible because the remote class is located in a single place); problem-

based and project-based learning methods are included–students must individually work to design a robot module (e.g. an intelligent gripping system) and in cooperation to solve one or two real mini-problems taken from industry (they have to spend this time in companies).

2. **Course support unit:** Consideration of .ppt and .avi files for on-line lectures; consideration of .doc, .pdf, .jpg and .avi files for off-line information delivery; guiding manual to support the project work is provided in .pdf format; during seminar and lecture classes, supplementary files could be generated (see the use of whiteboards and simulation tools)–these files can be saved in .jpg, .htm, .avi and other formats for later review; materials with detailed information on how to use files are necessary; supplementary case studies and exercises should be included for the complex parts of the course–thus, off-line learning is facilitated; two local assistants are taken into account (students that support teacher in initializing and turning off the system); each session is recorded by the videoconference system for later use, both by students and teacher; a specific software tool for remote computer monitoring can be activated to "record" any action performed by each student on his/her terminal.

3. **Simulation unit:** Use educational licenses for specialized web-based robot simulation and programming software–in this case RobotStudio and Robot HandlingPro are included in the training program because their producers provide free licenses for educational purposes; other CAD/CAR/CAE/CAP software tools and PLC simulation tools can be accessed by students via the remote desktop control application (e.g. VNC).

4. **Course management unit:** An open source repository system is implemented (e.g. Moodle); Moodle offers sufficient features for a good management of the web-based course (e.g. upload/download of files, forums, course scheduling, role assignment, and many other facilities which practically cover all usual actions for managing a course remotely); both teachers and students are trained in advance on how to use the technology for communication and interactive/cooperative work (training manuals and training sessions have to be taken into account); students can contribute to the adaptation of the course for the web-based delivery.

5. **Communication and interactive/cooperative work unit:** A videoconference system is installed–a standard videoconference system between point A and point B uses two video cameras: one, at point A, broadcasts person(s) A (e.g. the teacher) and displays the image that camera B broadcasts; camera B broadcasts person(s) B (e.g. the students) while displaying image received from point A; two multi-media projectors are installed in the classroom, one connected with the computer where the remote desktop control application is installed (e.g. VNC) and another one connected to a videoconference camera (e.g. Polycom); in order to increase system's robustness, additional communication means are integrated–they comprise e-mail, Skype and Yahoo messenger, with web camera included on both sites; interaction between students and teacher is increased using the tablet PC and an interactive whiteboard (e.g. CorelGrafigo); in order to make different annotations on the work performed by students, specific software applications, like Windows Journal, are integrated; projects and homework can be remotely analyzed by the teacher using the virtual whiteboard and the tablet PC, as well as the remote desktop control server application and the videoconference system.

The scenario for communication and interactive/cooperative work between teacher and students looks like this: (a) teacher and students can see each other by using the standard video-conference system based on two video cameras; each camera needs to have a direct internet connection and a real IP address; a broadband internet connection is recommended; (b) the broadcast quality is usually a standard TV broadcast quality (for example a PAL video system has a 720 x 576 pixel resolution); (c) teacher watches students on a computer monitor that is directly connected to his/her videoconference camera; (d) teacher works on his tablet PC which needs to have a direct internet access; it needs to run a remote desktop control server application (e.g. VNC); teacher runs the course presentation or any support applications on his/her tablet PC; (e) students watch teacher on a projection screen; their videoconference camera is connected to the multimedia projector; (f) students are connected to the teacher's tablet PC by using a remote desktop control client application like VNC and are able to see teacher's desktop, i.e. the presentation or any other application (students need to have internet access); they may even remotely work on teacher's computer (even an on-line exam may be performed in this way); (g) teacher controls who or if anyone can remotely control his/her desktop; teacher may accept connections in read-only mode when presenting his/her course and then allows connections with control rights for any interactive tasks. Figure 6 suggestively illustrates this scenario.

The architecture of the web-based course revealed in this case study has a high degree of versatility, meaning that with low effort and in a very short time it can be adapted to many other web-based courses in engineering. The system described in this case study has been effectively tested in practice, providing a very good satisfaction both to students and teachers. After one year of implementation, it was adopted by other ten engineering courses (i.e. Computer Aided Design, Creativity and Innovation in Engineering, Quality Engineering and Management, Machine Elements Design, Mechanisms Design, Hydraulic and Pneumatic Systems, Objected Oriented Programming, Integrative Project, Process Monitoring and Control, Machine Tool Design).

FURTHER RESEARCH DIRECTIONS

Research presented in this material could be continued by including in the planning scheme additional target-functions. Two of the most chal-

Figure 6. Communication between teacher and students by means of a videoconference system

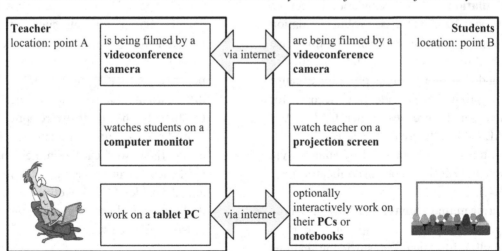

lenging ones are "design for web-based course content quality" and "design for fast web-based course customization." Both design for course content quality and course customization lead to a significant number of conflicting problems. As a consequence, adequate planning and innovation is required to overpass crisis points in web-based course system conceptualization, development and implementation.

Performance planning could be further deployed to the level of teaching and learning processes, as well as to the pedagogical methods. Here, there is a huge space of action, where competitive design methods and methodologies would be extremely useful to handle less tangible dimensions and incomplete data.

Development of more realistic and intuitive simulation tools for web-based courses should be another area where competitive design would play an important role. Actually, this is a generous area for future developments, because every producing organization will have to become conscious that provision to universities of simulation tools related to their products will represent a competitive weapon. These tools will have to operate effectively via internet. In this respect, future developments of web technologies will support this process. Moreover, new organizational models of the educational providers will have to be considered.

Universities will have to enhance their partnerships with producing organizations, acting as "extended open universities." In this area, competitive design will provide a reliable framework to ensure a superior balance between effort and effect in web-based course conceptualization, development and delivery.

Further challenges in web-based course development will occur. They will be related to the development of a new generation of web-based courses in engineering. Such courses will consider remote interactions with hardware systems, also including capabilities for "sensing" the remote environment. In this area, competitive design will

play an important role for setting up ingenious and commercial sustainable solutions, too.

CONCLUSION

The consideration of competitive design philosophy for setting up successful web-based courses in engineering is introduced in this material. Web-based courses in engineering are complex systems, which have to face with various challenges from technical, technological, financial and organizational points of view. From this perspective, such courses have to be designed and developed in strong connection with the context where they are going to be implemented, thus no "best-of-the-world" solution exists, and only context-related optimal solutions could be proposed at the best. Therefore, application of performance planning and innovative problem solving tools is necessary during the conceptualization and design of such courses.

This chapter introduces a novel framework for concurrent planning of a web-based course considering its multiple facets. The planning process starts from the incipient phases of web-based course development. The goal is to formulate a competitive solution in an effective and efficient way, with lower development costs and in a reasonable period of time. Several concluding remarks derive from the application of methodology on the case study.

First, designing a successful web-based course in engineering requires simultaneous consideration of multiple sides characterizing the respective course. Second, the complex nature of the course necessitates an innovative approach of solution formulation in order to overpass inevitable conflicting problems of technical, technological, organizational and financial nature. Third, a web-based course in engineering should be seen as a dynamic system, whose evolution in time is strongly dependent by the correlations between the life-cycles of its major sub-systems (e.g. communication unit, management unit, simulation unit, support unit, etc.).

REFERENCES

Altshuller, G. (2000). *The innovation algorithm. TRIZ, systematic innovation and technical creativity*. Worcester, MA: Technical Innovation Center.

Barros, B., Read, T., & Verdejo, M. F. (2008). Virtual collaborative experimentation: An approach combining remote and local labs. *IEEE Transactions on Education*, *51*(2), 242–250. doi:10.1109/TE.2007.908071

Bhatt, R., Tang, C. P., Lee, L. F., & Knovi, V. (2009). A case for scaffolded virtual prototyping tutorial case-studies in engineering education. *International Journal of Engineering Education*, *25*(1), 84–92.

Bier, I.D., & Cornesky, R. (2001). Using QFD to construct a higher education curriculum. *Quality Progress*, April, 64-68.

Brackin, P. (2002). Assessing engineering education: an industrial analogy. *International Journal of Engineering Education*, *18*(2), 151–156.

Brad, S. (2005). Need for radical innovation to ensure high quality of continuous education in engineering. In C. Oprean, & C. Kifor (Ed.), *3rd Balkan Region Conference on Engineering Education* (pp. 77-80). Sibiu: Lucian Blaga University Press.

Brad, S. (2008). Complex system design technique. *International Journal of Production Research*, *46*(21), 5979–6008. doi:10.1080/00207540701361475

Brad, S. (2009). Concurrent multifunction deployment. *International Journal of Production Research*, *47*(19), 5343–5376. doi:10.1080/00207540701564599

Callaghan, M. J., Harkin, J., McGinnity, T. M., & Maguire, L. P. (2008). Intelligent user support in autonomous remote experimentation environments. *IEEE Transactions on Industrial Electronics*, *55*(6), 2355–2367. doi:10.1109/TIE.2008.922411

Du, J., Li, F., & Li, B. (2008). The case of interactive web-based course development. In E. Leung, F. Wang, L. Miao, J. Zao, & J. He (Ed.), *2nd Workshop on Blended Learning* (pp. 65-73). Jinhua: Springer-Verlag Berlin.

Ebner, M., & Walder, U. (2008). E-education in civil engineering–a promise for the future? In P. Vainiunas & L. Juknevicius (Ed.), *6th AECEF Symposium on Civil Engineering Education in Changing Europe* (pp. 16-26). Vilnius: Vilnius Gediminas Technical Univ. Press.

Finger, S., Gelman, D., Fay, A., & Szczerban, M. (2005). Supporting collaborative learning in engineering design. In W. Shen, A. James, K. Chao, M. Younas, & J. Barthes (Ed.), *9th International Conference on Computer Supported Cooperative Work in Design* (pp. 990-995). Coventry: Coventry Univ. Press.

Hamada, M. (2008). Web-based environment for active computing learners. In O. Gervasi, & B. Murgante (Ed.), *International Conference on Computational Sciences and Its Applications* (pp. 516-529). Perugia: Spriger-Verlag Berlin.

Helander, M. G., & Emami, M. R. (2008). Engineering e-laboratories: Integration of remote access and e-collaboration. *International Journal of Engineering Education*, *24*(3), 466–479.

Hutchings, M., Hadfield, M., Horvath, G., & Lewarne, S. (2007). Meeting the challenges of active learning in web-based case studies for sustainable development. *Innovations in Education and Teaching International*, *44*(3), 331–343. doi:10.1080/14703290701486779

Jou, M., Chuang, C. P., Wu, D. W., & Yang, S. C. (2008). Learning robotics in interactive web-based environments by PBL. *IEEE Workshop on Advanced Robotics and Its Social Impacts* (pp. 206-211). Taipei: Nat. Taiwan Normal Univ. Press.

Kaminski, P. C., Ferreira, E. P. F., & Theuer, S. L. (2004). Evaluating and improving the quality of an engineering specialization program through the QFD methodology. *International Journal of Engineering Education, 20*(6), 1034–1041.

Koksal, G., & Egitman, A. (1998). Planning and design of industrial engineering education quality. *Computers & Industrial Engineering, 35*(3-4), 639–642. doi:10.1016/S0360-8352(98)00178-8

Kolmos, A., & Du, X. (2008). Innovation for engineering education–problem and project based learning (PBL) as an example. In Aung, W., Mecsi, J., Moscinski, J., Rouse, I., & Willmot, P. (Eds.), *Innovations 2008: World Innovations in Engineering Education and Research* (pp. 119–128). Arlington, VA: Begell House Publishing.

Kumar, A., & Labib, A. W. (2004). Applying quality function deployment for the design of a next-generation manufacturing simulation game. *International Journal of Engineering Education, 20*(5), 787–800.

Lau, H., Mak, K., & Ma, H. (2006). IMELS: An e-learning platform for industrial engineering. *Computer Applications in Engineering Education, 14*(1), 53–63. doi:10.1002/cae.20067

Li, L. Y., & Wang, H. J. (2007). A new method for building web-based virtual laboratory. In H. Liu, B. Hu, X.W. Zheng, & H. Zhang (Ed.). *1st International Symposium on Information Technologies and Applications in Education* (pp. 555-558). Kunming: Shandong Normal Univ. Press.

Liang, J. S. (2009). Development for a web-based EDM laboratory in manufacturing engineering. *International Journal of Computer Integrated Manufacturing, 22*(2), 83–99. doi:10.1080/09511920801911019

Lindroos, K., Malmivuo, J., & Nousiainen, J. (2007). Web-based supporting material for biomedical engineering education. In T. Jarm, P. Kramar, & A. Zupanic (Ed.), *11th Mediterranean Conference on Medical and Biological Engineering and Computing* (pp. 1111-1114). Ljubljana: Springer-Verlag Berlin.

Mackey, T., & Ho, J. (2008). Exploring the relationship between web usability and students' perceived learning in web-based multimedia (WBMM) tutorials. *Computers & Education, 50*(1), 386–409. doi:10.1016/j.compedu.2006.08.006

Nedic, Z., & Machotka, J. (2006). Interactive electronic tutorials and web based approach in engineering courses. In V. Uskov (Ed.), *5th IASTED International Conference on Web-based Education* (pp. 243-248). Puerto Vallarta: Acta Press.

Ogot, M., & Okudan, G. (2007). A student-centered approach to improving course quality using quality function deployment. *International Journal of Engineering Education, 23*(5), 916–928.

Popescu, S., Brad, S., & Popescu, D. (2006). The engineering study program as a customized product: barriers and directions for intervention, In E. Westkaemper (Ed.), 1st CIRP International Seminar on Assembly Systems, (pp. 289-294). Stuttgart: IFF Univ. Stuttgart.

Raharjo, H., Xie, M., Goh, T. N., & Brombacher, A. C. (2007). A methodology to improve higher education quality using the quality function deployment and analytic hierarchy process. *Total Quality Management & Business Excellence, 18*(10), 1097–1115. doi:10.1080/14783360701595078

Rizzotti, S., & Burkhart, H. (2006). Web-based test and assessment system: Design principles and case study. In V. Uskov (Ed.), *5th IASTED International Conference on Web-based Education* (pp. 37-42). Puerto Vallarta: Acta Press.

Rojko, A., Hercoq, D., & Jezemik, K. (2008). Educational aspects of mechatronic control course design for collaborative remote laboratory. In IEEE (Ed.), *IEEE 13th International Power Electronics and Motion Control Conference* (pp. 2349-2353). Poznan: Poznan Univ. Press.

Saaty, T. L., & Vargas, L. G. (2001). *Models, methods, concepts & applications of the analytic hierarchy process*. Secaucus, NJ: Springer.

Salihbegovic, A., & Tanovic, O. (2008). Internet based laboratories in engineering education. In V. Luzar, V. Dobric, & Z. Bekic (Ed.), *30th International Conference on Information Technology Interfaces* (pp. 163-170). Cavtat: Zagreb Univ. Press.

Saygin, C., & Kahraman, F. (2004). A Web-based programmable logic controller laboratory for manufacturing engineering education. *International Journal of Advanced Manufacturing Technology*, *24*(7-8), 590–598. doi:10.1007/s00170-003-1787-7

Sessink, O., van der Schaaf, H., Beeftink, H., Hartog, R., & Tramper, J. (2007). Web-based education in bioprocess engineering. *Trends in Biotechnology*, *25*(1), 16–23. doi:10.1016/j.tibtech.2006.11.001

Suliman, S. M. A. (2006). Application of QFD in engineering education curriculum development and review. *International Journal of Continuing Engineering Education and Lifelong Learning*, *16*(6), 482–492. doi:10.1504/IJCEELL.2006.011892

Takago, D., Matsuishi, M., Goto, H., & Sakamoto, M. (2007). Requirements for a web 2.0 course management system of engineering education. In IEEE Computer Society (Ed.), *9th IEEE International Symposium on Multimedia* (pp. 435-440). Taichung: Asia Univ. Press.

Toral, S., Barrero, F., & Martinez-Torres, M. (2007). Analysis of utility and use of a web-based tool for digital signal processing teaching by means of a technological acceptance model. *Computers & Education*, *49*(4), 957–975. doi:10.1016/j.compedu.2005.12.003

Vargas, H., Sanchez, J., Duro, N., Dormido, R., Farias, G., & Dormido, S. (2008). A systematic two-layer approach to develop web-based experimentation environments for control engineering education. *Intelligent Automation and Soft Computing*, *14*(4), 505–524.

Wang, X., Dannenhoffer, J., Davidson, B., & Spector, M. (2005). Design issues in a cross-institutional collaboration on a distance education course. *Distance Education*, *26*(3), 405–423. doi:10.1080/01587910500291546

Xuelian, H. (2008). Research and practice on web-based distance education of master of engineering. In L. Maoqing (Ed.), *3rd International Conference on Computer Science & Education* (pp. 1511-1514). Kaifeng: Xiamen Univ. Press.

Yeo, R. (2008). Brewing service quality in higher education: Characteristics of ingredients that make up the receipe. *Quality Assurance in Education*, *16*(3), 266–286. doi:10.1108/09684880810886277

Chapter 11
WIRE:
A Highly Interactive Blended Learning for Engineering Education

Yih-Ruey Juang
Jinwen University of Science and Technology, Taiwan

ABSTRACT

Much research has shown that the blended learning can effectively enhance the motivation, communication skills, and learning achievement compared with teaching in a single form. However, a crucial issue in blended learning is how to integrate each blended format, media and experience into a coherent learning model, and then to keep interaction between teacher and students either in or outside the classroom. This study introduces a highly interactive strategy for blended learning that incorporates web-based and face-to-face learning environments into a semester course through answering the warm-up questions before class, interactive teaching in class, and review and exercise after class. By the empirical study in a 'Data Structure' class, most students made progress in learning achievement and gain more motivation and interaction within the class.

INTRODUCTION

Blended learning is a learning solution that incorporates instructor-led classroom events and any form of technology enabled instruction used outside the classroom (Pulichino, 2006). Many research findings show that the blended learning can more effectively enhance the motivation, communication skills, and learning achievement than the learning in a single form (Troha, 2002;

Barbian 2002; Zenger & Uehlein, 2001). However, a crucial issue in blended learning is how to integrate each blended format, media and experience into a coherent learning model, and then to keep interaction between teacher and students either in or outside the classroom. Learners will not busily deal with multiple learning styles and can absorb the knowledge delivered in a continuous learning path.

This study presents a blended learning model that integrates four activities into a continuous and complete learning experience, which comprises

DOI: 10.4018/978-1-61520-659-9.ch011

the warm-up before class, interaction in class, and review and exercise after class, WIRE for short. The WIRE model has been experienced in the Department of Information Management of a university in Taiwan for 18 weeks of a semester. For quasi-experimental research method, the participants were divided into two groups (classes). The experimental group adopts WIRE model while the control group adopts lecture-based activities with the encouragement of warm-up and review lessons. The preliminary findings from the learning achievement, questionnaire, and focus group interview show some conspicuous dissimilarities between the experimental group and control group on the degrees of the motivation, interaction between teacher and students, and collaboration among students The WIRE model can help enhance the learning effects.

THE REQUIREMENTS OF DESIGNING BLENDED LEARNING IN SCHOOLS

The Aspect of Cognitive Development

The type of a blended model can be various combinations of different environments, strategies, technologies, and medias for learning (Osguthorpe & Graham, 2003; Singh, 2003; Valiathan, 2003; Rossett, Douglis, & Frazee, 2003), which depends on the characteristics of audience, time, scale, application, content, and resource (Bersin, 2004). The designer of blended learning should consider how to gain the biggest benefits for the organization under the minimum cost of time consuming or resources. However, in the school setting, a teacher who adopts blended learning in a subject for a whole semester has to design an appropriate blended model that generally integrates eLearning and face-to-face classroom

learning formats to help students keep active on the scheduled instructional plan. Especially, the learning content and its objectives are different from the training courses for enterprises, which focused on the knowledge construction and the cultivation of learning abilities. Some learning outcomes cannot be evaluated immediately and explicitly, but will exert the influence on students' future development. Therefore, the instructional design for the blended learning in schools should take into account the students' cognitive development in learning the new knowledge and then adopt appropriate learning technologies.

By referring to the learning objectives in cognitive domain of Bloom's Taxonomy (1956), the instructional plan should be designed in a smooth process that can develop learners' knowledge and intellectual skills, starting from simplest behavior to the most complex, that is, from knowledge, comprehension, application, analysis, synthesis, to evaluation. The challenge in designing a blended learning model in the school setting is to gradually incorporate the cognitive skills, from basic to advanced, into a serious of learning activities with the use of appropriate technologies. This study divided the whole process of learning a new concept or topic into three stages, including the 'Warm-up' before class, the 'Interaction' with teacher and classmates in class, and the 'Review and Exercises' after class, that is WIRE model. In order to cultivate students' high-order thinking skills, students are requested to understand the assigned learning materials as far as they can before class. Then, in face-to-face class, the teacher can put the emphasis on the analysis and application of the new knowledge to inspire students to higher order thinking, such as the synthesis and evaluation of knowledge. The most important key principle in WIRE model is to keep effective interaction between teachers and students during the learning process in order to maximize the learning effect towards the abilities of synthesis and evaluation.

The Aspect of Technology Adoption

Modern technologies that support for teaching and learning generally consist of learning platform (course management system (CMS), learning management system (LMS), virtual learning environment (VLE) or instructional management system (IMS), such as Blackboard, Moodle, Sakai, etc.), learning content authoring and presentation tools (such as Articulate, Camtasia, Adobe Captivate, Microsoft PowerPoint, etc.), classroom technology products (such as interactive whiteboard, clickers, classroom management, etc.), web-served content (such as Wikipedia, Elsevier, Safari Montage, etc.), adaptive & gaming (such as adaptive tutoring, educational gaming, or other domain-specific learning environment products), content management (including course management system content management, digital repositories, and document management systems), mobile technology, etc. What technology the teacher will adopt depends on the instructional design and objective environment, but more consideration is focused on the usability and complexity. By referring to Rogers' model for the adoption and diffusion of innovations, the authors believe a new technology for education will be relatively accepted by most teachers when its diffusion, at least, has got into or has stepped across the 'early majority' stage. Teachers are interested in trying new technology with slight challenges but easy to use.

According to 2008 Snapshot Report on Learning Modalities released by eLearning Guide Research (2008), the first five popular learning modalities are:

1.   Printed-based materials,
2.   Classroom instruction,
3.   Asynchronous web-based learning,
4.   In-person mentoring and tutoring, and
5.   Conference calls.

And the first five popular learning modalities with technology support are:

1.   Asynchronous web-based learning,
2.   Email,
3.   Online assessment and testing,
4.   Online references, and
5.   Synchronous web-based learning.

Even in the 2006 report on the popular mixes of media in blended learning (Pulichino, 2006), the results are similar to the above except that the 4th and 5th items of popular learning modalities are replaced by 'Online assessment and testing' and 'Online reference' respectively, and the 5th item in the popular learning modalities with technology support is replaced by 'Learning content management systems.' The physical communication in a face-to-face context is still the main stream for human resource training in enterprises. The adoption of learning technologies concentrates on the traditional functions provided in general eLearning platforms.

In school setting, the authors referred to the study of the Second Information Technology in Education Studies (SITES 2006) launched by the IEA (International Association for the Evaluation of Educational Achievement). There are 23 education systems (22 countries), 400 schools per countries, join the study. The population of schools will be all schools in a country that have eighth grade students. According to the report of SITES 2006 (Law, Pelgrum, & Plomp, 2008), the first five ICT tools with high frequency of use in teaching are:

1.   Equipment and hands-on materials (e.g., laboratory equipment, musical instruments, art materials, overhead projectors, slide projectors, electronic calculators),
2.   General office suite (e.g., word-processing, database, spreadsheet, presentation software),

3. Digital resources (e.g., portal, dictionaries, encyclopedia),
4. Tutorial/exercise software, and
5. Multimedia production tools (e.g., media capture and editing equipment, drawing programs, webpage/multimedia production tools).

The traditional equipment and hand-on materials are still the popular delivery media. Except that, the adoption manner in school setting concentrates on the editing of digital learning content and the applications of Internet resources or tutorial/exercise software for delivery learning content. In higher education, the authors referred to the report on the latest survey of higher education teachers in Taiwan (Center for Educational Research and Evaluation, 2005) and found 67.1% of colleges or universities have purchased eLearning platforms, e.g. CMS, LMS, VLE or IMS, to support teaching and learning but only 27.0% of teachers that have personally used the platform for teaching. We are not surprised by the big gap because the adoption and diffusion of eLearning platform has not steadily stayed at 'early majority' stage. Even though the teachers who were in the 27% population have adopted it, how many of them who have really rethought and adjusted their teaching with the support of eLearning platform. To some extent, they just treat it as a bulletin board for course information or a space for digital learning materials.

Summarily, in this study, we believe the adoption of learning technologies in the design of blended learning should answer the following essential questions:

1. What technologies are easy to use?
2. What technologies nowadays the teachers are relatively familiar with?
3. What technologies can be closely integrated into the instructional strategy?
4. What technologies can enhance the learning effect?

Under this concept, the learning technologies adopted by this study may sound uncreative and nothing new, but the meaningful usage in the whole blended model will inspire teachers more innovative teaching ideas.

THE DESIGN OF WIRE MODEL FOR BLENDED LEARNING

Based on the concepts described above, this study designs a blended learning model, the WIRE model, that comprises three stages, the Warm-up before class, Interaction in class, and Review and Exercises after class. The learning activities of the three stages are arranged as a cyclic process that incorporates various ICT tools into face-to-face and online contexts. The ICT tools used in WIRE model are nothing new but focused on the meaningful use for each learning activity, and are in accord with the requirements mentioned in previous section. The detailed activities and the technology support in each stage of the WIRE model are briefly summarized in Table 1 and detailed in the following paragraphs.

In *Warm-up* stage, students are asked to read text or digital materials online and then to answer the warm-up questions submitted by the teacher. Teachers can browse those answers and abstract the misconceptions and those they have difficulties to comprehend. This stage put the emphasis on the initial reading of learning content and the preliminary understanding of students' knowledge background related to the content. Therefore, the main cognitive categories of warm-up questions should be offered based on the knowledge and comprehension according to Bloom's taxonomy. Teachers have not to mark, correct, or comment their answers and just roughly browse and refer them to adjust the lesson plan for the upcoming classroom activities. The technology support for this stage could be a threaded discussion tool like a forum generally provided in general eLearning platform (e.g. LMS, LCMS) or the blog-based tools.

Table 1. The activities and technology support in each stage of the WIRE model

Stages	Delivery mode	Students' activities	Teacher's activities	Categories of cognitive objectives	Technology support
Warm-up before class	Online	**Step 2**: Read the learning materials and answer the warm-up questions announced by the teacher.	**Step 1**: Announce warm-up questions to students. **Step 3**: Roughly browse the warm-up answers before entering classroom.	Knowledge and comprehension	Threaded discussion tools or blogs
Interaction in class	Face-to-face	Interact with the teacher and peers according to the lesson plan.	For each topic… **Step 4.1**: Give a simple test to check the warm-up state. **Step 4.2**: Give a mini-lecture based on the test results and give another deep question for peer discussion.	Analysis and application	Classroom Response Systems or group discussion tools
Review and **Exercises** after class	Online	**Step 6**: Revise the warm-up answers and do the assigned exercises individually or collaboratively	**Step 5**: Assign exercises or a small project for individuals or groups.	Synthesis and evaluation	Homework management tools, blogs, or PBL tools

In *Interaction* stage, the teacher implements the lesson plan designed beforehand in interactive form to attain better learning effect. This stage is conducted by a face-to-face format in classroom. First of all, the teacher gives students a simple test to check the warm-up state. Then, based on the test results, teachers give a mini-lecture to explain the answer with the introduction to its background knowledge, and ask another deep question for peer discussion. The deep questions can emphasize on the types of analysis and application to inspire high-order thinking about the concept. Finally, the teacher can go back to the first step for the next topic if most students are qualified to achieve the goals of current topic.

However, it is difficult to collect students' responses of the concept test since most university students are passive in responding to questions in classroom activities. Therefore, it is suggested to use the Classroom Response System (sometime called Personal Response System, Electronic Response System, Audience Response System, etc. e.g. clicker) to enable lectures to become

more interactive and measure student responses to any variety of instructor questions in real time. Lot of literature showed that CRSs can motivate and engage student learning more effectively than traditional lecture-oriented teaching (Fies & Marshall, 2006; Siau, Sheng, & Nah, 2006; Bunz, 2005; Caldwell, 2007). Each student holds a remote control to answer teacher's concept tests interwoven with each lecture, and the teacher hence can rapidly gain the correctness rate to make a decision what to do next, such as to teach again, to lower the difficulty, to provide more hint, peer discussion, or to go to next topic. Sometimes, the concept tests could be open-ended questions that can be responded from group representatives. Students can use the group discussion tools provided in eLearning platform, or blogs to abstract the discussion results for further discussion or sharing.

In the stage of *Review and Exercises*, students are asked to modify their pervious answers of warm-up questions in the original discussion threads, and to supplement what they have learned in classroom. Teacher then browses the review and compare with the previous answers to assess how effective the classroom activities are. In order to get the learning outcome into students' long-term memory, the teacher can offers one to three exercises to ensure their understanding. In addition, to cultivate the abilities of synthesis and evaluation, the exercises could be a small project for individuals or groups. To a certain extent, it could be a form of problem-based learning that encourages student to cooperatively accomplish the assigned mission. The supporting technology should provide a common workspace with adequate scaffolding tools for each group to initiate the project.

Each cyclic process of the WIRE model, called a scenario, could be conducted for a lesson, chapter, unit, or any scale of learning content. When designing instructional plan for a WIRE scenario, the teacher can classify the learning content into three parts for each stage according to the cognitive objectives, and design corresponding concept questions, deep questions, and high-order thinking questions or small projects for each stage respectively. A scenario could be a combination of the three stages but should keep the sequence from W, I, to R & E. That is, a scenario may consist of multiple warm-up, interactions, reviews, or exercises based on the attributes of learning content. For example, a common scenario that frequently occurs in freshman year can arrange multiple combinations of warm-up and interaction if most learning content emphasized on the comprehension of basic knowledge, such as W >> I >> W >> I >> R & E. Another example, if the focus of learning activity is put on the synthesis and evaluation, the WIRE cycle can be designed with more time for review and exercises, such as W >> I >> R & E >> R & E.

RESEARCH DESIGN

The WIRE model was experimented upon a class of the Department of Information Management of a university of science and technology in Taiwan for 18 weeks a semester. The course was originally offered by distance learning, but the researchers changed it to adopt the blended mode which interwove with some face-to-face classes. The subject for this study is 'Data Structure' which is a fundamental course for data processing. Therefore, the WIRE scenario designed for each chapter of the learning material consists of twice warm-ups and interactions, and then twice stages for review and exercise (see Figure 1). The whole process of this scenario was conducted for three weeks on a chapter of the textbook, and therefore there are five chapters provided in the experiment.

By following the general class scheduling rule in universities, this study supposes each week has one day for interactive activity either online or in face-to-face classroom, such as the day 4, 11, and 18 in Figure 1. Each WIRE cycle has only one face-to-face classroom interaction (Interaction2 on day 11), and the other two interactive activities (Interaction1 and Interaction3) were conducted by the synchronous text chat on eLearning platform. For warm-ups, the learning content for Warm-up1

Figure 1. The WIRE scenario spread over three weeks

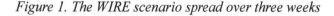

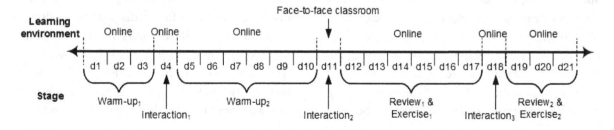

stage concentrates on the comprehension of the new knowledge, so the intention of Interaction1 stage is to clarify the confusion about the new knowledge; and then the Warm-up2 stage spends six days to prepare the learning content which will be taught in Interaction2 stage face-to-face. The rest days after the Interaction2 stage are allocated for the stage of Review and Exercise, which includes Interaction3 for consulting about the exercises or mini-PBL.

Participants

The experiment has 88 participants which were divided into two groups (classes). The experimental group has 45 students while the control group has 43 students. The experimental group adopts the WIRE model while the control group adopts lecture-based instruction but is verbally encouraged to warm-up and review lessons. Since the participants have been divided into two classes when they entered the department, the researchers cannot reorganize and randomly assign them into two new classes just for the experiment. Therefore, all participants were asked to take a prior-ability test which comprises basic Java programming and algebra. The test results revealed no significant by the Levene's test of homogeneity of variance (F=.824, p=.429>.05). That means the discrete distribution of the two classes have no significant difference, and they are in equal background knowledge and abilities to learn the subject 'Data Structure.' In addition, the students of each class have been divided into some heterogeneous groups where each group comprises 3 to 5 students for the group works.

Research Questions, Tools and Procedure

The researchers attempt to answer the following questions in this study by learning achievement test, questionnaire, and focus-group interview.

1. Can the WIRE model effectively motivate students to learn?
2. Can the WIRE model effectively facilitate interaction either between teacher and students or among students?
3. Can the WIRE model effectively increase students learning achievement?

The examination paper for evaluating the difference between the groups in learning achievement was made by the teacher according to the learning content, and was used after the experiment. In addition, the researchers made a questionnaire to survey the view of the experimental group on the WIRE model. Finally, 10 students of the experimental group were invited to the focus-group interview. The selection principle is the heterogeneity, that is, the interviewees have different attributes in learning achievement, participation frequency of learning activities, and different level of agreement to the WIRE model.

The questionnaire was made by five-point Likert scale. The pretest samples are 32 students who come from another class of the same department and grade and have adopted the WIRE model for one semester. After the pretest, the researchers kept 16 items for the formal questionnaire by item analysis and factor analysis. It contains three factors, the strategy of warm-up lessons, interactive activities in class, and the strategy of review and exercise after class and their reliability estimates of Cronbach's Alphas are 0.627, 0.870, and 0.911 respectively. Taken as a whole, the Cronbach's Alpha reliability coefficient is 0.892.

The research procedure is listed as below.

- Step 1: Give a test for both experimental and control groups in prior ability of basic Java programming and algebra, so as to evaluate sample representativeness and then to adjust the lesson plan and question difficulty for each instructional stage.
- Step 2: Treat experimental group with the WIRE scenario showed as Figure 1 while

the control group was treated with lecturing and collective discussions. The main educational technology involves the forum tool, homework tool, chat tool, and group tool, collected in Blackboard, an eLearning platform.

- Step 3: Give a test for both groups in learning achievement. The experimental group also took the questionnaire and a few selected participants were invited to focus-group interview.

RESULTS AND DISCUSSIONS

By analyzing the score of learning achievement test, the average score of experimental group is higher than the control group at about 11.2 points (100 points as full marks). The t-test for the difference between the means of the two groups has reached significant level (t(88)=3.116, p<.05). The result reveals that the WIRE model can effectively enhance the learning achievement of students. The detailed results of the group statistics for learning achievement test and its independent samples test are shown in Table 2 and 3.

In the aspect of questionnaire survey, the research group conducted a questionnaire (using a five-point Likert scale, in which 1 = strongly disagree or poor and 5 = strongly agree or excellent) for the participants to evaluate the motivation, interaction, and learning achievement under the adoption of WIRE model. In 45 responses, those have fixed responses, too many unanswered questions, and unreasonable responses were eliminated, and finally 41 valid responses (validation rate=91.1%) were used for statistics.

To answer the research questions, the questions in the questionnaire were divided into three categories, the motivation, interaction, and learning achievement. From the statistics, the three weighted averages for the three categories are 3.78, 3.72, and 3.73 respectively (see Table 4). That reveals that most participants agree the course that adopted WIRE model can effectively increase the learning motivation, interaction between teacher and students, and learning achievement than the other courses without adopting WIRE model. Since all students were asked to answer the warm-up questions, they have to read the text and digital content provided by the teacher

Table 2. Group statistics for learning achievement test

	Group	N	Mean	Std. Deviation	Std. Error Mean
Score of learning achievement test	Experiment group	45	63.870	19.500	2.907
	Control group	43	52.691	13.467	2.054

Table 3. Independent samples test for the results of learning achievement test

	Levene's Test for Equality of Variances		t-test for Equality of Means						
	F	Sig.	t	df	Sig. (2-tailed)	Mean Difference	Std. Error Difference	95% Confidence Interval of the Difference	
								Lower	Upper
Equal variances assumed	5.962	0.017	3.116	86	0.0025	11.179	3.588	4.046	18.313
Equal variances not assumed			3.141	78.419	0.0024	11.179	3.559	4.094	18.264

so that they had prepared when they entered the classroom. In contrast with traditional instructional methods, students would pay much attention on classroom activities (avg.=3.9) and were more responsible to answer teacher's questions, because the teacher's lecture can arouse students' resonance in their brain.

From the results of focus-group interview, most interviewees generally accepted that the WIRE model can enhance their attention, interaction, and motivation in the classroom activities, and therefore the learning achievement was promoted. They were interested in trying the novel delivery model. Some interviewee said:

I like the new learning style. When I encountered the questions that I had no idea to solve, I take seriously to discuss with my classmates instead of giving up. And then I felt the sense of accomplishment when I solved the questions. Also, I relatively understand the lessons more than the other courses.

However, some problems and suggestions were discussed. First of all, some interviewees hoped they have more time for face-to-face classroom learning, although they appreciate the distance learning with WIRE model. They want to learn from teachers for solving the questions the encountered in the warm-up stage, and are familiar with the face-to-face discussion. Like most research results of online learning, the learning strategies that adopt communication online focuses on the affordance of provided tools to support learning motivation, interactions, and cooperation. Most learners preferred to discuss lessons with others in face-to-face settings.

Second, some warm-up questions are difficult to solve. Since the teacher just only provided one face-to-face teaching opportunity for each one book chapter, the students' real abilities cannot be precisely predicted. It is a challenge for a teacher to ask an appropriate question by the observation from the answers of previous warm-up questions. The questions must not only reflect the concepts of the learning materials but be moderately difficult, so that the learners' self-confidence can be inspired. In the future study, those questions used in WIRE model should be collected for quantitative difficulty analysis every time when used. The experience will be gradually accumulated for asking a good question.

Table 4. The questionnaire results

Categories	Questions (N=41)	Average
Motivation	The warm-ups increased my learning motivation on the course.	3.66
	I was engaged in the classroom activities because of the warm-up before class.	3.73
	In contrast with other courses, I paid much attention on classroom learning.	3.90
	In contrast with other courses, I had much motivation on classroom learning.	3.83
Interaction	The warm-ups increased the interaction of me with my teacher.	3.61
	The warm-ups increased the interaction of me with my classmates.	3.80
	In contrast with other courses, I have more interaction of me with my teacher in classroom activities.	3.44
	In contrast with other courses, I got more interaction of me with my classmates in classroom activities.	3.73
Learning achievement	The warm-ups can increase my learning achievement of this course.	3.66
	Revising the warm-up questions as the lesson review after class can help me connect the knowledge acquired before and after classroom learning.	3.63
	The exercises assigned after class can enhance my learning achievement.	4.02

Finally, some interviewees expressed they expected to have more exercises for practice, so that they had more confidence to be confronted with the test. The researchers also found most of them who commented that suggestion have relatively high learning achievement and are highly active in all learning activities. On the other hand, maybe some students did not like to receive heavy loading for exercises. Therefore, it is also the challenge for teachers to arrange the time and quantity for exercises either in classroom teaching or online activities, and to be in accord with all students' demand as possible as the teachers can do.

CONCLUSION AND FUTURE WORKS

The concept of WIRE model sounds nothing new and is the effective approach generally accepted, but most teachers and students cannot put it into practice on a regular and sustained basis. The crucial problem is that students cannot urge themselves to warm-up lessons before class and review lessons after class, and teachers have no enough time to examine the students' works. The WIRE model engages students' willingness to follow a series of learning activities that can enhance interactivity between the teacher and students. The technology support also reduces teachers' burden during the activities. Nevertheless, the instructional strategy that adopts WIRE model in practice has some challenges for further research, including the difficulty of warm-up questions, the quantity of exercises, and the instructor's burden.

REFERENCES

Barbian, J. (2002). Blended works: Here's proof! *Online Learning*, 6(6), 26–31.

Bersin, J. (2004). *The blended learning book: Best practices, proven methodologies, and lessons learned*. San Francisco, CA: Pfeiffer.

Bloom, B. S. (1956). *Taxonomy of educational objectives, Handbook I: The cognitive domain*. New York: David McKay Co Inc.

Bunz, U. (2005). Using scantron versus an audience response system for survey research: Does methodology matter when measuring computer-mediated communication competence? *Computers in Human Behavior*, 21, 343–359. doi:10.1016/j.chb.2004.02.009

Caldwell, J. E. (2007). Clickers in the large classroom: Current research and best-practice tips. *CBE - Life Sciences Education, 6*(spring), 9-20.

Center for Educational Research and Evaluation. (2005). Report on the survey of higher education teachers in 2004. Retrieved June 13, 2009 from http://www.cher.ntnu.edu.tw/analyze/data/edu92_teacher/pdf/2/2-2.pdf, (the Integrated Higher Education Database in Taiwan)

Eklund, J., & Kay, M. (2001). *Strategy 2001: Evaluation of the Usage of National Flexible Learning Toolboxes - (Series 2)*. Commissioned Report for the Australian National Training Authority.

Eklund, J., & Kay, M. (2003). Evaluation of the usage of National Flexible Learning Toolboxes (Series 3). *Commissioned Report for the Australian National Training Authority*. eLearning Guide Research (2008). Snapshot Report on Learning Modalities. Retrieved June 13, 2009 at http://www.elearningguild.com/research/archives/index.cfm?action=viewonly2&id=129&referer=http%3A%2F%2Fwww%2Eelearningguild%2Ecom%2F

Fies, C., & Marshall, J. (2006). Classroom response systems: A review of the literature. *Journal of Science Education and Technology*, 15(1), 101–109. doi:10.1007/s10956-006-0360-1

Law, N., Pelgrum, W. J., & Plomp, T. (Eds.). (2008). *Pedagogy and ICT use in schools around the world: Findings from the IEA SITES 2006 study*. Hong Kong: CERC-Springer.

Morrison, Don. (2003) The Search for the Holy Recipe. Retrieved June 13, 2009 from http://www. elearningpost. com/archives/2003_04. asp

Osguthorpe, R. T., & Graham, C. R. (2003). Blended learning environments: Definitions and directions. *The Quarterly Review of Distance Education, 4*(3), 227–233.

Pulichino, (2006). The trends in blended learning research report. Santa Rosa, CA: The eLearning Guild.

Rogers, E. M. (2005). *Diffusion of innovations.* Glencoe: Free Press.

Rossett, A., Douglis, F., & Frazee, R. V. (2003). Strategies for building blended learning. *Learning Circuits.* Retrieved June 13, 2009 from http://www.astd.org/LC/2003/0703_rossett.htm

Siau, K., Sheng, H., & Nah, F. F.-H. (2006). Use of a classroom response system to enhance classroom interactivity. *IEEE Transactions on Education, 49*(3), 398–403. doi:10.1109/TE.2006.879802

Singh, H. (2003). Building effective blended learning programs. *Educational Technology, 43*(6), 51–54.

Smith, J. M. (2001) Blended Learning An old friend gets a new name. Retrieved June 13, 2009 from http://www.gwsae. org/Executiveup-date/2001/March/blended

Troha, F. J. (2002) Bulletproof instructional design: A model for blended learning. *USDLA Journal, 16*(5).

Valiathan, P. (2003) Blended learning models. *Learning Circuits, ASDT Online Magazine.* Retrieved June 13, 2009 from http://www. learn-ingcircuits.com/2002/aug2002/valiathan.html

Zenger, J., & Uehlein, C. (2001) Why Blended Will Win. *T+D, 55*(8), 54-62.

Chapter 12
Sights Inside the Virtual Engineering Education

Giancarlo Anzelotti
University of Parma, Italy

Masoumeh Valizadeh
University of Guilan, Iran

ABSTRACT

Engineering is the area of creation and the creation has to be motivated in engineering education. The development of fast Internet access and Virtual environment have facilitated this field of education. considering time and cost and the critical position of engineer training, this chapter is introducing purpose-oriented small software as a more focused and cost-effective tool that can be implemented in universities and engineering education institutes. The chapter contains some examples of such software in mechanic and material engineering fields.

INTRODUCTION

The critical tenet of engineering education reform is the integral role of virtual environment capabilities which is provided by fabulous advances in information technology. Current technological progresses combined with changes in engineering content and instructional method require engineering instructors to be able to design intensive and concentrated lessons for exploration and discovery of the engineering concepts through appropriate computer applications. In actual practice, however, most computer applications provided for engi-

neering education consist of software designed for a specific educational purpose. Furthermore, economical constraints often stand in the way of incorporating such special purpose software into an instructional setting.

Owing to all the revolutions made and is being made in information transfer and computer capabilities and speed, recent decades have been witnessed many changes in the methods of engineering education and there are many new researches and methods for this task (Willis 1993; Koksal 1998; Kaufman 2000; Cloete 2001; Wankat 2002; Baker 2005; Hutten 2005; Yu 2006; Burnely 2007; Moreno 2007; Haghi 2008; Monahan 2008; Shih 2008; Toth 2008;

DOI: 10.4018/978-1-61520-659-9.ch012

Vadillo 2009). This paper discusses an alternative to the traditional approach which shifts the instructional focus from specific computer applications to more sophisticated uses of general purpose software. In particular, educational uses of purpose-oriented small software which can be implemented in multi purpose software are exampled as an introduction to this approach. In another vision, purpose-oriented small software can be designed as a cost-effective approach to provide practical experience to undergraduate students. The concept which is called '*Virtual Laboratory*' and is being widely developed in engineering schools and educational programs (Gervasi 2004; Ramasundaram 2005; Kukreti 2008; Tan 2008; Vadillo 2009).

The software and approaches described here have been presented to graduate and undergraduate engineering education majors in a continuing education course in universities or factories.

ENGINEERING EDUCATION IS EVOLVING

Since the first years of its emergence in mid 20th century (1940-1945), the computers have been welcomed and started to be employed by governments, companies, factories and moreover in education. By this new invention of human being the computations needed the handling of big amount of data and calculations have been "translated" in computer language and little by little have been automatized and implemented by this powerful contrivance. It changed the approach to design radically and convinced the engineers, mathematicians, economists, industrialists all over the world to employ it as a potent tool in their activities. Since that time we have been witnessing significant jumps in the science of computing and design. Nowadays, thanks to the enormous power of today's computers respect to what was available just few years ago, we are able to solve complex, multifarious tasks

involving different science fields and may need weeks or months of work if they would have been done by hands.

Along with the significant changes in designing and calculating methods, occurred through the application of electronic devices like calculating machines as a simple example, but mostly the computers, the methods of teaching the design and modeling techniques have changed remarkably. However comparing to all the present branches of science and technology, the engineering and engineering education have evolved more and have progressed fabulously. Nowadays the use of computer-aided design software (CAD) or finite element systems (FEA) have become *de facto* a standard and the essential topics in engineering courses, both at school and university levels. In the next step, further improvements in the field of computer languages has made the use of custom-made software more widely possible, and has empowered the engineers to solve different requirements of the industries.

Professional programming languages like Python™ or Visual Basic® have been developed with a reasonably user friendly, easy learning curve and practical performances, providing lower investment costs for users. On the other hand to confront with more complicated calculations middle-level languages like C++ or High-level languages like FORTRAN represent a fairly better performance in a more admissible time. Furthermore, concerning to the fact that the software may be used in different operating systems and to aid the students to use them in laboratory or at home then a cross-platform programming approach consisting of languages, tools, libraries, etc. that let the programmer to port easily the code from one operating system to another one can be more favorable.

However, the field of education is usually concerned about the understanding of concepts as well as how to speed-up the learning process, where the ease of knowledge transfer from the teacher to the students is the key issue.

In this area, one of the problems is the lack of suitable thinking approach. The students are used to solve design problem still based on analytical approach, not based on synthetic approach. They have difficulty in applying the acquired specific knowledge to their design task, although they have already learned that from related nature science or engineering science courses. However, the engineers of 21st century need not only innovative talents but also interdisciplinary abilities to develop new technologies and products. Appropriate teaching aids are therefore necessary for the challenge of engineering design education. In addition to teaching the knowledge of engineering science, today the design and simulation methodology play more and more an important role in engineering. Therefore it seems that new requirements on the curriculum of engineering design courses are necessary to enforce the students to use the present facilities to actuate and accelerate their learning progressions (Tsai 2008).

Concerning to this task, a wide range of professional software have been made that can be used to increase the effectiveness of learning as well as teaching such as SolidWorks®, AutoCAD®, ArchiCAD®, Abaqus®, LabView®, Matlab® *etc*. Although the professional software provide noteworthy facilities in both industrial and educational fields, but for many of the educational centers it is not affordable to pay for their licenses and to employ them for educational affairs. However the fact that there are always possibilities to enhance students' cognition through some proper ploys such as developing e-tools and freeware or small software sounds encouraging.

Relevant visual tools ensure quick and comprehensive transfer of information and convey data, coding of which in written or spoken form takes longer time and more facilities, or in many situations even it is impossible. The main forms of visualization can be counted as images, animations and simulations (Upitis 2008).

DEVELOPING SMALL SOFTWARE IN ENGINEERING EDUCATION

The authors believe that even with purposeful small software especially with a plain, but intelligible graphic interface, the students will be able to reach a better understanding of the design and simulation courses, admitting that there is no other branch of science as objective, tangible and virtualizable as engineering.

Thus, in this field the application of script languages, embedded inside general purpose[1] software is preferable like Python™ that has embedded inside Abaqus®. According to the authors' experience, the use of Visual Basic® for Application (VBA) for the beginner level of programming inside an engineering course, aimed to solve simple design problems and not directly to programming is usually a good choice.

Programming itself may sometimes help the students to clarify the complex mathematical environment needed to solve engineering problems. Regarding to this fact that math is just a powerful tool in the hands of the engineers it is the essential term of every designing or simulation, but should not be considered as the main task of the engineering.

Focusing on mechanical engineering as the branch of the engineering which is sorely involved in design and simulation, the next paragraphs will present some examples of software aimed to help the students to understand better the design of mechanical parts. The design of a camshaft or simply of a single cam is usually a difficult task to solve. Because it deals not only with the design of the shape and to make the lever in a way that follows a certain displacement law, but also it needs to take into account the speed and more important the acceleration law for the lever (and sometimes also the jerk, first derivative of the acceleration). When the general geometry of the cam and the lever are simple and the applied displacements and loading system are not complex, the design and its calculations can be solved manually. But to design more complicated systems, the use of a computer aided approach is unavoidable. A problem-oriented software is able to define different geometries using a trial-and-error algorithm in few minutes. While the number of input and output variables is high, a congruous computer software gives the possibility to the user to focus on just few of them a time and to draw graphs of the variation of the output respect to the change of input. Through an automatic algorithm based on trial-and-error or more complex optimization schemes like genetic algorithm (GA),the software may be even able to determine the best design itself.

However the programmer has to consider that for the educational purposes, graphical user interface (GUI), its ease of use, comprehensiveness and visual aspects, plays a very important role, because it is the main object attracts the learners and incite them to focus on the lesson. Here the virtual environments may offer better facilities. Through applying this approach, more clear and exact illustrations of the objects and the problems will be available. Therefore every sketch, drawing or schemes, but moreover the simulated animations and movies may accelerate the learning process.

As an example Figure 1 shows a graphical user interface designed for the definition of the input needed to design a desmodromic cam. In Figure 1a the software asks for the general properties of the cam so it provides the opportunity to import the geometrical information. Figure 1b introduces a simple 7-steps acceleration law, so it asks for loading system.

Programming for such specialized software needs the deep and comprehensive understanding of the programmer about the designing of the desmodromic cam, to consider every kind of mistake the user may make and prevents them by providing appropriate warnings or error messages.

Figure 2 shows how the software can easily present different representing graphs, such as the displacement (on the top of Figure 2a) to the acceleration (on the bottom of Figure 2a).

Figure 1. Graphical user interface (GUI) of CamCreator

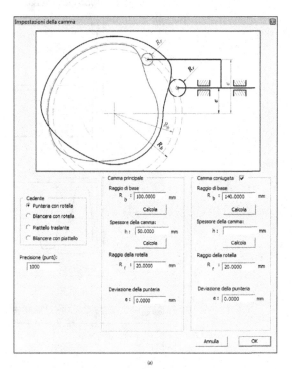

(a)

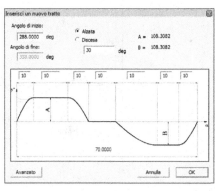

(b)

Such software can be equipped by virtual 3D configuration which is the forte of computer aided engineering education. This appliance supplies a live performance of the machine and its kinematical mechanisms or its dynamics in a 3D world, without building the real part or prototyping it.

To make this scenario cost-effective there are freeware utilities such as OpenGL, to animate the final results. This application has been shown in Figure 3.

Figure 2. A small software like CamCreator can lead to precious statements

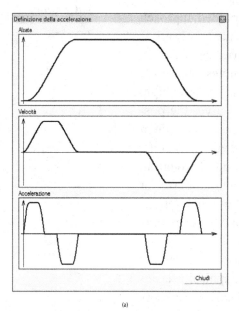

(a)

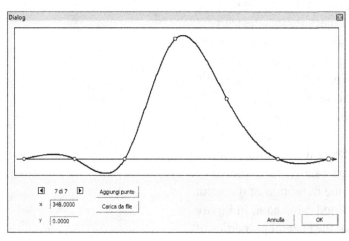

(b)

Figure 3. 3D visualization: a key to deep understanding

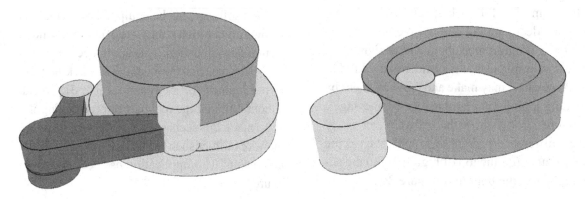

The second example is a software to teach the gears and the interaction between the gears to the students. This subject usually it is not deeply clear to the students. The influence of corrections on the change of the geometry or the undercut phenomenon is most of the times a complicated subject. For this purpose laboratory sessions may help them to figure out the mechanism, but this needs the laboratory being equipped with a milling machine or some kind of milling system able to cut the gear profile out of a cylindrical piece of metal. Then, according to the

course time and the disbarments to the course, just few specimens may be cut to show the students, or the same specimens should be showed at each year.

To solve these limits, purpose-oriented software has been developed by the authors in order to represent different gears geometry to the students.

In Figure 4a a simple approach to gears-cut is presented. The virtual rotation without slipping of a rack on a smooth circle is creating the shape of the gears. As shown in Figure 4b, several possible position of the rack can be visualized.

Figure 4. Gear simulation by Gears-Cut software

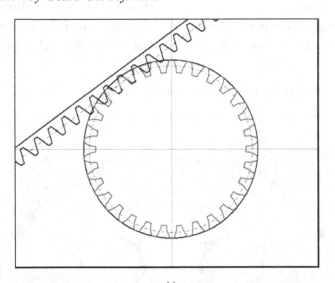

(a)

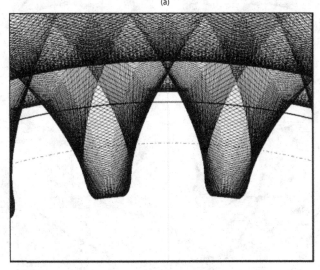

(b)

Testing of strange combinations of the input variables makes the lessons more interesting to the students and more over it also introduces concepts like the minimum number of teeth. Usually this is introduced as the result of an equation aimed to obtain geometry of the gear in order to avoid the teeth interference.

Through the visual approach of such software the student can also easily realize how the shape of teeth is changing from a high number of teeth to the low number of teeth, as shown in Figure 5; where we have just 5 teeth (meanwhile the classical minimum number of teeth is around 17 for a pressure angle of 20 degrees).

The use of a well-designed user interface and a real-time changing of the visual results may help students to "play" with the inputs and to see the result of their combinations directly.

For example the use of a slide may easily teach the influence of a parameter to a student watching directly the changing of the geometry from a very low value to a very high value of that parameter.

The authors believe that an effective educational approach is not just the use of one particular tool or software even if it is very powerful, but make the students used to system of tools as the best choice. This may be sometimes time consuming and they may require a level of knowledge which is not easily accessible in every school or factory. However the use of different general purpose software, smartly connected to each other, is a fast and flexible technique which can be adapted

Figure 5. An example of 5 teeth gear designed by Gear-Cut

to several problems with small changes.

The example which is presenting here is the parametric design of a loft. This problem involves the design skills of the students. It is also a good example of how to combine the masteries of CAD and programming to design an educational appliance.

The design of pillars, beams, stairs and the parapet requires particular knowledge of different courses and push the students on the use of local or international norms, as a structure addresses many people and high safety standards have to be concerned. To design such structure weight and properties of different materials (steel, wood, plastic, etc.) used to build it have to be considered. On the other hand taking in to account different input variables needs a parametric design.

Excel®, Calc, Numbers, Matlab®, Octave and much other software can be used. Even some freeware software may provide enough professional capabilities. Considering the necessity of application of a spreadsheet Microsoft® Excel® can be a good choice as it is a standardized utility in many factories and universities. The use of a spreadsheet let the students to design easily a user interface where inputs and outputs are well presented and also where all the materials are written in tables, equipped by all their properties needed for the design.

Once introduced the inputs, a series of well-linked cells may lead to a series of outputs giving us the opportunity to draw the complete structure. However this is possible only if simple approaches are to be used. The task shows itself more strongly concerning to the fact that the iterative design is not always accessible (especially if the design presents several branches). In this case the connection between the cells may become incredibly complex or even impossible. Moreover the debug of the algorithm will be a very complex task as well. Thus the use of Visual Basic® for Application may solve these limits and let the student to embed the design algorithm inside the spreadsheet, but using a simple programming language.

The students have to study all kinds of simple structures that constitute the loft such as beams and pillars and draw the designs by hands. The algorithms have to be solved leaving parametric input variables. They have to be able to distinguish all the limits in the calculations and designing in order to encounter the warning and error messages that they will face with during using the e-tool approaches caused by applying improper combinations of values of the parameters.

The present spreadsheet is able to design even each small bolt. To avoid the time consuming step of updating all the dimensions by the designer after making each kind of changes and moreover, to prevent the errors due to the accidental forgetfulness the worksheet has been connected directly to a CAD system. Thus it allows the drawings to be updated automatically at each change of the input parameters. Finally thanks to realistic rendering software like Blender or 3DStudio it is possible to convert the engineering drawing of the structure to a more desirable "realistic" representation (Figure 6).

In this example all calculations are considered and solved by VBA (some of them in closed form, others by iterative calculations). The cells of the spreadsheet are organized in a way that the user finds easily the series of necessary input and the series of important output. Using a 3D parametric CAD software like SolidEdge®, the user can connect the output cells of the spreadsheet (representing all the dimensions of all the part of the loft) to the CAD system using OLE (Object Linking and Embedding) approach. To do this the CAD model of the loft has to be drawn and each dimension has to be linked to the spreadsheet.

In different fields of engineering, the study of the influence of pores on the fatigue life of mechanical parts is a very important task. In fact the pores are defects of empty volumes presenting in material due to the production process and are very expensive to reduce or control.

Recent studies on this area are more focused on a better understanding of the influence of the

Figure 6. Employing a system of software to present a virtual reality

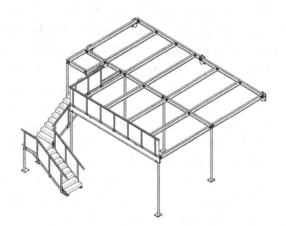 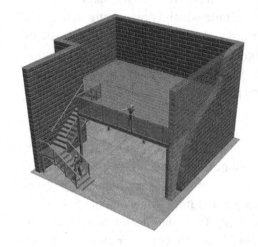

pores in order to predict the fatigue life or to induce special morphology distribution of pores during the producing process and to control the mechanical properties. To get closer to the natural morphology of the pores, now the researchers are more interested in three dimensional characterizations of pores which present the actual shape and volume of the pores.

Computed tomography (CT) is a technique to identify the volumetric distribution of the "density" of the material, describing the pores three dimensionally. CT is a precise digital geometry process able to generate three dimensional volumes of the inside of an object using a large series of two dimensional images and joining them together. The information held in the volumetric pixel, the voxel, is the absorption coefficient of the material, which is somehow related to the density of the material itself.

The student may be interested in the three-dimensional description of a pore, starting from the volumetric information of a specimen portion. This is useful to understand the real shape of a complex phenomenon that can be for example the casting of metal materials, and it may be also useful for a subsequent finite element analysis based on real morphology and not an approximation.

In Figure 7a it is possible to see the distribution of pores inside a small cylindrical specimen

of 4mm diameter this model has been made in 3D-Pores using the measured data of 3D X-ray. Figure 7b is a typical portion of cross-section of such a specimen. However the 2D pore cross-section would not help the student to understand the correct shape as the real three-dimensional reconstruction. The three-dimensional pore shown in Figure 8 is an example of a real pore reconstructed by 3D-Pores.

The other example which is related to several deviations of engineering is measuring the material properties and mechanical behaviors. These are usually the main subject of many engineering courses. In addition to many theoretical lessons, the students have to make the experiments in laboratory and gain a deep understanding of different properties and different behaviors.

Experiment design, making the measurements, collecting the data, statistical investigation and interpretation of data are different part of real experiments that is more productive if are done by the students themselves rather than being based on just the theoretical sessions.

Many mechanical processes are strain-driven, thus it is very important to understand its behavior due to loading. In order to experience different approaches to strain measurements, there are different devices used to help the students to understand the meaning of deformation and

Figure 7. Pore 3D approach vs. 2D

(a)

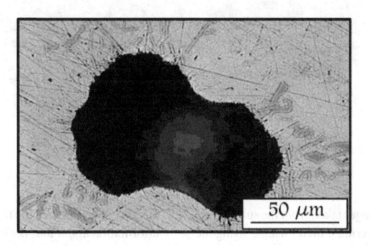

(b)

heterogeneity. The classical and older method to teach strain measurements to the students is measuring the strain by clamping the specimen and applying the load. This method is usually the less precise and it needs some specific cares such as finding the adopted gage-length. Extensometer is a more accurate method which needs a higher attention during the installation and implementation processes Extensimetry is a desired method to measure the local strains and for most of engineering materials with a small level of heterogeneity or homogeneous materials. In this case the theory behind the strain-gage working principle and the attention to the implementing procedure is taught to the students.

A relatively new and progressing technique to have full-field strain information of the specimen surface is the digital image correlation (DIC). This is an optical non-contact method able to identify the displacement field of a specimen, employing the observation of the specimen surface by a digital camera. The algorithm is based mainly on the

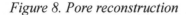

Figure 8. Pore reconstruction

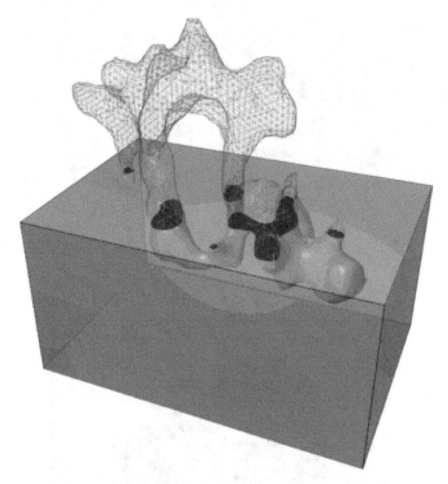

analysis of the images taken from the surface of a specimen and describing their changes using the constitutive law of the solid continuous mechanics. The image taken at the undeformed state is called *master picture* and each of the following images, describing a special deformed state, are called *slave images*. As it is shown in Figure 9 the algorithm starts with identifying a number of sub-windows on the master picture. Each of them is representing an area of the specimen using a characteristic matrix of grey-level pixels. The aim is to follow the displacement and deformation of these sub-windows in each of the slave picture. At the end of the analysis the algorithm leads to a series of time-changing displacements for each sub-window[2]. According to the magnification of

the acquisition device the user may have general information about the specimen or quite detailed information at the meso- or micro-level (Anzelotti 2008; Nicoletto 2008; Nicoletto 2009; Nicoletto in Press).

DIC is practically based on visual concepts. It introduces the image processing algorithm to the students, make them familiar with the explanation of the displacements calculations and moreover it is a good approach to activate the students interests in the topic of strain measurement. DIC is a measuring method mostly used by the researchers and as an upcoming and growing method has been less used for educational purposes. However it is potentially capable of being used in engineering education as it easily provides the visual require-

Figure 9. Applying DIC approach for strain measurement

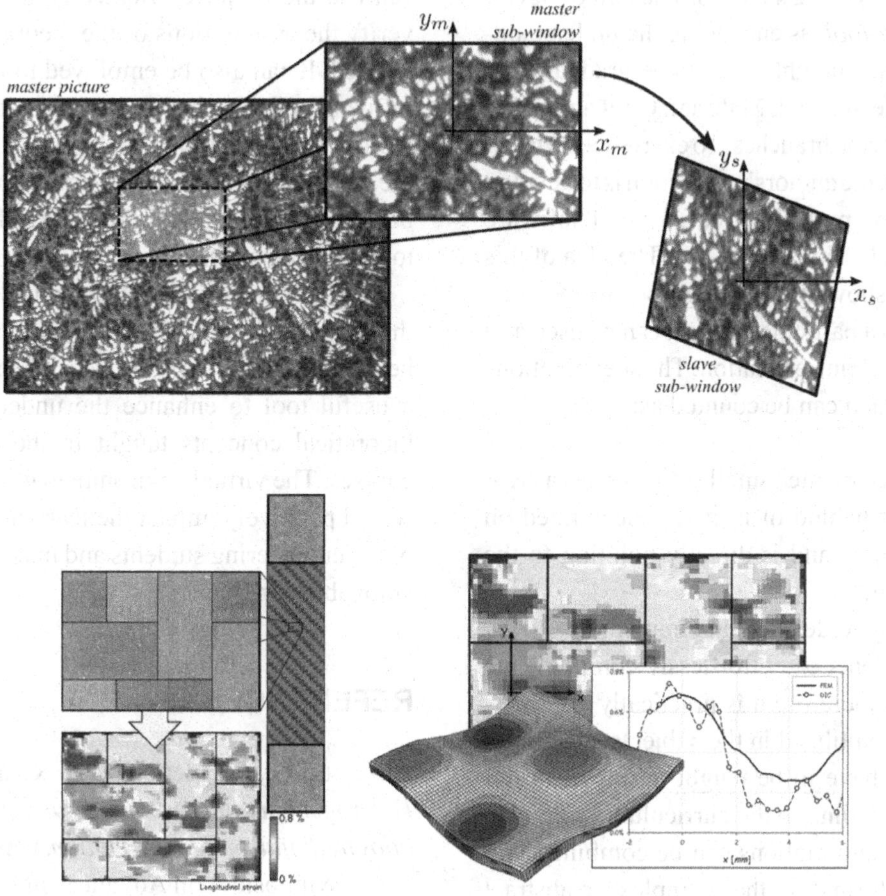

ments to motivate the students effectively. The fact that DIC approach can be implemented for different kinds of materials such as metals, ceramics, composites and even more heterogeneous structures such as textiles empowers it to be employed in several engineering fields.

CONCLUSION REMARKS

Recent experience in different engineering education programs have shown that only systematic, research based, approach could be used for engineer training and education. The engineering

education designers have to consider the engineering market demands. The industries would like to employ well educated skilled engineers and a skilled engineer is the engineer who has experience. Time and costs as two essential parameters do not let the engineering educational centers to train enough skilled engineers. To cope with this task and in order to reach an effective engineering education approach, a comprehensive educational design is required; starting from the schools and continuing to universities or research centers. An effective educational design is based on adequate and appropriate tools connected to each other through definite standards.

This chapter emphasizes on the power of virtual tools for engineering subjects. The aim of developing these *e-tools* is enhancing the understanding of concepts taught in the undergraduate and graduated Mechanical, Material, Civil and many other engineering branches. To exert such a virtual environment, the authors' suggestion is to develop and/or employ purpose-oriented small software along the professional software. The idea of this chapter is to show that even light software which is focused on a particular subject can be used as a key for engineering education. The specifications of this approach can be counted as:

- purpose-oriented small software has a clear thought behind of it, it is concentrated on the subject and is directly pointing to the problem;
- it is easy to develop and implement;
- the students can effortlessly communicate with it, therefore it is practically appropriate to be utilized in the subjects and courses that have to be taught in one semester and according to the curriculum;
- visuals and motions can be combined in a single format so that complex or abstract concepts can be illustrated through visual simulation (the old apothegm: "a picture is worth a thousand words");
- it is easy to make it adapted with different operating systems, so that the students can use it with no trouble even at home or at school;
- it can be joined and employed by novel education systems such as e-learning or generally distance education;
- it is cost effective;
- it is effectively a motivative, inspiring tool.

The virtual environment is expected to enhance the understanding of how and why one phenomenon happens, why a material behavior differs from the other, what is the optimum design for a device or a structure and help the students to real-ize the material properties and behavior. Students can use the purpose-oriented small software to verify the assumptions of the theories they have learned. It can also be employed to illustrate the influence of the change in assumptions and to realize the limitations and boundaries. Hence the students will be able to simulate mechanical behavior connecting their theoretical knowledge to the virtual experiments.

With the same idea of the virtual laboratories, the purpose-oriented software discussed above is being developed with the goal of making them a useful tool to enhance the understanding of theoretical concepts taught in the engineering courses. The virtual environment of these e-tools would positively impact the learning experience of the engineering students and make learning an enjoyable experience.

REFERENCES

Anzelotti, G., Nicoletto, G., & Riva, E. (2008). *Heterogeneous microscopic strains in a cast Al-Si alloy by digital image correlation.* Paper presented at the Symposium on Advances in Experimental Mechanics, České Budějovice.

Baker, A., Navarro, E. O., & van der Hoek, A. (2005). An experimental card game for teaching software engineering processes. *Journal of Systems and Software, 75,* 3–16. doi:10.1016/j.jss.2004.02.033

Burnely, S. (2007). The use of virtual reality technology in teaching environmental engineering. *Engineering Education . Journal of the Higher Education Academy Engineering Subject Center, 2*(2).

Cloete, E. (2001). Electronic education system model. *Computers & Education, 36,* 171–182. doi:10.1016/S0360-1315(00)00058-0

Gervasi, O., Rignelli, A., Pacifici, L., & Lagana, A. (2004). VMSLab-G: a virtual laboratory prototype for molecular science on the Grid. *Future Generation Computer Systems, 20,* 717–726. doi:10.1016/j.future.2003.11.015

Haghi, A. K., Mottaghitalab, V., & Akbari, M. (2008). *The Scholarship of Teaching Engineering: Some fundamental issues.* New York: IGI Global.

Hutten, H., Stiegmaier, W., & Rauchegger, G. (2005). KISS—A new approach to self-controlled e-learning of selected chapters in Medical Engineering and other fields at bachelor and master course level. *Medical Engineering & Physics, 27,* 611–616. doi:10.1016/j.medengphy.2005.05.003

Kaufman, D. B., Felder, R. M., & Fuller, H. (2000). Accounting for individual effort in cooperative learning teams. *Journal of Engineering Education, 89*(2), 133–140.

Koksal, G., & Egitman, A. (1998). Planning and design of industrial engineering education quality. *Computers & Industrial Engineering, 35,* 639–642. doi:10.1016/S0360-8352(98)00178-8

Kukreti, A., & Zaman, M. (2008). *Virtual laboratory modules for undergraduate strength of materials course.* Paper presented at the Engineering Education, Loughbrough, London.

Monahan, T., McArdle, G., & Bertolotto, M. (2008). Virtual reality for collaborative e-learning. *Computers & Education, 50,* 1339–1353. doi:10.1016/j.compedu.2006.12.008

Moreno, L., Gonzalez, C., Castilla, I., Gonzalea, E., & Sigut, J. (2007). Applying a constructivist and collaborative methodological approach in engineering education. *Computers & Education, 49,* 891–915. doi:10.1016/j.compedu.2005.12.004

Nicoletto, G., Anzelotti, G., & Riva, E. (2008). Mesomechanic strain analysis of twill-weave composite lamina under unidirectional in-plane tension. *Composites. Part A, Applied Science and Manufacturing, 39*(8), 1294–1301. doi:10.1016/j.compositesa.2008.01.006

Nicoletto, G., Anzelotti, G., & Riva, E. (2009). Mesoscopic strain fields in woven composites: Experiments vs. finite element modeling. *Optics and Lasers in Engineering, 47*(3-4), 352–359. doi:10.1016/j.optlaseng.2008.07.009

Nicoletto, G., Marin, T., Anzelotti, G., & Roncella, R. (in press). Application of high magnification digital image correlation technique to micromechanical strain analysis. *Strain.*

Ramasundaram, V., Grunwald, S., Mangeot, A., Comerford, N. B., & Bliss, C. M. (2005). Development of an environmental virtual field laboratory. *Computers & Education, 45,* 21–34. doi:10.1016/j.compedu.2004.03.002

Shih, W. C., Tseng, S. S., & Yang, C. T. (2008). Wiki-based rapid prototyping for teaching-material design in e-Learning grids. *Computers & Education, 51,* 1037–1057. doi:10.1016/j.compedu.2007.10.007

Tan, A. C. C., Tang, T., & Paterson, G. (2008). *Web-based Remote Vibration Experimental Laboratory.* Paper presented at the Engineering Education, Loughbrough, London.

Toth, P. (2008). *New Method in Quality Assurance of Electronic-based Teaching Materials.* Paper presented at the Engineering Education, Loughbrough, London.

Tsai, S., Yeh, T. L., Lee, C. K., & Wang, N. C. (2008). *Teaching Strategies for Design Realization in Engineering Design Education.* Paper presented at the Engineering Education, Loughbrough, England.

Upitis, G., Maziais, J., & Rudnevs, J. (2008). *Problem-Oriented E-tools for Studies in Mechanical Engineering*. Paper presented at the Engineering Education, Loughborough, England.

Vadillo, M. A., & Matute, H. (2009). Learning in virtual environments: Some discrepancies between laboratory- and Internet-based research on associative learning. *Computers in Human Behavior*, *25*, 402–406. doi:10.1016/j.chb.2008.08.009

Wankat, P. C., Felder, R. m., Smith, K. S., & Oreovicz, F. S. (2002). *The scholarship of teaching and learning in engineering*. Washangton: AAHE/Carnegie Foundation for the Advancement of Teaching.

Willis, B. D. (1993). *Distance Education- A Practical Guide*. Englewood Cliffs: Educational Technology Publications.

Yu, C. Y., & Shaw, D. T. (2006). *Fostering Creativity and Innovation in Engineering Students*. Paper presented at the 2006 International Mechanical Engineering Education Conference, joined Conference by ASME and CMES, Beijing, China.

ENDNOTES

[1] Here the concept of "general purpose" is referred to the possibility to use the software to solve different problems in different courses

[2] This is not the only existing algorithm but it is quite used by researchers and the authors do not want to present an exhaustive treatise about DIC, but just introduce it as teaching experience in engineering courses.

Chapter 13
Effective Design and Delivery of Learning Materials in Learning Management Systems

Mehregan Mahdavi
University of Guilan, Iran

Mohammad H. Khoobkar
Islamic Azad University of Lahijan, Iran

ABSTRACT

Learning Management Systems (LMS) enable effective design and delivery of learning materials. They are Web-based software applications used to plan, implement, and assess a specific learning process. LMSs allow learners to connect to and interact with the educational material through the Internet. They enable tools for authors (instructors) to design learning materials that include text, html, audio, video, etc. They also enable learner activity management in the learning process. Moreover, they provide tools for effective and efficient assessment of the learners. This chapter explores learning management systems and their key components that enable instructors organize and monitor learning activities of the learners. It also introduces the authoring features provided by such systems for preparing learning material. Moreover, it presents assessment methods and tools that enable evaluation of the learners in the learning process. Furthermore, existing challenges and issues in this field are explored.

INTRODUCTION

Learning Management Systems (LMS) are Web-based software applications used to plan, implement, and assess a specific learning process. They enable learners to use the learning material at a time and place of their choosing. Typically, a learning management system provides an instructor with a way to create and deliver content, monitor learners' participation, and assess their performance.

Constructive learning is enhanced by interaction with instructors and classmates, rather than simply interacting with content (Alexander, 2008). As a result, learners build new thoughts, ideas and concepts making use of their knowledge and experience (Beatty, 2003). An important aspect of constructive learning is that it gives responsibility and control over the learnt material to the learner.

DOI: 10.4018/978-1-61520-659-9.ch013

Learning Activity Management (LAM) systems are flexible learning design tools that enable instructors to organize and monitor learning activities of the learners. These activities include assignments, quizzes, and also collaboration. One of the benefits of LAM systems is that they can reduce staff uptake. Additionally, higher levels of pupil motivation are expected using the coherent, integrated and structured LAM systems, compared to traditional courses. Moreover, the self-paced LAM environment encourages students with anonymous favors develop their confidence, autonomous learning, and meta-cognitive skills. As a result, the users of such systems become more inclusive from the traditional ones. Assessment is also an integral part of LAM systems, that enable instructors effectively evaluate learners' activities in the learning process.

In this chapter, we study learning management systems and the main components that such systems should provide in order for instructors and learners to effectively participate in the learning process. We present the functionalities that such systems can provide for organizing and monitoring learning activities of the learners. We also present the authoring tools used for preparing learning material, as well as the assessment methods and tools that enable effective evaluation of the learners in the learning process.

MANAGING LEARNING ACTIVITIES

Traditional e-learning systems have focused on content delivery and individual interaction with this content. LAM systems extend this by combining content delivery with collaboration. They aim to combine the benefits of e-learning with the collaborative aspects of traditional (classroom-based) education, thus resulting in a more effective on-line learning environment. Some LAM systems have already been built to realize the above ideas. We describe some of the more prominent examples and then discuss the challenges in developing such systems.

Dalziel (2003) has developed a system called LAMS, which is perhaps the most complete LAM system currently available. LAMS provides authoring, learning, and monitoring modules (which we describe in more detail below). It has achieved widespread acceptance, due in part to its release as open-source software (Alexander (2008) notes that LAMS users number roughly 3200 in 80 countries). An additional factor in its adoption is that LAMS has been designed based on Learning Design standards so that designs may be shared, re-used, and re-purposed. Also, it has been designed so that it can work either as a stand-alone system or in conjunction with other Virtual Learning Environments (VLEs) and Learning Management Systems (LMSs).

The following example (from Dalziel (2003)) gives some idea of the capabilities of LAMS. It was initially designed for a class of 20-30 high-school history students, potentially located in more than one physical location, around the topic "What is Greatness?", and implemented using the LAMS system. The activity lasts for four weeks. In the first week, all students discuss their views on the topic in an online forum. In the second week, students are given access to a range of material on the topic, and asked to find an example web-site on the topic, which they then share and comment on to the whole class. In week 3, students are put into small discussion groups, where they chat interactively on-line to deal with questions provided by the teacher; one student acts as a "scribe" to record the discussion. The conclusions reached by each group are then posted for the whole class. Finally, in the fourth week, students individually write a report on the initial question ("What is Greatness?") which they submit via the system for marking. The activity concludes when the students receive marks and feedback from the teacher.

One of the powerful features of the learning activity approach is that the content of a sequence can be easily changed to suit a different discipline, while leaving the activity structure unchanged (the above example has been re-used to deal with the question "What is Jazz?"). The learning activity sequence can provide a "pedagogical template" which may be useful in many contexts by changing the "content" to suit different discipline areas. The focus on easy re-use with LAMS means that these changes can be implemented and ready to run with a new student group. Moreover, the pedagogical template itself should be modifiable, so that if a teacher wishes to change the order of the tasks, or add/remove activities from the template, this can be easily achieved. In LAMS, this is possible within the authoring environment using a simple drag-and-drop system which makes explicit the teaching and learning processes as a series of discrete activities.

Nor Azan (2007) has developed a LAM system named SPAP, based on ideas from the IMS Learning Design framework. In particular, it used the Conceptual Model to provide a containment framework that can describe the design of teaching and learning processes in a formal way. SPAP allows teachers to plan, manage and monitor learning activities, and enables learners to carry out these learning activities. SPAP provides a set of seven activity tools based on teaching methods such as discussion, problem solving and simulation. SPAP includes five modules, the most important of which are the authoring and learner modules. The authoring module provides a graphical interface that allows teachers to describe a sequence of learning activities (a learning design) and save the design. The learner module guides students through the learning designs specified by the teacher, using a graphical interface to clearly indicate progress. It also gives them access to a course overview developed by the teacher in a synopsis module. A monitoring module allows the teacher to track the progress of all students through the learning activities.

Mann (2008) considers the impact of online collaboration on the learning process. Collaborative learning is a process during which learners collaborate with each other for solving problems.

Collaborative learning is particularly useful in promoting team work in the learning process. Mann (2008) notes advantages of online collaboration as the free sharing of ideas and freedom from physical and time constraints. Despite the advantages, he also identifies a disadvantage: learners involved in unstructured online collaboration may find themselves in a situation similar to chatting on "MSN", which is more in line with socializing instead of learning. In other words, learners may misuse the online environment so that its effectiveness as a learning environment is significantly reduced. As a result, monitoring tools may be useful in order to lead learners to learn when they perform a learning activity. LAM environments may also provide a live monitoring tool for educators in order to monitor the progress of students so that inappropriate behaviors of misusing the collaboration tool in a destructive way can be minimized.

Fan and Jiaheng (2008) describe sequencing of activities based on ontology and activity graphs in personalized learning environments. As a result, the creation and sequencing of content-based, single learner, and self-paced learning objects would be enabled.

The most important activities in a LAM system can be categorized in four groups "informative", "evaluative", "collaborative", and "reflective" activities.

Informative activities include those that play an informative role for users. The most important activities in this category include:

- **Notice board:** It is a straightforward way of providing content to the learners (as a bulletin of the professors). It can be implemented in two methods. Firstly, a brief link-based legend index of items can be presented. This index points to different targets. Secondly, a larger area can provide as much content as possible. It might also use links. The content can be rich text, html, links, word documents, audio, video, etc.

- **Share resources:** The learners can share their resources using this activity.
- **Task list:** This activity allows learners to inform the monitors and the authors the completed tasks. It may be an individual's activity to manage the activities too.
- **Meeting with the instructor:** This activity enables the learners to participate in meeting sessions with their instructor(s) and resolve their possible problems.

Evaluative activities include those that are used in the evaluation process. Some of the most important activities in this category include:

- **Multiple choices:** This activity is a quiz consisting of several questions each of which containing several items to choose from. The learners should select the correct answer of each question in the specified time. The multiple choices are designed for each part of the course according to the needs of that part.
- **True/False questions:** True/False questions are used to assess conceptual knowledge of students in a quick way. The students' answers to these questions can show if they have understood the essence of the learning material.
- **Drag and drop:** Drag and drop activities are those in which students require to place text or images by dragging in the correct positions on the page.
- **Hotspot:** This is an activity in which students should select the correct area within an image.
- **Drop down list:** It is a selection list in which a set of answers is revealed so that students can select the answer to a question from the list.
- **Freeform text entry:** In this activity, students should answer a question by writing a text.

- **Sequence:** In this activity, students should arrange a series of options by dragging them in a certain order.
- **Submit files:** The learner's work or answers to exercises can be received by the system in different ways. One of these ways is submitting files. Learners can create files from their work and submit them in the specified area for assessment.
- **Research activity report:** The learners should inform the instructors the research activities/results of their current tasks.
- **Screen capture:** For some disciplines such as computer science, the result of students' activities should be evaluated carefully. The details should be considered by both the students and the professors. As a result, capturing a video from the steps of the students' activities could be an effective and powerful evaluation tool.

Collaborative activities enable collaboration between learners. Some of the most important activities in this category include:

- **Chat:** It is a popular activity for the learners. They can chat together, to the monitor/instructor based on their needs.
- **Scribe:** It is the reports that the learners write.
- **Forum:** It is an asynchronous discussion environment for the learners. The discussion threads usually are created by the authors.
- **Group communications:** the learners can join together as a group and communicate with each other. For example, file and resource sharing between them can be done.
- **Group survey:** The learners can complete a survey about a specified topic together. This makes both learning course materials and problem solving easier. This activity can also be considered as a reflective activity.

Reflective activities reflect the learners' ideas or thoughts during the learning process. Examples of these activities include:

- **Question and Answer:** The learners can ask questions and the instructors answer them. The questions and answers are normally visible to all.
- **Notebook:** It is an activity for recording learners' thoughts during a sequence.
- **Survey:** It collects the learners' surveys and opinions about a topic.
- **Voting:** It allows the learners to vote on a specific idea in the sequence, topic, or learning object.
- **Learner bulletin:** It is a place for the learners to inform the others about anything in the learning space.

In LAM systems, there are different roles participating simultaneously such as educational roles, controlling roles, advertising roles and so on. But in all of them, there are three common roles constituting the main LAM foundation: Learner (Student), Monitor (Assistant) and Author (Instructor). In the followings we describe these roles.

Learner (Student)

The students can study and learn what an instructor wants in a controlled manner. In fact, a LAM system works just like the instructor is asking questions, assigning tasks, teaching and overlooking the answers to the tasks/assignments to be done. These tasks include reading the provided lessons (in the form of plain text, html, and multimedia), chat, scribe, voting, and other previously mentioned activities. However, these tasks are not limited and can be extended to any other digital contributions.

The overall tasks/assignments can be shown and classified as progress bar or sequence of tasks. Each task in the sequence is represented by a specified name, shape and color. For example,

as in Alexander (2008), the completed tasks are shown with blue circles. The in-progress tasks are indicated with red squares and the tasks that haven't been reached are shown with green triangles. However, this is only an assumption and can include more comprehensive states and situations. The task progress bar (or task sequence) navigation can be facilitated with interactive user-friendly navigation tools including text, tables, images, video, and audio.

In the lower level, after selecting each task, the learner can contribute to the learning process specified by the instructor. For example, if the activity is an html scribe task, the learner should write something specific. If the activity is a multiple choice, then the learner should choose between the multiple choice(s). A notebook tool is also available for the students to write statements to be viewed entirety or collated into a set of entries available to be viewed by the instructor. An example of different possible areas for the learners is shown in Figure 1.

Monitor (Assistant)

A key feature in LAM systems is monitoring. By this feature, the instructors (or their assistants) can control the students' learning. There are three major parts for this feature: Lesson, Sequence, and Learner.

LESSON-RELATED MONITORING ACTIVITIES

The instructor (or his assistants) can control the overall properties of each lesson and the students that use them. He can also disable lessons, archive them, etc. The important properties of this activity may include:

- Viewing the lessons' status
- Viewing the statistics of each class, its learners and their selected lessons

Figure 1. An example of different possible learner areas

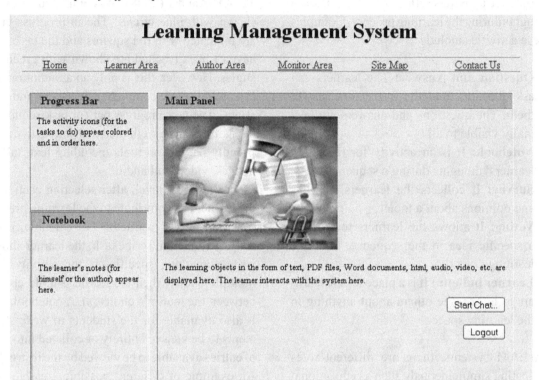

- Viewing the number of online learners for the lesson at each time
- Viewing the total number of learner logins for a lesson in a period of time
- Viewing the sequence status of the learners for the lesson.

SEQUENCE-RELATED MONITORING ACTIVITIES

A sequence is a chain of tasks or activities for the learners as defined by the instructors. The monitoring tasks for a particular sequence are:

- Grouping the learners (performed by the monitors)
- Assigning activities to the groups
- Choosing scribes
- Viewing the overall status of the learners in each sequence

- Viewing the total number of learners doing each task (online or offline)
- Refreshing the system to be synchronized with the sequence status of the current learner
- Controlling the learners such as opening or closing gates[1], forcing the learners to move to a specific activity (by moving their icon to the desired activity) or moving out the learners from an activity or a sequence
- Making minor sequence content changes by adding or removing particular activities.

LEARNER-RELATED MONITORING ACTIVITIES

Monitoring the learners' progress is an important aspect of LAM systems. In the traditional systems, it is constituted from several face-to-face contacts during the learning period. The learner-related monitoring tasks include:

- Viewing the overall progress of learners
- Viewing the sequence progress of learners (completed, current and new activities that haven't been undertaken)
- Viewing the detailed status of the learners in each sequence (according to the selected learner)
- Editing permissions of the learners to sequences (or parts of them) created by the authors (instructors)
- Exporting portfolios for a learner or a group of learners in groupware activities.

Author (Instructor)

Each activity in a LAM system should be started by or under supervision of the author. In fact, creating multiple fascinating pathways for the learners is not a troublesome task today, but having a meaningful consistent set of controlled structures with unified educational goals in such systems makes it hard to be implemented, considering the collaboration between different parts of the system. In such a system, an author can control the system in these ways:

- Creating, storing, modifying and reusing the learner activity sequences
- Viewing a preview of their authored sequences
- File operations (new, open, save, and save as)
- Editing operations (copy–paste and cut–paste)
- Sequencing tools (transition, optional activity, flow, gate, grouping, branching, and preview)
- Importing and exporting sequences to be shared with colleagues

AUTHORING TOOLS

Authoring tools[2] enable creating effective and perhaps interactive learning material which may include multimedia. Using such tools, existing learning objects can be adapted for use in different courses or for different purposes. The provided learning material can be used by various learners with different skills. One of the important criteria for selecting authoring tools is their ease for use and creative freedom so that a powerful authoring tool can lead to a richer course content. In general, authoring tools cause cost reduction and help in investment return. Authoring tools can be categorized in three generations: Template-based Tools, General Authoring Tools, and High-end Simulation Tools[3]. In the following, we briefly describe these categories.

- **Template-Based Tools**: Courses designed by these tools may contain a sequence or a combination of dialog boxes or similar objects. There is no need for any skill in order for learners to use them. These tools work well where ease of use has higher priority than creative freedom and flexibility, since these tools have low flexibility.
- **General Authoring Tools**: These tools are more useful for creating general purpose learning material and are widely used in the market for current courses. There is a need for more skills in order for users to use them. They also have more flexibility than the first generation as they give more creative freedom. In general, these tools are considered as general purpose tools.
- **High-end Simulation Tools**: They are extended tools of high-end multimedia simulations. There is a need for high skills in order to use them. They have high flexibility and they give more creative freedom than the previous generations. Better quality of learning can be achieved by using these tools compared to the previous ones.

ASSESSMENT

Due to the increased number of students and classes in universities and other educational institutions, traditional assessment can be time consuming, difficult and error prone. Computer Aided Assessment (CAA) refers to the use of computer technology as part of the assessment process and is a rapidly developing research area as new technologies are harnessed. Assessing students' answers to a set of questions in order to either evaluate their ability or giving them instant feedback has become an important part of learning management systems. Assessment can be categorized as summative and formative assessment. Summative assessment is used to evaluate students and their ability over the learning material. Formative assessment is used to enhance students' learning by giving them immediate feedback. Formative assessment does not normally contribute in the marking process.

Assessment can be done in different areas such as Mathematical capabilities of learners, programming skills, language abilities and etc. There are software systems that enable automatic assessment, such as Web Work[4]. Its goal is to make homework more effective and efficient. Each Web Work problem set is personalized. That is, each student has a different version of each problem; for example the numerical values in the formulas may be slightly different. Students complete the assignments, log onto the Internet and enter their answers using a Web browser. Web Work system responds students by telling them whether an answer (or a set of answers) is correct or not and also records whether the student answered the question correctly or not. The students are free to try a problem as many times as they wish until the due date. A key educational benefit of this system is that if a student gets a wrong answer, he gets immediate feedback while the problem is still fresh in his mind. The student can then correct a careless mistake, review the relevant material before approaching the problem again, or seek help (frequently via e-mail) from his friends, the

TA or the instructor. CourseMarker[5] is also another software system that enables automatic assessment of learners. According to Higgins & Hegazy & Symeonidis & Tsintsifas (2003), CourseMarker is an extended version of Ceilidh system so that it addresses issues and limitations of its assessment system. In other words, CourseMarker not only has Ceilidh features but also adds new features such as scalability, performance, maintainability, extensibility, and usability. CourseMarker can assess various types of student works such as multiple choice questionnaires, essays, computer programs in different languages and etc. It also supports object-oriented languages. In Course-Marker, feedback to students is not limited to a mark but also a set of details is given in order to improve the performance of students' learning process. In other words, CourseMarker gives an instant detailed feedback to students regarding their missing marks and then the students can correct their mistakes using the feedbacks.

The most important benefit of CAA is the ability of setting a number of topic-related assessments at the end of each topic in order to engage the students. Students are required to answer the questions in order to pass each topic. Compared to the traditional methods, more activity of the learners is required since a test is necessary for each topic. However, this will make sure that students have a fair grasp of the topic before they move on to the next topic. The next advantage is the quick feedback given to students so that they get an immediate warning if their answer is wrong. As a result, students have the opportunity to correct their answer which in turn results in a quicker learning process. The other benefit of CAA is that unlike traditional methods, increasing the number of students does not reduce the performance of the assessment process. It makes it possible to perform assessments for a large number of students efficiently. Saving time for is another advantage of CAA and increasing the number of students does not really affect the amount of time needed for assessment. Moreover, the process is independent from time and location and can be performed at any time and any place. Unlike traditional methods that may assign different marks for an answer at different times, CAA leads to consistency in marking and makes sure that the same mark is given to an answer at all times.

One of the disadvantages of CAA is the potential risk of hardware and software failure during assessment. A serious system crash that leaves the server out of action prevents students to complete their assessments. In addition, students are required to have IT skills. Thus, some students may be discriminated against in the assessment because of their poor IT skills. On the other hand, some types of questions cannot be marked automatically (such as descriptive questions). Assessment of higher order skills can be difficult as well, as preparing suitable questions which assess higher order skills can be troublesome and time consuming.

Bennett (1998) has categorized CAA in three generations. In the first generation of CAA, questions are an electronic version of traditional questions. Computer technology is used for converting these questions into an electronic format, propagating them among students and collecting them. The second generation takes advantage of interactive multimedia such as image, audio and video. Using multimedia technology enables assessing learners' skills, such as formatting a document using a word processor or preparing a spreadsheet document. In the third generation, the learners are assessed during the learning process. In other words, learning and assessment are inseparable. Learners get feedback during learning which in turn enhances the quality of the learning process.

There are two types of questions used in CAA: selective and constructive (descriptive). Selective questions are those that include a set of predefined answers, so that learners can choose from. Multiple choices, true/false questions, and sequences are examples. Constructive questions are those that an answer should be made for them by the learners. Freeform text entry, single word or phrase questions, and Mathematical expressions are examples. Unlike selective questions, the marking process of constructive questions is more difficult. However, constructive questions are more useful for assessing higher order skills.

CONCLUSION

In this chapter, we studied learning activity management systems as a tool for designing, managing and delivering online learning material. We also described components and key features of e-learning systems. We also studied assessment as an integral part of any learning management system. Assessment can be performed during the learning process. As a result, instant feedback can be given to the learners. This, in turn, enhances the quality of learning.

An overall comparison between e-learning and traditional learning systems shows that e-learing systems lead to lower cost, more motivation for learners, more effective education, and adaptability with different educational environments. These systems help instructors preserve the integrity of their course contents.

On the other hand, some people believe that these new learning systems encourage learners to individual activities and discourage social and group activities. Some controlling aspects also remain disputable due to the vast provision of resources and availability of the Internet for younger learners. These are challenging issues in such systems compared to traditional learning systems where face to face conversations, sessions, meetings, scientific tours, etc. are common.

REFERENCES

Alexander, C. (2008). An overview of LAMS (Learning Activity Management System). *Teaching English with Technology Journal, 8*(3).

Alexander, C. (2009). LAMS Revisited. *Teaching English with Technology Journal, 9*(1).

Beatty, K. (2003). *Teaching and Researching Computer-Assisted Language Learning.* London: Longman.

Bennett, R. (1998) Reinventing Assessment: Speculations on the Future of Large-Scale Educational Testing. *Educational Testing Service.* Retrieved from http://www.ets.org/research/pic/bennett.html

Cameron, L., & Dalziel, J. (2008), Perspectives on Learning Design. In *Proceedings of the 3rd International LAMS & Learning Design Conference* (pp. 81-86). Sydney: LAMS Foundation.

Dalziel, J. (2003). Implementing Learning Design: The Learning Activity Management System (LAMS). In *Proceedings of the 20th Annual Conference of the Australian Society for Computers in Learning in Tertiary Education* (ASCILITE).

Fan, Z., & Jiaheng, C. (2008). Learning activity sequencing in personalized education system. *Journal of Natural Science, 13*(4), 461–465.

Higgins, C., Hegazy, T., Symeonidis, P., & Tsintsifas, A. (2003). The CourseMarker CBA System: Improvements over Ceilidh. *Education and Information Technologies, 8*(3), 287–304. doi:10.1023/A:1026364126982

Lambert, G. (2004). *What is Computer Aided Assessment and how can I use it in my teaching? Learning & Teaching Enhancement Unit (LTEU).* Canterbury Christ Church University College.

Mann, S. (2008). The problems of online collaboration for junior high school students: Can the Learning Activity Management System (LAMS) benefit students to learn via online learning? In L. Cameron & J. Dalziel (Eds.), *Proceedings of 3rd International LAMS & Learning Design Conference 2008: Perspectives on Learning Design* (pp. 81-86). Sydney: LAMS Foundation.

Nor Azan, M. Z. (2007). Learning Activities Management System (SPAP). In *Proceedings of International Conference on Electrical Engineering and Informatics, Institut Teknologi Bandung, Indonesia.*

ADDITIONAL READING

Alexander, C. (2007a), Using the Internet in TESOL. In S. Walker, M. Ryan & R, Teed. (Eds.), Proceedings of Design for Learning (pp. 28-34). Greenwich University.

Alexander, C. (2007b). A case study of English language teaching using the Internet in Intercollege's language laboratory. *The International Journal of Technology . Knowledge and Society, 3*(1), 1–15.

Britain, S. (2004). A *Review of Learning Design: Concept, Specifications and Tools: A Report for the JISC E-Learning Pedagogy Programme.* Retrieved May 20, 2008, from http://www.jisc.ac.uk/uploaded_documents/ACF1ABB.doc

Cohen, R. F., Meacham, A., & Skaff, J. (2006). Teaching Graphs to Visually Impaired Students Using an Active Auditory Interface. In *Proceedings of the 37th SIGCSE technical symposium on Computer science education* (pp. 279-282), Houston, USA.

Crescenzi, P., & Nocentini, C. (2007). Fully integrating algorithm visualization into a CS2 course: a two-year experience. In *Proceedings of the 12th Annual Conference on Innovation and Technology in Computer Science Education* (pp. 296-300), Dundee, Scotland.

Curzon, P., & McOwan, P. W. (2008). Engaging with computer science through magic shows. *ACM SIGCSE Bulletin, 40*(3), 179–183. doi:10.1145/1597849.1384320

Dalziel, J. (2003a). *Discussion Paper for Learning Activities and Meta-data. Macquarie E-learning Centre of Excellence* (MELCOE). Retrieved May 10, 2008, from http://www.lamsinternational.com/documents/LearningActivities.Metadata.Dalziel.pdf

Erkan, A., VanSlyke, T. J., & Scaffidi, T. (2007). Data Structure Visualization with LaTeX and Prefuse. In *Proceedings of the 12th Annual Conference on Innovation and Technology in Computer Science Education, Dundee, Scotland.*

Lauer, T. (2006). Learner interaction with algorithm visualizations: viewing vs. changing vs. constructing. *ACM SIGCSE Bulletin, 38*(3).

Levy, P., Aiyegbayo, O., Little, S., Loasby, I., & Powell, A. (2008). *Designing and Sharing Inquiry-Based earning Activities: LAMS Evaluation Case Study.* (Centre for Inquiry-based Learning in the Arts and Social Sciences, University of Sheffield). Retrieved June 20, 2008, from http://www.jisc.ac.uk/media/documents/programmes/elearning-pedagogy/desilafinalreport.pdf

Myller, N., Laakso, M., & Korhonen, A. (2007). Analyzing engagement taxonomy in collaborative algorithm visualization. In *Proceedings of the 12th Annual Conference on Innovation and Technology in Computer Science Education* (pp. 251-255). Dundee, Scotland.

Russell, T., Varga-Atkins, T., & Roberts, D. (2005). Learning Activity Management System Specialist Schools Trust pilot. *A Review for Becta and the Specialist Schools and Academies Trust by CRIPSAT, Centre for Lifelong Learning, University of Liverpool.* Retrieved March 10, 2008, from http://www.cripsat.org.uk/current/elearn/bectalam.htm

Schweitzer, D., & Brown, W. (2007). Interactive visualization for the active learning classroom. *ACM SIGCSE Bulletin, 39*(1).

Subramanian, K. R., & Cassen, T. (2008). A cross-domain visual learning engine for interactive generation of instructional materials. *ACM SIGCSE Bulletin, 40*(1).

ENDNOTES

[1] Professors can generate stop points somewhere in a sequence by gates. Gates halt the process of learners through a sequence until a specified condition is met.

[2] http://www.trivantis.com/resources/articles/2007/whatisauthtool.html

[3] http://www.trivantis.com/resources/articles/2007/selectingatool.html

[4] http://webwork.csis.pace.edu/webwork

[5] http://www.cs.nott.ac.uk/

Chapter 14
Web–Based Training:
An Applicable Tool for Engineering Education

Masoumeh Valizadeh
University of Guilan, Iran

Giancarlo Anzelotti
University of Parma, Italy

Sedigheh Salehi
Bekaert, Kortrijk, Belgium

ABSTRACT

Today, the World Wide Web and Internet technologies are the main support for modern communication. Hence, a vast amount of information is available for students at any level. In this chapter, Web based training (WBT) is described as an efficient tool for engineering education. The benefits of using WBT as a replacement or complementary tool for traditional educational systems and its limitation are discussed. The education process is supposed to shape a professional who possesses not only scientific knowledge, but also exhibits a special set of personal features which includes being curious, innovative, and capable of teamwork. The question is whether distance education is capable of doing that without personal contact. The chapter argues that WBT can provide efficient tools such as forums, discussion boards, and video conference facilities.

1. INTRODUCTION

Distance learning is a progressive method of pedagogy employing the electronic technology to transfer the education to the students who are not on site and allows them to participate in educational activities on their own time. Since 1728 which the first ideas of distance learning has flickered through application of postal services, this technique has passed through several transition states like radio, television, telephone, computer and internet. Emersion of internet made a revolution in distance learning leading to a new generation called web-based learning. Web-based learning is practically the milestone of electronic learning or e-learning, profiting the popularity of the internet as a fast information source.

DOI: 10.4018/978-1-61520-659-9.ch014

Connolly and Stansfield have categorized the historical progress of web-based training through three distinct generations. The first generation took place from 1994-1999 and was marked by a passive use of the Internet where traditional materials were simply repurposed to an online format. The second generation appeared through 2000-2003. This spectrum is marked by the transition to higher bandwidths, rich streaming media, increased resources, and the move to create virtual learning environments that integrated access to course materials, communications, and student services. Internet, Intranets, local area networks (LAN) and wide area networks (WAN) have offered the learners the opportunity to use distance learning beyond pre-recorded classes, educational software and virtual laboratories. The third generation is currently in progress and it is marked by the incorporation of greater collaboration, socialization, project based learning, and reflective practices, through e-portfolios, wikis, blogs, social bookmarking and networking and online simulations. Online forums, blogs and discussion boards have become a precious resource in Learning Management Systems (LMSs). They allow learners to communicate with their compeers and tutors, empowering them to socialize and learn together online. Furthermore, the third generation is progressively being influenced by advances in mobile communicating (Harper 2004; Connolly 2007; Monahan 2008).

Due to its singular capabilities, web-based learning has entered and is widely used in every field of science and technology. It is not only warmly welcome at schools and universities, but also in factories and houses. However utilization of web-based learning technique requires tender and comprehensive attentions in designing, applying and assessing configurations, directed by when, where and for which purpose it is being employed.

Among different branches of human knowledge and sciences, engineering as well as medicine is more involved in practical and daily-life aspects, where the virtual utilities and educational soft-

ware can be utilized to consummate the practical features of engineering education. Furthermore the virtual environment of e-learning courses can provide cheaper, safer, more comprehensive and more inclusive approaches to engineering educational material. what usually the students need to learn in laboratories and workshops and it is costive and demanding for the universities and the schools. The aim of this chapter is to count the requirements of engineering education and to accord the facilities and inadequacies of e-learning as training technique in engineering instruction.

2. ENGINEERING

Engineering is the discipline of implementing of science and mathematics to develop explanation for the problems and to find practical solutions that have an applicable outcome. Engineering is the knowledge of creating a new world based on the rules which science has already discovered. Engineers suppose to be inventors. They design and manufacture machines, processes, systems and even economical constitutions. Utilizing science and mathematics, they observe the world, find the troubles, dream up new ideas and realize the dreams to improve the quality of life and make our world a more comfortable place to live.

Based on scientific principles, every engineering discipline has two main tasks (Wintermantel 1999):

1. To model subsystems using the theoretical and methodological scientific knowledge. In this respect, engineering is not different from the natural sciences.
2. To develop methods and procedures, which allow real systems in all their complexity to be designed and constructed even if not all subsystems have been precisely modeled due to a lack of a thorough knowledge of the underlying physics and chemistry.

The Engineers Council for Professional Development (ECPD), in USA has defined the term of engineering as: "the creative application of scientific principles to design or develop structures, machines, apparatus, or manufacturing processes, or works utilizing them singly or in combination; or to construct or operate the same with full cognizance of their design; or to forecast their behavior under specific operating conditions; all as respects an intended function, economics of operation and safety to life and property" (Science 1941).

3. REQUIREMENTS TO MAKE THE BEST ENGINEER

An engineer supposes to solve complex problems by the simplest solutions. They are often responsible for directly creating a new product or service. The main functions of an engineer have been defined through 7 terms by Encyclopedia Britannica (Smith 2009):

1. *Research*, Using mathematical and scientific concepts, experimental techniques, and inductive reasoning, the research engineer seeks new principles and processes.
2. *Development*, Development engineers apply the results of research to useful purposes. Creative application of new knowledge may result in a working model of a new electrical circuit, a chemical process, or an industrial machine.
3. *Design*, in designing a structure or a product, the engineer selects methods, specifies materials, and determines shapes to satisfy technical requirements and to meet performance specifications.
4. *Construction*, The construction engineer is responsible for preparing the site, determining procedures that will economically and safely yield the desired quality, directing the placement of materials, and organizing the personnel and equipment.
5. *Production*, Plant layout and equipment selection are the responsibility of the production engineer, who chooses processes and tools, integrates the flow of materials and components, and provides for testing and inspection.
6. *Operation*, The operating engineer controls machines, plants, and organizations providing power, transportation, and communication; determines procedures; and supervises personnel to obtain reliable and economic operation of complex equipment.
7. *Management and other functions*, in some countries and industries, engineers analyze customers' requirements, recommend units to satisfy needs economically, and resolve related problems.

By integrating all above mentioned functions, the required skills for all engineering disciplines can be sorted in 3 categories: mathematics and science, team working and curiosity, creativity and innovation (Figure 1). Every educational curriculum for engineering education has to fulfill these themes comprehensively.

Mathematics and Science

Math and science are the basis of engineering. Physics, chemistry, biology, economics, psychology as the laws, forces and resources of nature are the primal necessities of different engineering disciplines. The physical sciences are concerned with the laws that govern the world and the realities of nature. Mathematic however is a comprehensive concept that, in spite of being inspired by the rules of nature, does not bound up with it. It can exist and grow within itself. Mathematics and basic sciences are pure. Engineering, in contrast, is not pure. The process of training an engineer to apply the basic rules of science and to design a machine

Figure 1. The essential requirements of being an engineer

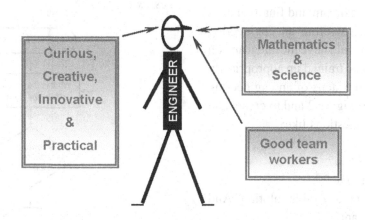

or a process requires a study of engineering models and deep understanding of basic science and the mathematical techniques. Those models require correspondence with reality in their conception, and precision in their description. And those mathematical techniques, like all mathematical techniques, require practice, sophistication and rigor. Consequently, the technological world of an engineer has been constructed on the pure disciplines of mathematics and basic sciences, but it is not contained in them (Chatterjee, 2005).

Teamwork

Most engineers work as a team. Therefore they should be good team players and acquire reasonable skills in communicating with others. This is highly important when planning and creating new projects. Teamwork and group work are increasingly common elements of engineering learning. When group activities and projects are assigned, two types of learning outcomes will emerge:

*   *Product outcomes*: outcomes related to the content of the course;
*   *Process outcomes*: outcomes related to team skills and contribution in group work activities.

The aims of teamwork development are that the members work interdependently and work towards both personal and team goals. They understand these goals are accomplished best by communal support. They feel a sense of ownership towards their role in the group because they committed themselves to goals. Members work in partnership together and use their talent and experience to supply the success of the team's purposes. As a part of every mutual work, members try to be honest, respectful, and listen to others point of view. They see conflict as a part of human nature and they react to it by treating it as an occasion to hear about new ideas and opinions. Members participate equally in decision-making, but each member understands that the leader might need to make the final decision if the team cannot come to a common consent (Kaufman 2000).

Practical, Innovative Personality

Conceptual advances have always been the driving force behind scientific progresses. This in turn relies on creativity and the ability to continue the production of new insights and innovative ideas. The history has given us many discoveries (especially in engineering and the sciences) linked to a sudden recognition, which are the outcomes of creative, dreamer mind of human beings. An engineer is who dreams, then assesses the realization possibilities of the dream and finally makes the dream a reality. Therefore an engineering student on one hand should have the talent and on the other hand has to be trained up appropriately to be able to follow the steps of an engineering decision according to Figure 2 and to create and find the answers for questions like:

1. What is the problem?
2. What do I want to do?
3. What are the benefits of this solution? And are they significant?
4. How can I implement the solution?
5. How can I know that if I have reached the benefits?
6. How can I represent a better solution with my new experiences?

4. ENGINEERING EDUCATION REQUIREMENTS

In a continuously changing world, many challenges and opportunities face the engineering profession and engineering education. Based on various technological and economic developments, fundamental changes have taken place over the last few years that affect the methods in engineering education. In consequence, the actual training trends like the trend to virtual and classical Instructional methods, the global availability of information on one hand and of information resources and accessibility on the other hand, the rise of learning demands and the particularities of engineering education require a strategic streamlining of each of these options. The basis for this strategic planning should be the instructional principal abilities and the use of the general available enabling technologies.

Engineering professionals working in industry are in general unsatisfied with the level of real-world preparedness acquired by recent university graduates entering the work market. Their dissatisfaction is understandable. In order for these

Figure 2. Decision making in engineering processes

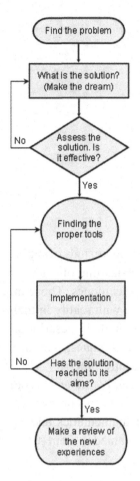

graduates to be productive in an industrial setting, one idea is that the organizations which hire them must supplement their university education with extensive on-the-job training and preparation that provides them with the skills and knowledge they lack (Conn 2002; Baker 2005).

On the other hand, following the tremendous progress in science and technology, engineering teaching materials are currently under development by the multidisciplinary (production engineers, computer scientists, economists) project teams. The philosophy behind this approach is to give access to all information that is relevant to understand and to work on the topic of interest. The background of the didactic approach of these projects is teaching and learning by distance instructions. The central focus is the specification of event-oriented lessons and seminars such as computer supported role-plays for distributed problem-solving, gaming in an open environment and multinational web-based teaching, where students will learn to work on open problems in different fields of engineering, having an active connection with other universities and also the manufacturers. They will be guided from theoretical learning in lessons via interdisciplinary distributed games to international teamwork in cooperation with partner universities. Step by step, their capabilities and entrepreneurial spirit to apply theoretical knowledge will increase, as well as their social and managerial skills. This approach takes up the idea of e-learning to enhance traditional more or less passive reception of knowledge by students what we know as learning by listening, towards an active, self-directed learning and cooperative knowledge creation learning by trying (Hirsch 1998; Pyrz 2003; Baker 2005; Grunbacher 2007).

Concerning to the necessities of modern life, engineering education requires essential transitions in methods and tools to accord with modern societies. So that, engineering education designers endeavor to design and to employ the most consistent methods which fulfill this purpose.

As a potent approach web-based training technique is one of the modern methods of distance learning which plays an advancing role in every field of education. It can be applied solely or joined with the traditional education systems.

5. WEB-BASED TRAINING

In order to overcome the problems of incompatibility and employ a new way of linking made possible by computers, called hypertext, a group of scientists at the European Laboratory for Particle Physics (CERN) in Geneva, Switzerland, began developing an Internet tool in 1989 to link the information achieved by all of the CERN researchers. Rather than presenting information in a linear or hierarchical fashion, hypertext permits information to be linked in a web-like structure. The tool provided a way to link textual information on different computers and created by different scientists. Nodes of information could be linked to other nodes of information in multiple ways. As a result, users can dynamically interlace the information in order to make them most convenient. The CERN project resulted in an innovative front-end to the Internet, now called the World-Wide Web (WWW) (Gottschalk 2009).

For educators, the internet and HTML have provided a new occasion for distance training. The ability of putting together graphics, text, sound, animations and movies into a single tool has empowered the educators to build virtual classroom on their WebPages. By this mean the organizations or individuals can create their own home pages and link them to other home pages or share them with other computers. The educational home pages can be supported by virtual laboratories, virtual libraries, text lessons, virtual classrooms using video conferencing, exercises and exams. These are all the body builders of web-based training.

As a simple definition, Web Based Training (WBT) is a form of training that can be developed by utilizing a network. Web Based Training is the Delivery of educational content via a Web browser over the public Internet, a private intranet, or an extranet. Web-based training often provides links to other learning resources such as references, e-mail, bulletin boards, and discussion groups. WBT also may include a facilitator who can provide course guidelines, manage discussion boards, deliver lectures, and so forth. When used with a facilitator, WBT offers some advantages of instructor-led training while also retaining the advantages of computer-based training' (Imperial-College-London 2009).

WBT is the innovative approach to distance learning in which training material is transformed by the technologies and attitudes of the internet and intranets. The specification of Web-based training is the presenting of live contents, in a structure allowing self-directed, self-paced instruction in any topic (Kilby 2009).

The same as all other distance learning techniques, WBT needs some indispensable components which can be categorized in 3 groups:

- Instructional design
- Knowledge delivery techniques
- Standards

Specifications and examples of these components are tabled in Table 1 (Bergstrom 2001; Digitalthink 2003; Friesen 2004; IEEE 2005; Le@ rningFederation 2008; Ryder 2008; DCMI 2009; IMS 2009). These are general concepts, details of which need to be revised and upgraded according to the demands in different educational fields.

6. WEB-BASED TRAINING FOR ENGINEERING STUDENTS

Engineering education is in fact engineering training, explained as the way of improving a person's performance related to job environments by providing instructions. Intention, designing, tools and media and assessment are usually the components of training programs which can increase the effectiveness of the Engineering training. Though, intent is the motivation and willing to participate in parts of the training, the design is the systematic structure of the training. It determines each step while the learners go ahead with the new skills. It also includes instructional strategies and related measurement issues about these strategies. Tools and media are the means of instruction transfer. It could be a classroom, a web-based training system or combinations of different tools. To finalize the learning process, a formalized assessments and certification provide accountability of the training (YilDirim 2009).

Two basic components of web-based training technology are Ideal Learners and Ideal Instructions. These are the delicate concepts of WBT needing to be taken care for engineering education. As Horton states an Ideal Learner is who (Horton 2001):

- Has definite goals
- Opens to learn independently and view learning prohibitively
- Is self-disciplined and have good time management
- Is moderately capable of field and know basic concepts and facts
- Has enough good writing skills
- Has enough domination in computing skills
- Has positive attitude toward technology in business and learning
- Is so relaxed while there is any technical problem and can cope with these problems

An Ideal Engineering Learner in web-based training system, in addition to above mentioned items has to have notable skills in mathematics be able to adapt himself with the group working and the projects which are usually defined as part of the learners assessment plan and schedule of the course.

The most important and crucial aspect to reach to an advantageous Ideal course is determining learners needs. In an ideal course, teaching is efficient. Precise objects are well defined and the learner acquires new knowledge by spending less time. Development process of WBT is the other crucial concern. Using web-based utilities, the designers can use several models and modulus for instructional objects (see Table 1).

The nature of WBT provides a remarkable array of opportunities and technologies such as XML (eXtensible Markup Language), Web Services, Peer-to-Peer and many other approaches. On the other hand there are many items in

Table 1.

Concept	Aspects	Examples
Standards	**Resource detection**	• ADL SCORM Content Aggregation Model • IEEE P1484.11 Computer Managed Instruction • IMS Content Packaging Specification • IMS Content Packaging Specification • IMS Simple Sequencing Specification
	Content Management	• Dublin Core Metadata Element Set • UK LOM Core • IEEE 1484.12.1-2002 Learning Object Metadata standard • IMS Learning Resource Meta-data Specification • EdNA Metadata Standard • Gateway to Educational Materials Element Set • Canadian Core Learning Resource Metadata Application Profile • SingCore
	Accessibility	• Le@rning Federation Accessibility Specification • Web Content Accessibility Guidelines • IMS AccessForAll Meta-data Specification
	Interoperability	• ADL SCORM • IMS Resource List Interoperability Specification • IMS Enterprise Information Model • IMS Enterprise Services Specification • IMS Question & Test Interoperability Specification • IMS Shareable State Persistence Specification • IMS Digital Repositories • IMS Learner Information Package Specification • ISO/IEC JTC1 SC36/WG3 Learner Information
	Pedagogy Quality	• Le@rning Federation Educational Soundness Specification
	Digital Rights Management	• ISO/IEC JTC1 SC36/WG4 Digital Rights Expression Language • Le@rning Federation Rights Management Specification
Instructional design	**Analysis**	• ADDIE • ASSURE • ARCS • CSCL • CSILE
	Design	
	Development	
	Implementation	
	Evaluation	
Knowledge Delivery Techniques	**Synchronous Technologies:**	• Telephone and Mobile • Web-based conference • Video conference
	Asynchronous Technologies:	• Postal materials • Audio and video cassettes • Radio and Television • Fax and E-mail • Voice mail, MMS and SMS • Message boards and Forums • E-portfolios and blogs

WBT environment influencing the outcome. Therefore one can find the most and the least useful information simultaneously. Hence the designers of WBT courses are responsible of performing the course in the most comprehensive and specific system.

Michael Allen counts three priorities for successful Ideal course designing (Allen 2003):

- Motivation: motivation is essential to learning because it induces attention, persistence and participation in learning activities. This concept is very familiar to literature of learning and instructional designers, but unfortunately, explicit managing of motivation seems to be creeping into educational plans very slowly. The fact is that highly motivated learners will find the means and the tools to access the learning material. Motivation is definitely a very effective task in engineering education, because it is directly connected to creativity and curiosity. A designed approach of motivation in engineering courses assists team working approaches and can encourage the students to exchange their knowledge more readily.

- Directing: highly motivated learners are ready to use any tool helping them to reach the learning material. Therefore it is essentially important to design and provide appropriate educational materials upon a precise schedule. By this way the effectiveness of motivation credit will not be spoiled. Priming appropriate learning content includes providing excellent indexing helping the students to identify an appropriate learning material, or applying an assessment that specifies what the students (in more progressed educational courses, what every individual student) may need. For engineering aspects the relevant directions are even more important. The engineering aspects are widespread in different fields of science and technologies. Meanwhile being founded on basic sciences concepts; it has more connection with applied science. Engineering is in fact the most notable representative of applied science, and due to the broadness and variety of its subjects it requires thoroughly disciplined plan for courses.

- Purposeful and impressive: an ideal course has to be designed for the wide spectrum of students with different capabilities. It has to be enough understandable for less skilled learners meanwhile it is privileging talent students to grasp deeper concepts with satisfactory and appealing supplements. However from another point of view providing appealing and purposeful material for the courses impresses the students and helps them to memorize and remember the instructional contents more easily. For engineering courses designing this approach is even more desirable and helpful, since as objective concepts, the engineering perceptions can be significantly described using the virtual aids of WBT. The main advantage of this approach is the ease of remembering the taught objects in real job status. To facilitate the instructional course with such configuration, the WBT designer may use the following utilities:
    - Attractive context and new examples
    - Providing a charming environment using several means represented by virtual world advantages, such as computer simulated environments
    - Problem-solving scenarios
    - Engaging themes, media and user interface elements
    - Practices, that in WBT cases Virtual Laboratories may empower the instructional contents with virtual practical sessions.

To express the advantages of the above mentioned issues, some simple but very applicable examples are presented in the following paragraphs, revealing the application of WBT in some engineering subjects.

6.1. Cases

Thermodynamics

Thermodynamics is the fundamental knowledge of many branches of engineering such as chemical engineering, mechanical engineering, textile engineering, polymer engineering and many others. One of the fundamental lessons taught in this course is Carnot cycle to make an example of reversible heat engine. This cycle is defined by several adiabatic and isothermal steps including several changes in volume, pressure, enthalpy and entropy of the system. It is a fairly complicated introduction on second law of thermodynamic inaugurating the headwords for more elaborate concepts. As a pivotal subject, teachers usually have to put a lot of time on it and even in this case the students look still confused.

Professor Michael Flower from Department of physics, university of Virginia, has an online course on modern physics, entitled 'Galileo and Einstein' at http://galileoandeinstein.physics.virginia.edu, contenting different fundamental laws of physics such as Galileo's Compound Motion, Kepler's law, Doppler Effect, Carnot Cycle and many others. The advantage of his on-line course is simplification of subjects providing animations, graphs, definitions and states, and all these are connected to each other. Through the visual and virtual interface of this course, the student would be able to follow easily the instructions. They would have the opportunity to make the calculations on each point of cycle, and view the correspondent effects on states, graphs and the heat engine status which they have to learn. Moreover it is easier for the teachers to teach (Hwang 2004; Ion 2007; Flowers 2008).

Polymer Chemistry

Polymer chemistry is the fairly new subdivision of chemistry being taught in chemical engineering, polymer engineering, fiber production engineering, textile engineering, composite and material scientists. Polymer science which is in fact the chemistry, rheology and processing of macromolecules has intricate subjects for both students (to learn) and professors (to make sure that the taught subjects have been understood by the students). American chemical society has several short courses on chemistry and chemical engineering, including polymer courses. The ACS Short Courses in Polymer Chemistry targets industry professionals, chemists, materials scientists, and engineers with a B.S. or higher degree seeking greater understanding of principles governing polymer science and a comprehensive, up-to-date overview of polymer subjects. Since no assumptions are made regarding the background of the participants, the courses attempt to address a variety of learning styles and knowledge bases through lecture, demonstrations, laboratory exercises, and computer-based independent study (ACS 2009).

The department of polymer science of the University of Southern Mississippi has established an on-line center, called Polymer Science Learning Center (PSLC). This center contents polymer courses for different levels from beginners, even kids to polymer explorers. As a multi-level, multimedia polymer education resource, PCLS is an interactive web site in polymer education by pioneering an innovative hands-on, inquiry-based, multi-faceted learning environment for pre-kindergarten through researchers that will result in all polymer education ventures originating by choice with the PSLC. The mission is to teach polymer science to students of all levels in the most memorable manner possible to better prepare them for both chemical careers and life (Mathias 2009).

7. CONCLUSION

World-Wide Web and Internet technologies are the main support of modern communication. They have established foundations of modern educational systems which are offering an extraordinary volume of information to a wide spectrum of learners, even at schools, universities and factories, or house users which are looking for specific objects. Internet contains some of the vantage components of distance education naturally supplying time and place independency as the server computers are working 24 hours a day, all over the year.

With all the advantages happening in distance communication, the teaching and training technology is gravitating more and more toward web-based aspects. The main users of web-based training technologies are still the students of the universities and schools. However, due to its virtual supplements, web-based training can star more effectively in some particular branches of science such as engineering.

- Web-based training is a novel instructional technique offering many superior advantages over traditional methods. Potential capabilities of web-based training nominate it as a suitable method for engineering education, some of which can be counted as:
- WBT offers time-effective and cost-effective training services.
- WBT overcomes the disadvantages of traditional instructional methods based on printed course material. In WBT the educational content can be changed or added fast and easily.
- WBT provides standardized courses ensuring congruous training.
- The educational content can be easily designed to be consistent with different budgets and cultures.
- Course material can be easily adapted and updated for different learners and according to their needs.

- Virtual environment of WBT provides very powerful configurations that can replace the costive practical sessions required in engineering education.

In addition to all the advantages of WBT, it includes some disadvantages as well:

- Initial implementation of a WBT system can be more costive than the cost of publishing course material as textbooks.
- Bandwidth limitations and other limitations on required electronic devices such as the capabilities of computers are considered as the constrains of WBT development.
- WBT is more adapted with self-directed students. But the students who are more used to traditional techniques of teacher-classroom - printed textbooks or the students who do not have enough skills in using computer or internet may find it difficult to get adapted.
- WBT provides time and place independency for the students; however the students are more possible to get interrupted because of the freedom they gain because of the natural self- dependency of WBT method.
- Even with the existence of Virtual laboratories, the students will not get the idea of the real training, as well as they are needed to be trained for their coming careers.

Collecting all the presented information, WBT can be counted as a powerful method when it is designed and applied in its right position. It is certainly the most perfect method to enter in the course content and actually start it, especially for engineering courses where the provided virtual environment of WBT can help the students to figure out the subject. WBT would be very effective when the technical capabilities of the students and their skills have been depicted realistically. The main advantage of WBT is that it can be easily customized regarding to the needs. The needs can be the level of education, the required skills, job aids, the culture of users and many other issues.

REFERENCES

ACS. (2009). ACS Complete Short Course Listings. Retrieved from http://portal.acs.org/portal/acs/corg/content?_nfpb=true&_pageLabel=PP_ARTICLEMAIN&node_id=273&content_id=CTP_006408&use_sec=true&sec_url_var=region1&__uuid=f3138678-d8ec-4b90-83a7-781b4b80b9bf

Allen, M. W. (2003). *Michael Allen's guide to e-learning: Building Interactive, Fun, and Effective Learning Programs for Any Company*. Hoboken, NJ: John Wiley & Sons, Inc.

Baker, A., Navarro, E. O., & van der Hoek, A. (2005). An experimental card game for teaching software engineering processes. *Journal of Systems and Software*, *75*, 3–16. doi:10.1016/j.jss.2004.02.033

Bergstrom. (2001). CMI Guidelines for Interoperability AICC. Retrieved from http://www.aicc.org/docs/tech/cmi001v3-5.pdf

Chatterjee, A. (2005). Mathematics in engineering. *Current Science*, *88*(3), 405–414.

Conn, R. (2002). Developing software engineers at the C-130J software factory. *IEEE Software*, *19*(5), 25–29. doi:10.1109/MS.2002.1032849

Connolly, T., & Stansfield, M. (2007). *Principles of effective online teaching*. Santa Rosa, CA: Informing Science Press.

DCMI. (2009). *Dublin Core Metadata Element Set, Version 1.1*. Retrieved from http://dublincore.org/documents/dces/

Digitalthink. (2003). SCORM™: The E-Learning Standard. Retrieved from http://www.adlnet.org/index.cfm?fuseaction=scormabt

Flowers, M., & Ching, H. W. (2008). *Lecture on Carnot Cycle*. Retrieved from http://galileoandeinstein.physics.virginia.edu/more_stuff/flashlets/carnot.htm

Friesen, N. (2004). *The E-learning Standardization Landscape*. Retrieved from http://www.cancore.ca/docs/intro_e-learning_standardization.html

Gottschalk, T. H. (2009). *Distance education at a glance*. Retrieved from http://www.uiweb.uidaho.edu/eo/distglan

Grunbacher, P., Seyff, N., Briggs, R. O., In, H. P., Kitapci, H., & Port, D. (2007). Making every student a winner: The WinWin approach in software engineering education. *Journal of Systems and Software*, *80*, 1191–1200. doi:10.1016/j.jss.2006.09.049

Harper, K. C., Chen, K., & Yen, D. C. (2004). Distance learning, virtual classrooms, and teaching pedagogy in the Internet environment. *Technology in Society*, *26*, 585–598. doi:10.1016/S0160-791X(04)00054-5

Hirsch, B. E., Thoben, K. D., & Hoheisel, J. (1998). Requirements upon human competencies in globally distributed manufacturing. *Computers in Industry*, *36*, 49–54. doi:10.1016/S0166-3615(97)00097-3

Horton, W. (2001). *Designing Web-Based Training: How to Teach Anyone Anything Anywhere Anytime*. New York: John Wiley & Sons, Inc.

Hwang, F. (2004). NTNUJAVA Virtual Physics Laboratory Java Simulations in Physics: Enjoy the fun! Retrieved from http://www.phy.ntnu.edu.tw/ntnujava/index.php?PHPSESSID=61e16b595f51a79534ea86dd507e6fce&topic=23.msg156#msg156

IEEE. (2005). *1484.12.1: IEEE Standard for Learning Object Metadata*. Retrieved from http://ltsc.ieee.org/wg12/

Imperial-College-London. (2009). *Information & Communication Technologies ICT; An E-Learning Glossary*. Retrieved from http://www3.imperial.ac.uk/ict/services/teachingandresearchservices/elearning/aboutelearning/elearningglossary

IMS. (2009). *Innovation Adoption Learning*. Retrieved from http://www.imsglobal.org/specifications.html

Ion, D. (2007). Physics and Engineering. Retrieved from http://www.cs.sbcc.cc.ca.us/~physics/

Kaufman, D. B., Felder, R. M., & Fuller, H. (2000). Accounting for individual effort in cooperative learning teams. *Journal of Engineering Education, 89*(2), 133–140.

Kilby, T. (2009). *What is Web-Based Training?* Retrieved from http://www.webbasedtraining.com/primer_whatiswbt.aspx

Le@rningFederation (2008). *Metadata Elements Comparison: Vetadata and ANZ-LOM*. Retrieved from http://www.thelearningfederation.edu.au/verve/_resources/ANZLOM-VETADATA-comparison-v1-0.pdf

Mathias, L., Frantz, P., Brust, G., Montogery, J., Smith, V., Simmons, M., et al. (2009). *Polymer Science learning Center*. Retrieved from http://pslc.ws/macrog.htm.

Monahan, T., McArdle, G., & Bertolotto, M. (2008). Virtual reality for collaborative e-learning. *Computers & Education, 50*, 1339–1353. doi:10.1016/j.compedu.2006.12.008

Pyrz, R., Olhoff, N., Lund, E., & Thomsen, O. T. (2003). Mechanics for the future. Retrieved from http://me.aau.dk/GetAsset.action?contentId=1987778&assetId=3356989

Ryder, M. (2008). *Instructional design models*. Retrieved from http://carbon.cudenver.edu/~mryder/itc_data/idmodels.html#modern

Science (1941). The Engineers' Council for Professional Development. *Science, 94*(2446), 456.

Smith, R. J. (2009). *Engineering*. Retrieved from http://www.britannica.com/EBchecked/topic/187549/engineering

Wintermantel, K. (1999). Process and product engineering Ð achievements, present and future challenges. *Chemical Engineering Science, 54*, 1601–1620. doi:10.1016/S0009-2509(98)00412-6

Wintermantel, K. (1999). Process and product engineering Ð achievements, present and future challenges. *Chemical Engineering Science, 54*, 1601–1620. doi:10.1016/S0009-2509(98)00412-6

YilDirim. S. (2009). Web-Based Training: Design and Implementation Issues. Retrieved from http://ocw.metu.edu.tr/informatics-institute/web-based-training-design-and-implementation.

YilDirim. S. (2009). *Web-Based Training: Design and Implementation Issues*. Retrieved from http://ocw.metu.edu.tr/informatics-institute/web-based-training-design-and-implementation

Chapter 15
A Diagnostic System Created for Evaluation and Maintenance of Building Constructions

Attila Koppány
Széchenyi István University, Hungary

ABSTRACT

The successful diagnostic activity has an important role in the changes of the repair costs and the efficient elimination of the damages. The aim of the general building diagnostics is to determine the various visible or instrumentally observable alterations, to qualify the constructions from the suitability and personal safety (accidence) points of view. Our diagnostic system is primarily based on a visual examination on the spot, its method is suitable for the examination of almost all important structures and structure changes of the buildings. During the operation of the diagnostic system a large number of data–valuable for the professional practice–was collected and will be collected also in the future, the analysis of which data set is specially suitable for revaluing construction and the practical application of the experiences later during the building maintenance and reconstruction work. For using the system a so-called "morphological box" has been created, that contains the hierarchic system of constructions, which is connected with the construction components' thesaurus appointed by the correct structure codes of these constructions' place in the hierarchy. The thesaurus was not only necessary because of the easy surveillance of the system, but to exclude the usage of structure-name synonyms in the interest of unified handling. The analysis of which data set is specially suitable for revaluing earlier built constructions and can help to create knowledge based new constructions for the future.

DOI: 10.4018/978-1-61520-659-9.ch015

INTRODUCTION

It is important, that people in architecture science give a useful guide–especially concerning questions raised by new trends of the changing, transforming building activities and construction development–to the profession practice. Seeing that in the aspect of adequacy the appearance of new constructions and building materials always raises new problems to be solved, and the experts in practice busy with the daily tasks of the profession 'according to Möller (1945) cannot always pay enough attention to them'. The efficient diagnostic activity as it has been explained before plays a very important part in the formation of maintenance costs and elimination of damage. It has a just as important part in the preparation of a decision, as having a clear picture of the technological conditions of buildings or group of buildings can be of service at the preparation before making financial decisions of great significance.

DIAGNOSTIC SYSTEM

A faulty diagnosis can lead to incorrect decisions causing financial loss. The research group of the Széchenyi István University (Győr) worked out such a comprehensive diagnostic system (see Figure 1) which contains a common inspection method 'according to Molnárka (2000) for the vast majority of constructional components (for traditional and actually used constructions in Hungary), and can be used for computer data registration and analysis.

The morphological box (Zwicky 1966) is connected with the construction components' thesaurus denoted by the correct structure codes of these constructions' place in the hierarchy. The theory of using morphological box for data registration in the process of building diagnostic (Koppány 1977) was published in Hungary seven years ago. The "matrix" construction of our morphological box fits to the methodology of the visual examination and to the hierarchy of the common building

Figure 1.

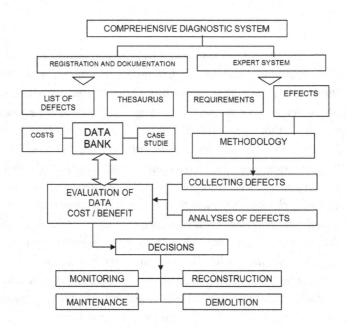

constructions in Hungary 'according to Koppány (2002) (see. Figure 2).

The main task is of our thesaurus (graph-version) to help the visual survey. It can be very useful to understand the hierarchy and the connections in the field of building constructions (see Figure 3). The thesaurus is not only necessary the easy surveillance of the system, but to exclude the usage of structure-name synonyms in the interest of unified handling. We have another tool too for the quick survey of the results of the visual examination, it is the hexagonal morphological box. The box shows the actual checked constructions or all constructions of the building. The various conditions of the building constructions can be marked with corresponding colours in the box-fields (see Figure 4).

The results of an examination are shown on Figure 5 and Figure 6. The diagnostic system has been successfully checked in the course of a course of an examination involving 60 buildings.

EXPERIENCES OF FLAT ROOFS EXAMINATION

The majority of the defects forming on *the flat roof water-proofing and heat insulation* can be observed on the conventional insulated roof. From the point of view of the protection of substance and providing of the proper application of the building it is very important to explore and repair professionally the failures as soon as possible. Concerning the costs it is very important to explore the failures in the beginning stage because the failures which can be repaired with relatively low costs at the beginning can be repaired only with much more expenditures and together with many other structures after the extension of the damage. It should be noted for example, that the water proofing should be checked by experts at least yearly because several extended and expensive damages can be prevented with this care. The typical failures of the roof can be sys-

Figure 2.

MAIN CONSTRUCTIONS	CONSTRUCTIONS AND THEIR SUPPLEMENTARY CONSTRUCTIONS			
CODE ⇒ A.	A.X	A.0X	A.00X	A.00X
1. FOUNDATIONS	1.1 Shallow foundations 1.2 Deep foundations		1.001 Insulations against ground-water	
2. VERTICAL LOAD BEARING CONSTRUCTIONS, NON LOAD BEARING CONSTRUCTIONS AND THEIR SUPPLEMENTARY STRUCTURES	2.1 Load bearing walls 2.2 Frames	2.01 Non load bearing walls 2.02 Partition walls 2.03 Doors and windows in internal walls	2.001 Wall-insulations 2.002 Plasters on facades 2.003 Plasters inside 2.004 Claddings inside	2.0001 Grates, balustrades, parapets
3. HORIZONTAL LOAD BEARING CONSTRUCTIONS, NON LOAD BEARING CONSTRUCTIONS AND THEIR SUPPLEMENTARY STRUCTURES	3.1 Ring beams 3.2 Arches, lintels 3.3 Floors, vaults	3.01 Suspended ceilings	3.001 Water-proofings to floors 3.002 Floorings	
4. STAIRCASES, BALCONIES, OPEN CORRIDORS, LOFTS AND THEIR SUPPLEMENTARY STRUCTURES	4.1 Staircases 4.2 Balconies, open corridors, lofts		4.001 Stair-tread covering 4.002 Balconies flooring 4.003 Open corridors flooring 4.004 Lofts flooring	4.0001 Balustrades, parapets, handrails to stairs 4.0002 Balustrades, parapets, handrails to balconies 4.0003 Balustrades, parapets, handrails to open corridors 4.0004 Balustrades, parapets, handrails to lofts
5. ROOFS, ROOF ACCESSORIES, CHIMNEIS, VENTILATION	5.1 Pitched roofs 5.2 Flat roofs 5.3 Roof superstructures 5.4 Chimneis 5.5 Air shafts 5.6 Vent pipes	5.01 Roofings	5.001 Water-proofing 5.002 Thermal insulation 5.003 Vapour-proofing 5.004 Flat roofs tiles and other functional layers 5.005 Roof edges and joining accessories	5.0001 Roof accessories
6. FACADES		6.01 Skirting board	6.001 Surfacings	6.0001 Grates, balustrades, parapets on the facades

Figure 3.

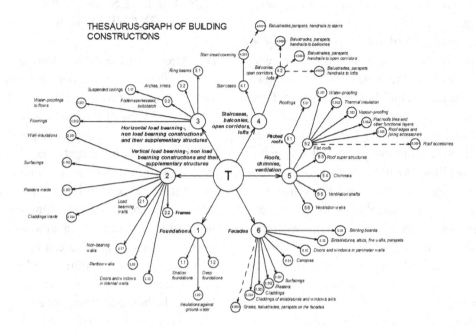

Figure 4.

Figure 5.

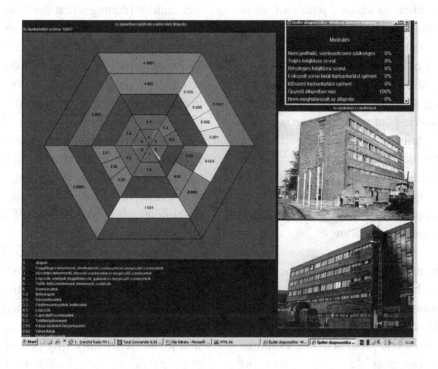

Figure 6.

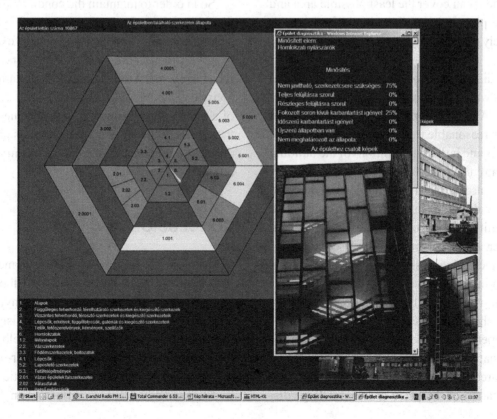

tematised according to various points of view. The analysis can be performed by the layers of the layer construction of the roof insulation and water-proofing or according to the contribution to the creation of the roof insulation (e.g. material manufacturing, planning, execution, operation), but it can be carried out according the so called weak points, details of the structural nodes.

Before introducing the diagnostic procedures and testing methods applied for the flat roof construction it is reasonable to determine some principles (Koppány & Graf 1985) in connection with the examinations as follows:

- the examination mustn't inhibit the proper use of the building;
- the examination should be quickly performable with easily usable tools;
- during the whole process of the examination the least possible damage can come out in the flat roof water-proofing;
- if the destructing examination is unavoidable it can cover the least possible area and the place of sampling should be immediately repairable (in a waterproof way).

The diagnostic work can consist of several phases - which are important from the end point of view. At first before the examination on the spot it is reasonable to inform on the basic data, structures and building conditions of the building to be examined. In case of old roof it is not always possible since the plans and other documents could get lost and they cannot be often reconstructed. In a significant part of the cases the structural character, layer construction, the used materials and the technologies should be identified during the examination on the spot.

During the visual examination the visual failures should be discovered then the analysis of the operation of the structure can lead to the determination of the more complex causes of the failures. During the visual examination the identification of the place of leak for large dis-

continuities (damages) on the water proofing of direct layer order has generally no difficulty. In case of quick examination the condition of the roof water-proofing is determined basically with visual examination, completed with a deteriorate free instrumental measurement if necessary (see Figure 7).

For comprehensive examinations the procedure covers the all structural and complementary elements of the roof. The condition, load capacity, deformation of the bare floor should be examined, the building physical properties of the floor structure, etc. should be evaluated.

There are a lot of interesting data from the examination of 60 flat roof constructions (see Figure 8). This figure shows the characteristics of frequency of the typical defects at the examined old flat roofs. The greatest number of defects were in the expansion joints area, nearly fifty percentage. The age of the examined roofs was on average 20 years. The roofs of the older industrial building had the worst condition, most of them were in bad repair.

So in order to maintain the condition, safe and durability of the flat roof water-proofing and insulation all significant factors should be examined even with a deterioration examination if necessary. It can happen that the diagnostics are started with a visual examination performed within the frame of a quick examination and to determine exactly the causes of the abnormalities, defects observed during the examination a complex examination is required.

CONCLUSION

In the paper is reported the development and structure of a field based survey methodology by the research group at the Széchenyi István University as practical diagnostic decision support tools. The new system was developed for the one of the biggest building holder-operator organisation in Hungary. The diagnostic system firstly based on a visual examination on the spot.

Figure 7.

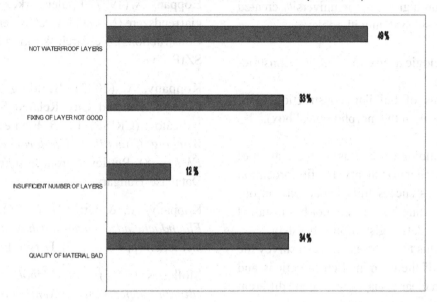

VISUAL EXAMINATION OF FLAT ROOF WATERPROOFING (60 BUILDING)

Figure 8.

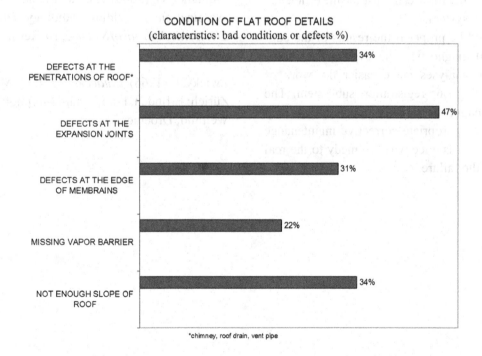

CONDITION OF FLAT ROOF DETAILS
(characteristics: bad conditions or defects %)

*chimney, roof drain, vent pipe

At the beginning of the developing work an important requirement was the visual demonstration of the examinations' results. It was expected from the tools of system to be able to show the general conditions of the buildings or the condition of selected constructions. In the process of visual examination the experts have a big amount of data. For the effective handling and using the data the research group of the university created a registration subsystem with tools:

- morphological box for building constructions and
- thesaurus of building construction (connections with the morphological box).

The morphological box as a diagnostic tool wasn't used earlier in Hungary. The first probation of this tool has successfully happened and our client, the building holder-operator has installed this tool in his data registration subsystem.

Using of this tools the experts can survey the connections of the examined constructions and can use also a clear visual survey of the different durability of the existing constructions. Some details from the results of the visual examination of 60 different flat roofs can illustrate the efficiency the new subsystem.

It would be proper if the results of the constructional diagnostics, the experiences of the pathologic analyses make easier the work of experts using our registration subsystem. The practical diagnostic decision support tool can help choose the appropriate corrective maintenance procedures, it is once which remedy to the real causes of the failure.

REFERENCES

Koppány, A. (1997). Épületszerkezetek morfológiai rendszere.(Morphological System of Building Constructions). Research Working Paper (p.40). SZIF, Győr

Koppány, A. (2002). Building Diagnostics. Construction and City Related Sustainability Indicators (CRISP). EU 5. Framework. *CRISP Working Committee Conference, Budapest. ÉMI* (p. 8). Retrieved from http://crisp.cstb.fr / Database-Hungary)

Koppány, A., & Graf, H. (1985). *Mängel an Flachdacheindeckungen und ihre Untersuchungsmethoden* (pp. 135–137). Berlin: Bauzeitung.

Möller, K. (1945). *Építési hibák és elkerülésük. (Building defects and prevention of defects)* (p. 32). Budapest: Egyetemi Nyomda.

Molnárka, G. (2000) The methodology in visual examination in building pathology. *Hungarian Electronic Journal of Science*, 10. Retrieved from www.sze.hu

Zwicky, F. (1966). *Entdecken* (p. 174). München, Zürich: Erfinden, Forschen im Morphologischen Weltbild, Droemer/Knaur.

Compilation of References

ACS. (2009). ACS Complete Short Course Listings. Retrieved from http://portal.acs.org/portal/acs/corg/content?_nfpb=true&_pageLabel=PP_ARTICLEMAIN&node_id=273&content_id=CTP_006408&use_sec=true&sec_url_var=region1&__uuid=f3138678-d8ec-4b90-83a7-781b4b80b9bf

Alexander, C. (2008). An overview of LAMS (Learning Activity Management System). *Teaching English with Technology Journal, 8*(3).

Alexander, C. (2009). LAMS Revisited. *Teaching English with Technology Journal, 9*(1).

Al-Jumaily, A., & Stonyer, H. (2000). Beyond Teaching and Research . *Changing Engineering Academic Work Global Journal of Engineering Education, 4*(1), 89–97.

Allen, M. W. (2003). *Michael Allen's guide to e-learning: Building Interactive, Fun, and Effective Learning Programs for Any Company.* Hoboken, NJ: John Wiley & Sons, Inc.

Altshuller, G. (2000). *The innovation algorithm. TRIZ, systematic innovation and technical creativity.* Worcester, MA: Technical Innovation Center.

Anohiina, A. (2005). Analysis of the terminology used in the field of virtual learning. *Educational Technology & Society, 8*, 91–102.

Anzelotti, G., Nicoletto, G., & Riva, E. (2008). *Heterogeneous microscopic strains in a cast Al-Si alloy by digital image correlation.* Paper presented at the Symposium on Advances in Experimental Mechanics, České Budějovice.

Astrom, A. (2008). *E-learning quality Aspects and criteria for evaluation of e-learning in higher education.* Stockholm: National Agency's Department of Evaluation.

Baker, A., Navarro, E. O., & van der Hoek, A. (2005). An experimental card game for teaching software engineering processes. *Journal of Systems and Software, 75*, 3–16. doi:10.1016/j.jss.2004.02.033

Barbian, J. (2002). Blended works: Here's proof! *Online Learning, 6*(6), 26–31.

Baron, R. (1998). *What Type Am I? Discover Who You Really Are.* New York: NY Penguin Putnam Inc.

Barros, B., Read, T., & Verdejo, M. F. (2008). Virtual collaborative experimentation: An approach combining remote and local labs. *IEEE Transactions on Education, 51*(2), 242–250. doi:10.1109/TE.2007.908071

Bártol, N. (1999). *Új segédanyag az épületszerkezettan hatékonyabb oktatásáért.* (New Education Aids for a More Effective Teaching of Building Construction Theory.) In *XXIV. Conference on Building Construction, "Széchenyi István" University, Győr* (pp. 54-62).

Beatty, K. (2003). *Teaching and Researching Computer-Assisted Language Learning.* London: Longman.

Beichner, R. J. (1990). The effect of simultaneous motion presentation and graph generation in a kinematics lab. *Journal of Research in Science Teaching, 27*(8), 803–815. doi:10.1002/tea.3660270809

Bennett, R. (1998) Reinventing Assessment: Speculations on the Future of Large-Scale Educational Testing. *Educational Testing Service.* Retrieved from http://www.ets.org/research/pic/bennett.html

Bergstrom. (2001). CMI Guidelines for Interoperability AICC. Retrieved from http://www.aicc.org/docs/tech/cmi001v3-5.pdf

Bersin, J. (2004). *The blended learning book: Best practices, proven methodologies, and lessons learned*. San Francisco, CA: Pfeiffer.

Bhatt, R., Tang, C. P., Lee, L. F., & Knovi, V. (2009). A case for scaffolded virtual prototyping tutorial case-studies in engineering education. *International Journal of Engineering Education, 25*(1), 84–92.

Bier, I.D., & Cornesky, R. (2001). Using QFD to construct a higher education curriculum. *Quality Progress*, April, 64-68.

Bloom, B. S. (1956). *Taxonomy of educational objectives, Handbook I: The cognitive domain*. New York: David McKay Co Inc.

Bloom, B. S. (1956). *Taxonomy of educational objectives: The classification of educational goals. Handbook I, Cognitive domain*. New York: Longmans, Green.

Boot, E. W., van Merrienboer, J. J. G., & Theunissen, N. C. M. (2008). Improving the development of instructional software: Three building-block solutions to interrelate design and production. *Computers in Human Behavior, 24*, 1275–1292. doi:10.1016/j.chb.2007.05.002

Borau, K., Ullrich, C., & Kroop, S. (2009). Mobile learning with Open Learning Environments at Shanghai Jiao Tong University, China. *Learning Technology publication of IEEE Computer Society Technical Committee on Learning Technology (TCLT) . Learning Technology Newsletter, 11*(1-2), 7–9.

Bouras, C., Philopoulos, A., & Tsiatsos, T. (2001). e-Learning through distributed virtual environments. *Journal of Network and Computer Applications, 24*, 175–199. doi:10.1006/jnca.2001.0131

Boyer, E. L. (1990). *Scholarship Reconsidered: Priorities of the Professoriate*. Stanford: Carnegie Foundation for the Advancement of Teaching.

Brackin, P. (2002). Assessing engineering education: an industrial analogy. *International Journal of Engineering Education, 18*(2), 151–156.

Brad, S. (2005). Need for radical innovation to ensure high quality of continuous education in engineering. In C. Oprean, & C. Kifor (Ed.), *3rd Balkan Region Conference on Engineering Education* (pp. 77-80). Sibiu: Lucian Blaga University Press.

Brad, S. (2008). Complex system design technique. *International Journal of Production Research, 46*(21), 5979–6008. doi:10.1080/00207540701361475

Brad, S. (2009). Concurrent multifunction deployment. *International Journal of Production Research, 47*(19), 5343–5376. doi:10.1080/00207540701564599

Bransford, J. B., Brown, A. L., & Cocking, R. R. (1999). *How People Learn*. Washington, D.C.: National Academy Press.

Bransford, J. B., Franks, J. J., Vye, N., & Sherwood, R. (1989). New approaches to instruction: Because wisdom can't be told . In Vosniadou, S., & Ortony, A. (Eds.), *Similarity and analogical reasoning* (pp. 470–497). Cambridge: Cambridge University Press. doi:10.1017/CBO9780511529863.022

Brant, G., Hooper, E., & Sugrue, B. (1991). Which comes first: The simulation or the lecture? *Journal of Educational Computing Research, 7*(4), 469–481.

Brown, T. (2005). Toward a model for m-learning in Africa. *International Journal on E-Learning, 4*(3), 299–315.

Bunz, U. (2005). Using scantron versus an audience response system for survey research: Does methodology matter when measuring computer-mediated communication competence? *Computers in Human Behavior, 21*, 343–359. doi:10.1016/j.chb.2004.02.009

Burnely, S. (2007). The use of virtual reality technology in teaching environmental engineering. *Engineering Education . Journal of the Higher Education Academy Engineering Subject Center, 2*(2).

Burstein, J., Kukich, K., Wolff, S., Chi, L., & Chodorow, M. (1998). Enriching automated essay scoring using discourse marking. In *Proceedings of the Workshop on Discourse Relations and Discourse Marking, Annual Meeting of the Association of Computational Linguistics, Montreal, Canada*.

Burstein, J., Leacock, C., & Swartz, R. (2001). Automated evaluation of essay and short answers. In M. Danson (Ed.), *Proceedings of the Sixth International Computer Assisted Assessment Conference, Loughborough University, Loughborough, UK*.

Butler, C. (2009). *The development of a human-centric fuzzy mathematical measure of human engagement in interactive multimedia systems and applications.* Unpublished doctoral dissertation, University of Central Florida.

Buzzetto-More, N. A. (2008). Student Perceptions of Various E-Learning Components. *Interdisciplinary Journal of E-Learning and Learning Objects, 4*, 113–135.

Caillot, M. (Ed.). (1991). *Learning Electricity and Electronics with Advanced Educational Technology.* New York: Springer-Verlag.

Caldwell, J. E. (2007). Clickers in the large classroom: Current research and best-practice tips. *CBE - Life Sciences Education, 6*(spring), 9-20.

Callaghan, M. J., Harkin, J., McGinnity, T. M., & Maguire, L. P. (2008). Intelligent user support in autonomous remote experimentation environments. *IEEE Transactions on Industrial Electronics, 55*(6), 2355–2367. doi:10.1109/TIE.2008.922411

Cameron, L., & Dalziel, J. (2008), Perspectives on Learning Design. In *Proceedings of the 3rd International LAMS & Learning Design Conference* (pp. 81-86). Sydney: LAMS Foundation.

Campanella, S., Dimauro, G., Ferrante, A., Impedovo, D., Impedovo, S., & Lucchese, M. G. (2007). Engineering e-learning surveys: a new approach. *International Journal of Education and Information Technologies, 1*(2), 105–113.

Campione, J. C., & Brown, A. L. (1985). Dynamic assessment: One approach and some initial data. Tech Rep. No. 361, Univ. of Illinois at Urbana-Champaign, Champaign, IL.

Campione, J. C., & Brown, A. L. (1987). Linking dynamic assessment with school achievement . In Lidz, C. S. (Ed.), *Dynamic Assessment: An Interactional Approach to Evaluating Learning Potential* (pp. 479–495). New York: Guilford Press.

Center for Educational Research and Evaluation. (2005). Report on the survey of higher education teachers in 2004. Retrieved June 13, 2009 from http://www.cher.ntnu.edu.tw/analyze/data/edu92_teacher/pdf/2/2-2.pdf, (the Integrated Higher Education Database in Taiwan)

Chang, B., Wang, H., & Lin, Y. (2009). Enhancement of mobile learning using wireless sensor network. *Learning Technology publication of IEEE Computer Society Technical Committee on Learning Technology (TCLT) . Learning Technology Newsletter, 11*(1-2), 22–25.

Chatterjee, A. (2005). Mathematics in engineering. *Current Science, 88*(3), 405–414.

Chen, R., & Hsiang, C. (2007). A study on the critical success factors for corporations embarking on knowledge community-based e-learning. *Information Sciences, 177*, 570–586. doi:10.1016/j.ins.2006.06.005

Chittaro, L., & Ranon, R. (2007). Web3D technologies in learning, education and training: Motivations, issues, opportunities. *Computers & Education, 49*, 3–18. doi:10.1016/j.compedu.2005.06.002

Chorpothong, N., & Charmonman, S. (2004). An eLearning project for 100,000 students per year in Thailand. *International Journal of The Computer, the Internet and Management, 12*, 111-118.

Christie, J. R. (1999). Automated essay marking-for both style and content. In M. Danson (Ed.), *Proceedings of the Third Annual Computer Assisted Assessment Conference, Loughborough University, Loughborough, UK*.

Christie, J. R. (2003). Email communication with author. 14th April.

Clark, D. (2004). ADDIE - 1975. from http://www.nwlink.com/%7Edonclark/history_isd/addie.html

Clement, J. J. (1993). Using bridging analogies and anchoring intuitions to deal with students' preconceptions in physics. *Journal of Research in Science Teaching, 30*(10), 1241–1257. doi:10.1002/tea.3660301007

Clement, J. J. (1994). Use of physical intuition and imagistic simulation in expert problem solving . In Tirosh, D. (Ed.), *Implicit and Explicit Knowledge*. Norwood, NJ: Ablex Publishing Corp.

Cloete, E. (2001). Electronic education system model. *Computers & Education, 36*, 171–182. doi:10.1016/S0360-1315(00)00058-0

CNN (1999). *Recovering from the Exxon Valdez oil spill in Alaska*.

Cockburn, A., & Mckenzie, B. (2004). Evaluating spatial memory in two and three dimensions. *International Journal of Human-Computer Studies, 61*, 359–373. doi:10.1016/j.ijhcs.2004.01.005

Coleman, D. J. (1998). *Applied and academic geomatics into the 21st Century*. Paper presented at the FIG Commission 2, XXI Inter. FIG Congress, Brighton, England.

Colston, R. (2008). *ADDIE model*. Retrieved from http://www.learning-theories.com/addie-model.html

Conn, R. (2002). Developing software engineers at the C-130J software factory. *IEEE Software, 19*(5), 25–29. doi:10.1109/MS.2002.1032849

Connolly, T., & Stansfield, M. (2007). *Principles of effective online teaching*. Santa Rosa, CA: Informing Science Press.

Corbeil, J., & Valdes-Corbeil, M. (2007). *Are You Ready for Mobile Learning? Frequent use of mobile devices does not mean that students or instructors are ready for mobile learning and teaching*. Retrieved July 17, 2009, from http://www.educause.edu/EDUCAUSE+Quarterly/EDUCAUSEQuarterlyMagazineVolum/AreYouReadyforMobileLearning/157455

Crosby, M., Auernheimer, B., Aschwanden, C., & Ikehara, C. (2001). Physiological data feedback for application in distance education. In *Proceedings of the 2001 Workshop on Perceptive User Interfaces*. Orlando, Florida.

CSU. (2004). *Chico Distance & Online Education*. Retrieved from http://rce.csuchico.edu/ online/site.asp

Cucchiarelli, A., Faggioli, E., & Velardi, P. (2000). Will very large corpora play for semantic disambiguation the role that massive computing power is playing for other AI-hard problems? *2nd Conference on Language Resources and Evaluation (LREC), Athens, Greece*.

Dale, E. (1969). *Audiovisual Methods in Teaching* (3rd ed.). New York, NY: Dryden Press.

Dalgarno, B., & Hedberg, J. (2001). *3D Learning environments in tertiary education*. Paper presented at the 18th annual conference of the Australasian Society for Computers in Learning in Tertiary Education, Melbourne, Australia.

Dalziel, J. (2003). Implementing Learning Design: The Learning Activity Management System (LAMS). In *Proceedings of the 20th Annual Conference of the Australian Society for Computers in Learning in Tertiary Education* (ASCILITE).

Davis, M. L., & Cornwell, D. A. (2001). *Introduction to Environmental Engineering*. McGraw Hill.

DCMI. (2009). *Dublin Core Metadata Element Set, Version 1.1*. Retrieved from http://dublincore.org/documents/dces/

De Lucia, A., Francese, R., Passero, I., & Tortora, G. (2008). *Development and evaluation of a virtual campus on Second Life: The case of SecondDMI*. Computers & Education.

de Oliveira, P. C. F., Ahmad, K., & Gillam, L. (2002). A financial news summarization system based on lexical cohesion. In *Proceedings of the International Conference on Terminology and Knowledge Engineering, Nancy, France*. Breck, E.J. et al. (2000). How to Evaluate Your Question Answering System Every Day… and Still Get Real Work Done, In *Proc. LREC-2000, Linguistic Resources in Education Conf., Athens, Greece*.

De Regt, H. W., & Dieks, D. (2002). A contextual approach to scientific understanding. Retrieved from March 15th, 2003 from http://philsci-archive.pitt.edu/documents/disk0/00/00/05/53/index.html

Deerwester, S. C., Dumais, S. T., Landauer, T. K., Furnas, G. W., & Harshman, R. A. (1990). Indexing by latent semantic analysis. *Journal of the American Society for Information Science American Society for Information Science, 41*(6), 391–407. doi:10.1002/(SICI)1097-4571(199009)41:6<391::AID-ASI1>3.0.CO;2-9

Derry, S. J., Wilsman, M. J., & Hackbarth, A. J. (2007). Using contrasting case activities to deepen teacher understanding of algebraic thinking and teaching. *Mathematical Thinking and Learning, 9*(3), 305–329.

Dick, W., & Carey, L. (1996). *The systematic design of instruction* (4th ed.). New York: Harper Collins.

Digitalthink. (2003). SCORM™: The E-Learning Standard. Retrieved from http://www.adlnet.org/index.cfm?fuseaction=scormabt

Dilmaghani, M. (2003). *National providence and virtual education development capabilities in higher education.* Paper presented at the Virtual University Conference Kashan, Iran.

Dr. Fátrai, G. (1998). *Építéskivitelezés online-jegyzet* (On-Line Lecture Notes on Building Effectuation). Retrieved March 24, 2009, from www.arc.sze.hu/kivitelea

Dr. Koppány, Attila (1998 and 2006). *Épületszerkezettan online-jegyzet* (On-Line Lecture Notes for Building Construction). Retrieved March 24, 2009, from www.arc.sze.hu/epszerkea

Dr. Orbán, J. (2000-2007). *Orisoft Építőanyagipari Katalógus* (Orisoft Catalogue of Building Materials). Retrieved March 24, 2009, from www.orisoft.pmmf.hu

Dr. Somfai, A. (2006). *Számítógépes Építészeti Modellezés online-jegyzet* (On-Line Lecture Notes on Computer Aided Architectural Modelling). Retrieved March 24, 2009, from www.arc.sze.hu/cad

Du, J., Li, F., & Li, B. (2008). The case of interactive web-based course development. In E. Leung, F. Wang, L. Miao, J. Zao, & J. He (Ed.), *2nd Workshop on Blended Learning* (pp. 65-73). Jinhua: Springer-Verlag Berlin.

Duit, R., Jung, W., & von Rhoneck, C. (Eds.). (1984). *Aspects of Understanding Electricity.* Kiel, Germany: Verlag, Schmidt, & Klaunig.

Dunn, R. (1990). Understanding the Dunn and Dunn Learning Styles Model and the Need for Individual Diagnosis and Prescription. *Reading. Writing and Learning Disabilities, 6*, 223–247.

Dunn, R. (1992). Learning Styles Network Mission and Belief Statements Adopted. *Learning Styles Network Newsletter, 13*(2), 1.

Dunn, R., Beaudry, J. S., & Klavas, A. (1989). Survey of Research on Learning Styles. *Educational Leadership, 46*(6), 50–58.

Ebner, M., & Walder, U. (2008). E-education in civil engineering–a promise for the future? In P. Vainiunas & L. Juknevicius (Ed.), *6th AECEF Symposium on Civil Engineering Education in Changing Europe* (pp. 16-26). Vilnius: Vilnius Gediminas Technical Univ. Press.

Eklund, J., & Kay, M. (2001). *Strategy 2001: Evaluation of the Usage of National Flexible Learning Toolboxes - (Series 2).* Commissioned Report for the Australian National Training Authority.

Eklund, J., & Kay, M. (2003). Evaluation of the usage of National Flexible Learning Toolboxes (Series 3). *Commissioned Report for the Australian National Training Authority.* eLearning Guide Research (2008). Snapshot Report on Learning Modalities. Retrieved June 13, 2009 at http://www.elearningguild.com/research/archives/index.cfm?action=viewonly2&id=129&referer=http%3A%2F%2Fwww%2Eelearningguild%2Ecom%2F

E-learning. (2006). *Evaluation management.* Retrieved from http://www.elearning-engineering.com/evaluation.htm

EmptyOceans Empty Nets (2003). Bullfrog Films.

Engelhart, P. V., & Beichner, R. J. (2004). Students' understanding of direct current resistive electrical circuits. *American Journal of Physics*, *72*(1), 98–115. doi:10.1119/1.1614813

Fan, Z., & Jiaheng, C. (2008). Learning activity sequencing in personalized education system. *Journal of Natural Science*, *13*(4), 461–465.

Farrell, S., Hesketh, R. P., Newell, J. A., & Slater, C. S. (2001). Introducing freshmen to reverse process engineering and design through investigation of the brewing process. *I. J. E. E (Norwalk, Conn.)*, *17*(6), 2001.

Felder, R. M. (1993). Reaching the Second Tier: Learning and Teaching Styles in College Science Education. *Journal of College Science Teaching*, *23*(5), 286–290.

Fies, C., & Marshall, J. (2006). Classroom response systems: A review of the literature. *Journal of Science Education and Technology*, *15*(1), 101–109. doi:10.1007/s10956-006-0360-1

Finelli, C. J., Klinger, A., & Budny, D. D. (2001). Strategies for improving the classroom environment . *Journal of Engineering Education*, *90*(4), 491–501.

Finger, S., Gelman, D., Fay, A., & Szczerban, M. (2005). Supporting collaborative learning in engineering design. In W. Shen, A. James, K. Chao, M. Younas, & J. Barthes (Ed.), *9th International Conference on Computer Supported Cooperative Work in Design* (pp. 990-995). Coventry: Coventry Univ. Press.

Flowers, M., & Ching, H. W. (2008). *Lecture on Carnot Cycle*. Retrieved from http://galileoandeinstein.physics.virginia.edu/more_stuff/flashlets/carnot.htm

Foertsch, J., Moses, G., Strikwerda, J., & Litzkow, M. (2002). Reversing the lecture / homework paradigm using e-TEACH Web-based streaming video software. *Journal of Engineering Education*, *91*(3), 267–275.

Forks, G. (2009). *UND Online & Distance Education*. Retrieved from http://distance.und.edu/

Fozdar, B., & Kumar, L. (2007). Mobile learning and student retention. *The International Review of Research in Open and Distance Learning, 8*(2), from http://www.irrodl.org/index.php/irrodl

Frances, C., Pumerantz, R., & Caplan, J. (1999). Planning for instructional technology. What you thought you knew could lead you astray. *Change, 31*(4), 24–33. doi:10.1080/00091389909602697

Freudenthal, H. (1993). Thoughts on teaching mechanics: Didactical phenomenology of the concept of force. *Educational Studies in Mathematics*, *25*(1/2), 71–88. doi:10.1007/BF01274103

Friesen, N. (2004). *The E-learning Standardization Landscape*. Retrieved from http://www.cancore.ca/docs/intro_e-learning_standardization.html

Gentner, D., & Gentner, D. R. (1983). Flowing waters or teeming crowds: Mental models of electricity . In Gentner, D., & Stevens, A. L. (Eds.), *Mental Models*. Erlbaum.

Gervasi, O., Riganelli, A., Pacifici, L., & Laganà, A. (2004). VMSLab-G: a virtual laboratory prototype for molecular science on the Grid. *Future Generation Computer Systems*, *20*, 717–726. doi:10.1016/j.future.2003.11.015

Gervasi, O., Rignelli, A., Pacifici, L., & Lagana, A. (2004). VMSLab-G: a virtual laboratory prototype for molecular science on the Grid. *Future Generation Computer Systems*, *20*, 717–726. doi:10.1016/j.future.2003.11.015

Ghosh, S., & Fatima, S. S. (2007b). *Retrieval of XML data to support NLP applications*. Paper presented at ICAI'07- The 2007 International Conference on Artificial Intelligence Monte Carlo Resort, Las Vegas, Nevada, USA, June 25-28, 2007.

Ghosh, S., & Fatima, S. S. (2007c). *A Web Based English to Bengali Text Converter*. Paper presented at the 3rd Indian International Conference on Artificial Intelligence (IICAI-07), Pune, India, December 17-19, 2007.

Ghosh, S., & Fatima, S.S. (2007a). *Use of local languages in Indian portals. CSI Communication*, 4-12.

Gibbs, R. W. (2006). *Embodiment and Cognitive Science*. Cambridge University Press.

Gibson, J. J. (1979). *The Ecological Approach to Visual Perception*. Boston: Houghton Mifflin.

Giveki, F. (2003). *Learning New Methods in Distance Higher Education*. Paper presented at the Virtual University Conference, Kashan.

Gladun, A., Rogushina, J., Garcıa-Sanchez, F., Martınez-Bejar, R., & Fernandez-Breis, J. T. (2009). An application of intelligent techniques and semantic web technologies in e-learning environments. *Expert Systems with Applications, 36*(2), 1922–1931. doi:10.1016/j.eswa.2007.12.019

Glenberg, A. M., Gutierrez, T., Levin, J. R., Japuntich, S., & Kaschak, M. P. (2004). Activity and Imagined Activity Can Enhance Young Children's Reading Comprehension. *Journal of Educational Psychology, 96*(3), 424–436. doi:10.1037/0022-0663.96.3.424

Goldin-Meadow, S. (1999). The role of gesture in communication and thinking. *Trends in Cognitive Sciences, 3*(11), 419–429. doi:10.1016/S1364-6613(99)01397-2

Goldin-Meadow, S. (2005). *Hearing Gesture: How Our Hands Help Us Think*. Belknap Press.

Gooding, D. (1992). Putting agency back into experiment . In Pickering, A. (Ed.), *Science as Practice and Culture*. University of Chicago Press.

Gottschalk, T. H. (2009). *Distance education at a glance*. Retrieved from http://www.uiweb.uidaho.edu/eo/distglan

Grondlund, N. E. (1985). *Measurement and evaluation in teaching*. New York: Macmillan.

Grunbacher, P., Seyff, N., Briggs, R. O., In, H. P., Kitapci, H., & Port, D. (2007). Making every student a winner: The WinWin approach in software engineering education. *Journal of Systems and Software, 80*, 1191–1200. doi:10.1016/j.jss.2006.09.049

Guertin, L., Bodek, M., Zappe, S., & Kim, H. (2007). Questioning the student use of and desire for lecture podcasts. *Journal of Online Learning and Teaching, 3*(2), 133–141.

Gutwill, J. P., Fredericksen, J. R., & White, B. Y. (1999). Making their own connections: Students' understanding of multiple models in basic electricity. *Cognition and Instruction, 17*(3), 249–282. doi:10.1207/S1532690X-CI1703_2

Haghi, A. K., Mottaghitalab, V., & Akbari, M. (2008). *The Scholarship of Teaching Engineering: Some fundamental issues*. New York: IGI Global.

Hamada, M. (2008). Web-based environment for active computing learners. In O. Gervasi, & B. Murgante (Ed.), *International Conference on Computational Sciences and Its Applications* (pp. 516-529). Perugia: Spriger-Verlag Berlin.

Hamza-Lup, F. G., & Stefan, V. (2007). *Web 3D & Virtual Reality - Based Applications for Simulation and e-Learning*. Paper presented at the 2nd International Conference on Virtual Learning, ICVL, Constanta, Romania.

Harper, K. C., Chen, K., & Yen, D. C. (2004). Distance learning, virtual classrooms and teaching pedagogy in the Internet environment. *Technology in Society, 26*, 585–598. doi:10.1016/S0160-791X(04)00054-5

Harper, K. C., Chen, K., & Yen, D. C. (2004). Distance learning,virtual classrooms,and teaching pedagogy in the Internet environment. *Technology in Society, 26*, 585–598. doi:10.1016/S0160-791X(04)00054-5

Härtel, H. (1982). The electric circuit as a system: A new approach. *European Journal of Science Education, 4*(1), 45–55.

Harvey, R., Johnson, F., Marchese, A. J., Newell, J. A., Ramachandran, R. P., & Sukumaran, B. (1999). Improving the Engineering and Writing Interface: An Assessment of a Team-Taught Integrated Course. *ASEE Annual Meeting*, St. Louis, MO.

Hearst, M. (2000). The debate on automated essay grading. *IEEE Intelligent Systems, 15*(5), 22–37. doi:10.1109/5254.889104

Hejazi, A. (2007). *New wine-old bottles*. Retrieved from http://ictarticles.blogspot.com/2007/05/new-wine-old-bottles.html

Helander, M. G., & Emami, M. R. (2008). Engineering e-laboratories: Integration of remote access and e-collaboration. *International Journal of Engineering Education, 24*(3), 466–479.

Hesketh, R. P., Farrell, S., & Slater, C. S. (2003). An Inductive Approach to Teaching Courses in Engineering. 2003 ASEE Annual Conference, Session 2531, June 2003.

Hesketh, R. P., Jahan, K., & Marchese, A. J. (1997) Integrating Hands-on Education to Freshman Engineers at Rowan College. *1997 ASEE Zone 1 Spring Meeting.* West Point, NY, April, 1997.

Higgins, C., Hegazy, T., Symeonidis, P., & Tsintsifas, A. (2003). The CourseMarker CBA System: Improvements over Ceilidh. *Education and Information Technologies, 8*(3), 287–304. doi:10.1023/A:1026364126982

Hinchcliff, J. (2000). Overcoming the Anachronistic Divide: Integrating the Why into the What in Engineering Education. *Global Journal of Engineering Education, 4*(1), 13–18.

Hirsch, B. E., Thoben, K. D., & Hoheisel, J. (1998). Requirements upon human competencies in globally distributed manufacturing. *Computers in Industry, 36,* 49–54. doi:10.1016/S0166-3615(97)00097-3

Holton, D. L., Verma, A., & Biswas, G. (2008). Assessing student difficulties in understanding the behavior of AC and DC circuits. In *Proceedings of the 2008 ASEE Annual Conference.* Pittsburgh, PA.

Holton, D.L. (in press). How people learn with computer simulations. *Handbook of Research on Human Performance and Instructional Technology.* IGI Global.

Honan, W. (1999, January 27). High tech comes to the classroom: Machines that grade essay. *New York Times.*

Horton, W. (2001). *Designing Web-Based Training: How to Teach Anyone Anything Anywhere Anytime.* New York: John Wiley & Sons, Inc.

Hung, T. C., Wang, S. K., Tai, S. W., & Hung, C. T. (2007). An innovative improvement of engineering learning system using computational fluid dynamics concept. *Computers & Education, 48,* 44–58. doi:10.1016/j.compedu.2004.11.003

Hung, T., Liu, C., Hung, C., Ku, H., & Lin, Y. (2007, September 3-7). *The Establishment of an Interactive E-learning System for Engineering Fluid Flow and Heat Transfer.* Paper presented at the International Conference on Engineering Education – ICEE 2007, Coimbra, Portugal.

Hutchings, M., Hadfield, M., Horvath, G., & Lewarne, S. (2007). Meeting the challenges of active learning in web-based case studies for sustainable development. *Innovations in Education and Teaching International, 44*(3), 331–343. doi:10.1080/14703290701486779

Hutten, H., Stiegmaier, W., & Rauchegger, G. (2005). KISS—A new approach to self-controlled e-learning of selected chapters in Medical Engineering and other fields at bachelor and master course level. *Medical Engineering & Physics, 27,* 611–616. doi:10.1016/j.medengphy.2005.05.003

Hwang, F. (2004). NTNUJAVA Virtual Physics Laboratory Java Simulations in Physics: Enjoy the fun! Retrieved from http://www.phy.ntnu.edu.tw/ntnujava/index.php?PHPSESSID=61e16b595f51a79534ea86dd507e6fce&topic=23.msg156#msg156

IEEE. (2005). *1484.12.1: IEEE Standard for Learning Object Metadata.* Retrieved from http://ltsc.ieee.org/wg12/

Imperial-College-London. (2009). *Information & Communication Technologies ICT; An E-Learning Glossary.* Retrieved from http://www3.imperial.ac.uk/ict/services/teachingandresearchservices/elearning/aboutelearning/elearningglossary

IMS. (2009). *Innovation Adoption Learning.* Retrieved from http://www.imsglobal.org/specifications.html

Informa Telecoms & Media. (2008). 3G Americas: Number of mobile devices passes 4 billion mark. Retrieved June 15, 2009 from http://www.fiercewireless.com/story/3g-americas-number-mobile-devices-passes-4-billion-mark/2008-12-23?utm_medium=rss&utm_source=rss&cmp-id=OTC-RSS-FW0

Ion, D. (2007). Physics and Engineering. Retrieved from http://www.cs.sbcc.cc.ca.us/~physics/

Jahan, K., & Dusseau, R. A. (1998). Teaching Civil Engineering Measurements through Bridges. In *Proceedings of the 1998 Annual Conference of ASEE*, Seattle, Washington, June, 1998.

Jahan, K., & Dusseau, R. A. (1998). Water Treatment through Reverse Engineering. In *Proceedings of the Middle Atlantic Section Fall 1998 Regional Conference*, Washington D.C., November 6-7, 1998.

Jahan, K., Hesketh, R. P., Schmalzel, J. L., & Marchese, A. J. (2001). Design and Research Across the Curriculum: The Rowan Engineering Clinics. *International Conference on Engineering Education. August, 6 – 10, 2001 Oslo, Norway.*

Jahan, K., Marchese, A. J., Hesketh, R. P., Slater, C. S., Schmalzel, J. L., Chandrupatla, T. R., & Dusseau, R. A. (1998). Engineering Measurements and Instrumentation for a Freshman Class. In *Proceedings of the 1998 Annual Conference of ASEE*, Seattle, Washington, June, 1998.

Jara, C. A., Candelas, F. A., Torres, F., Dormido, S., Esquembre, F., & Reinoso, O. (2009). Real-time collaboration of virtual laboratories through the Internet. *Computers & Education, 52*(1), 126–140. doi:10.1016/j.compedu.2008.07.007

Jerrams-Smith, J., Soh, V., & Callear, D. (2001). Bridging gaps in computerized assessment of texts. In *Proceedings of the International Conference on Advanced Learning Technologies* (pp. 139-140). IEEE.

Johnson, D. W., Johnson, R. T., & Smith, K. A. (1991). *Active Learning: Cooperation in the College Classroom.* Edina, MN: Interaction Book Company.

Johnson, D. W., Johnson, R. T., & Smith, K. A. (1992). *Cooperative Learning: Increasing College Faculty Instructional Productivity. ERIC Digest.* Washington, D.C: The George Washington University, School of Education and Human Development.

Jonasson, J. (2005). *3D learning environment. Will it add the 3rd dimension to e-learning and teaching?* Paper presented at the 3rd International Conference on Multimedia and ICTs in Education, Cáceres, Spain.

Jones, R. (2009). *Physical ergonomic and mental workload factors of mobile learning affecting performance of adult distance learners: Student perspective.* Doctoral dissertation, University of Central Florida, 2009.

Jou, M., Chuang, C. P., Wu, D. W., & Yang, S. C. (2008). Learning robotics in interactive web-based environments by PBL. *IEEE Workshop on Advanced Robotics and Its Social Impacts* (pp. 206-211). Taipei: Nat. Taiwan Normal Univ. Press.

Jung, C. G. (1971). *Psychological Types.* Princeton, NJ: UA University Press.

Kaminski, P. C., Ferreira, E. P. F., & Theuer, S. L. (2004). Evaluating and improving the quality of an engineering specialization program through the QFD methodology. *International Journal of Engineering Education, 20*(6), 1034–1041.

Kao, W., Ye, J., Chu, M., & Su, C. (2009). Image quality improvement for electrophoretic displays by combining contrast enhancement and halftoning techniques. *IEEE Transactions on Consumer Electronics, 55*(1), 15–19. doi:10.1109/TCE.2009.4814408

Kaufman, D. B., Felder, R. M., & Fuller, H. (2000). Accounting for individual effort in cooperative learning teams. *Journal of Engineering Education, 89*(2), 133–140.

Kilby, T. (2009). *What is Web-Based Training?* Retrieved from http://www.webbasedtraining.com/primer_whatiswbt.aspx

Klem, A., & Connell, J. (2004). Relationships matter: Linking teacher support to student engagement and achievement. *The Journal of School Health, 74*(7), 262–273. doi:10.1111/j.1746-1561.2004.tb08283.x

Koksal, G., & Egitman, A. (1998). Planning and design of industrial engineering education quality. *Computers & Industrial Engineering, 35*(3-4), 639–642. doi:10.1016/S0360-8352(98)00178-8

Kolb, D. A. (1981). *Learning styles and disciplinary differences.* San Francisco, CA: Jossey-Bass.

Kolb, D. A. (1984). *Experiential Learning: Experience as the Source of Learning and Development* (1st ed.). Englewood Cliffs, NJ: Prentice-Hall.

Kolmos, A. (1996). Reflections on project work and Problem-Based Learning. *European Journal of Engineering Education, 21*(2), 141–148. doi:10.1080/03043799608923397

Kolmos, A., & Du, X. (2008). Innovation for engineering education–problem and project based learning (PBL) as an example . In Aung, W., Mecsi, J., Moscinski, J., Rouse, I., & Willmot, P. (Eds.), *Innovations 2008: World Innovations in Engineering Education and Research* (pp. 119–128). Arlington, VA: Begell House Publishing.

Koppány, A. (1997). Épületszerkezetek morfológiai rendszere.(Morphological System of Building Constructions). Research Working Paper (p.40). SZIF, Győr

Koppány, A. (2002). Building Diagnostics. Construction and City Related Sustainability Indicators (CRISP). EU 5. Framework. *CRISP Working Committee Conference, Budapest. ÉMI* (p. 8). Retrieved from http://crisp.cstb.fr /Database-Hungary)

Koppány, A., & Graf, H. (1985). *Mängel an Flachdacheindeckungen und ihre Untersuchungs-methoden* (pp. 135–137). Berlin: Bauzeitung.

Kozma, R., Russell, J., Jones, T., Marx, N., & Davis, J. (1996). The use of multiple, linked representations to facilitate science understanding . In Vosniadou, S., Glaser, R., De Corte, E., & Mandl, H. (Eds.), *International Perspectives on the Psychological Foundations of Technology-Based Learning Environments* (pp. 41–60). Hillsdale, NJ: Erlbaum.

Kukreti, A., & Zaman, M. (2008). *Virtual laboratory modules for undergraduate strength of materials course.* Paper presented at the Engineering Education, Loughbrough, London.

Kukulska-Hulme, A. (2002). Cognitive ergonomic and affective aspects of PDA use for learning. In

Kumar, A., & Labib, A. W. (2004). Applying quality function deployment for the design of a next-generation manufacturing simulation game. *International Journal of Engineering Education, 20*(5), 787–800.

Laham, D., & Foltz, P. W. (2000). The intelligent essay assessor . In Landauer, T. K. (Ed.), *IEEE Intelligent Systems.*

Lambert, G. (2004). *What is Computer Aided Assessment and how can I use it in my teaching? Learning & Teaching Enhancement Unit (LTEU).* Canterbury Christ Church University College.

Landauer, T. K., Foltz, P. W., & Laham, D. (1998). An introduction to latent semantic analysis. *Discourse Processes, 25.* Retrieved from http://lsa.colorado.edu/ papers/ dp1.LSAintro.pdf. doi:10.1080/01638539809545028

Larkey, L. S. (1998). Automatic essay grading using text categorization techniques. In *Proceedings of the 21st ACM/SIGIR (SIGIR-98)* (pp. 90-96). ACM.

Larkey, L. S. (2003). Email communication with author. 15th April. Mason, O. & Grove-Stephenson, I. (2002). Automated free text marking with paperless school. In M. Danson (Ed.), *Proceedings of the Sixth International Computer Assisted Assessment Conference, Loughborough University, Loughborough, UK.*

Lau, H., Mak, K., & Ma, H. (2006). IMELS: An e-learning platform for industrial engineering. *Computer Applications in Engineering Education, 14*(1), 53–63. doi:10.1002/cae.20067

Law, N., Pelgrum, W. J., & Plomp, T. (Eds.). (2008). *Pedagogy and ICT use in schools around the world: Findings from the IEA SITES 2006 study.* Hong Kong: CERC-Springer.

Le@rningFederation (2008). *Metadata Elements Comparison: Vetadata and ANZ-LOM.* Retrieved from http:// www.thelearningfederation.edu.au/verve/_resources/ ANZLOM-VETADATA-comparison-v1-0.pdf

Lee, J., & Lee, W. (2008). The relationship of e-Learner's self-regulatory efficacy and perception of e-Learning environmental quality. *Computers in Human Behavior, 24,* 32–47. doi:10.1016/j.chb.2006.12.001

Lee, J., Hong, N. L., & Ling, N. L. (2002). An analysis of students' preparation for the virtual learning environment. *The Internet and Higher Education, 4,* 231–242. doi:10.1016/S1096-7516(01)00063-X

Leonard, W. J., Dufresne, R. J., & Mestre, J. P. (1996). Using qualitative problem-solving strategies to highlight the role of conceptual knowledge in solving problems. *American Journal of Physics, 64*(12), 1495–1503. doi:10.1119/1.18409

Li, L. Y., & Wang, H. J. (2007). A new method for building web-based virtual laboratory. In H. Liu, B. Hu, X.W. Zheng, & H. Zhang (Ed.). *1st International Symposium on Information Technologies and Applications in Education* (pp. 555-558). Kunming: Shandong Normal Univ. Press.

Liang, J. S. (2009). Development for a web-based EDM laboratory in manufacturing engineering. *International Journal of Computer Integrated Manufacturing, 22*(2), 83–99. doi:10.1080/09511920801911019

Lindroos, K., Malmivuo, J., & Nousiainen, J. (2007). Web-based supporting material for biomedical engineering education. In T. Jarm, P. Kramar, & A. Zupanic (Ed.), *11th Mediterranean Conference on Medical and Biological Engineering and Computing* (pp. 1111-1114). Ljubljana: Springer-Verlag Berlin.

Ling, C., Gen-Cai, C., Chen-Guang, Y., & Chuen, C. (2003). *International Conference on Communication Technology proceeding, 2*, 1655-1661.

Littlejohn, A., Falconer, I., & Mcgill, L. (2008). Characterising effective eLearning resources. *Computers & Education, 50*, 757–771. doi:10.1016/j.compedu.2006.08.004

Liu, C. C., Hung, C. T., Hung, T. C., Pei, B. S., & Zhang, L. (2006). *The development of an innovative interactive e-learning system in computational thermal-hydraulics for engineers.* Paper presented at the EDMEDIA.

Liu, Y. (2009). Exploring Students' Perceptions toward Using Interactive Response System. *Learning Technology publication of IEEE Computer Society Technical Committee on Learning Technology (TCLT) . Learning Technology Newsletter, 11*(1-2), 29–32.

Mackey, T., & Ho, J. (2008). Exploring the relationship between web usability and students' perceived learning in web-based multimedia (WBMM) tutorials. *Computers & Education, 50*(1), 386–409. doi:10.1016/j.compedu.2006.08.006

Magnusson, S. J., Templin, M., & Boyle, R. A. (1997). Dynamic science assessment: A new approach for investigating conceptual change. *Journal of the Learning Sciences, 6*, 91–142. doi:10.1207/s15327809jls0601_5

Magoha, P. W., & Andrew, W. O. (2004). The global perspectives of transitioning to e-learning in engineering education. *World Transactions on Engineering and Technology Education, 3*(2), 205–210.

Malachowski, M. J. (2002). *ADDIE based five-step method towards instructional design.* Retrieved from http://fog.ccsf.cc.ca.us/~mmalacho/OnLine/ADDIE.html

Mamchur, C. (1990). *Cognitive Type Theory and Learning Style, Association for Supervision and Curriculum Development. A teacher's guide to cognitive type theory and learning style.* Alexandria, VA: Association for Supervision and Curriculum Development.

Mangina, E., & Kilbride, J. (2008). Utilizing vector space models for user modeling within e-learning environments. *Computers & Education, 51*, 493–505. doi:10.1016/j.compedu.2007.06.008

Mann, S. (2008). The problems of online collaboration for junior high school students: Can the Learning Activity Management System (LAMS) benefit students to learn via online learning? In L. Cameron & J. Dalziel (Eds.), *Proceedings of 3rd International LAMS & Learning Design Conference 2008: Perspectives on Learning Design* (pp. 81-86). Sydney: LAMS Foundation.

Marchese, A. J., Newell, J., Ramachandran, R. P., Sukumaran, B., & Schmalzel, J. L & Maraiappan, J. L. (1999). The Sophomore Engineering Clinic: An Introduction to the Design Process through a Series of Open Ended Projects. In *Proc. Conf. Amer. Soc. Eng. Edu*, Charlotte, NC.

Marchese, A. J., Ramachandran, R. P., Hesketh, R., Schmalzel, J., & Newell, H. L. (2003). The competitive assessment laboratory: Introducing engineering design via consumer product benchmarking. *IEEE Transactions on Education, 46*(1), 197–205. doi:10.1109/TE.2002.808216

Martins, L. L., & Kellermanns, F. W. (2004). A model of business school students' acceptance of a web-based course management system. *Academy of Management Learning & Education, 3*, 7–26.

Marton, F., & Pang, M. F. (2006). On some necessary conditions of learning. *Journal of the Learning Sciences, 15*(2), 192–220. doi:10.1207/s15327809jls1502_2

Mathias, L., Frantz, P., Brust, G., Montogery, J., Smith, V., Simmons, M., et al. (2009). *Polymer Science learning Center.* Retrieved from http://pslc.ws/macrog.htm.

McDermott, L. C., & van Zee, E. H. (1984). Identifying and addressing student difficulties with electric circuits. In Duit, R., Jung, W., & von Rhoneck, C. (Eds.), *Aspects of Understanding Electricity.* Kiel, Germany: Verlag, Schmidt, & Klaunig.

McQuiggan, S., Lee, S., & Lester, J. (2006). Predicting User Physiological Response for Interactive Environments: An Inductive Approach. In *Proceedings of the Second Conference on Artificial Intelligence and Interactive Entertainment.* Marina del Rey, California.

Meyer, S. A., & Morgenstern, M. A. (2005). Small scale biodiesel production: A laboratory experience for General Chemistry and Environmental Science students. *Chemical Educator, 10,* 1–3.

Ming, P. Y., Mikhailov, A. A., & Kuan, T. L. (2000). Intelligent essay marking system. In C. Cheers (Ed.), *Learners Together, Feb. 2000, NgeeANN Polytechnic, Singapore.* http://ipdweb.np.edu.sg/lt/feb00/ intelligent_essay_marking.pdf

Mitchell, T., Russel, T., Broomhead, P., & Aldridge, N. (2002). Towards robust computerized marking of free-text responses. In M. Danson (Ed.), *Proceedings of the Sixth International Computer Assisted Assessment Conference, Loughborough University, Loughborough, UK.*

Mokros, J. R., & Tinker, R. F. (1987). The impact of microcomputer-based labs on children's ability to interpret graphs. *Journal of Research in Science Teaching, 24*(4), 369–383. doi:10.1002/tea.3660240408

Möller, K. (1945). *Építési hibák és elkerülésük. (Building defects and prevention of defects)* (p. 32). Budapest: Egyetemi Nyomda.

Molnárka, G. (2000) The methodology in visual examination in building pathology. *Hungarian Electronic Journal of Science, 10.* Retrieved from www.sze.hu

Molnárka, G. (2006). *Hallgatói mobilitási portál* (Mobility Portal for Students). Retrieved March 24, 2009, from http://www.sze.hu/leonardo

Monaghan, J. M., & Clement, J. (2000). Algorithms, visualization, and mental models: High school students' interactions with a relative motion simulation. *Journal of Science Education and Technology, 9*(4), 311–325. doi:10.1023/A:1009480425377

Monahan, T., McArdle, G., & Bertolotto, M. (2008). Virtual reality for collaborative e-learning. *Computers & Education, 50,* 1339–1353. doi:10.1016/j.compedu.2006.12.008

Moreno, L., Gonzalez, C., Castilla, I., Gonzalea, E., & Sigut, J. (2007). Applying a constructivist and collaborative methodological approach in engineering education. *Computers & Education, 49,* 891–915. doi:10.1016/j.compedu.2005.12.004

Morrison, Don. (2003) The Search for the Holy Recipe. Retrieved June 13, 2009 from http://www.elearningpost.com/archives/2003_04. asp

Morse, A., & Suktrisul, S. (2006). Introducing eLearning into Secondary Schools in Thailand. *Special Issue of the International Journal of the Computer, the Internet and Management, 14,* 81-85.

Motiwalla, L. F. (2007). Mobile learning: A framework and evaluation. *Computers & Education, 49,* 581–596. doi:10.1016/j.compedu.2005.10.011

Myers, L. B., & McCaulley, M. H. (1985). *Manual-Guide to the Development and Use of the Myers-Briggs Indicator.* Palo Alto, CA: Consulting Psychologists Press.

Nedic, Z., & Machotka, J. (2006). Interactive electronic tutorials and web based approach in engineering courses. In V. Uskov (Ed.), *5th IASTED International Conference on Web-based Education* (pp. 243-248). Puerto Vallarta: Acta Press.

Nelasco, S., Arputtharaj, A. N., & Alwinson, E. G. (2007). ELearning for Higher Studies of India. *Special Issue of the International Journal of the Computer . The Internet and Management, 15,* 161–167.

Nelson, T. O. (1984). A comparison of current measures of accuracy of feeling-of-knowing predictions. *Psychological Bulletin, 95,* 109–133. doi:10.1037/0033-2909.95.1.109

Nemirovsky, R., & Monk, S. (2000). "If you look at it the other way…": An exploration into the nature of symbolizing . In Cobb, P. (Eds.), *Symbolizing and communicating in mathematics classrooms: Perspectives on Discourse, Tools, and Instrumental Design* (pp. 177–221). New Jersey: Erlbaum.

Nersessian, N. J. (2002). Maxwell and "the Method of Physical Analogy": Model-based reasoning, generic abstraction, and conceptual change. In D. Malament (Ed.), Essays in the History and Philosophy of Science and Mathematics (pp. 129-166). Lasalle, Il: Open Court.

Nicoletto, G., Anzelotti, G., & Riva, E. (2008). Mesomechanic strain analysis of twill-weave composite lamina under unidirectional in-plane tension. *Composites. Part A, Applied Science and Manufacturing, 39*(8), 1294–1301. doi:10.1016/j.compositesa.2008.01.006

Nicoletto, G., Anzelotti, G., & Riva, E. (2009). Mesoscopic strain fields in woven composites: Experiments vs. finite element modeling. *Optics and Lasers in Engineering, 47*(3-4), 352–359. doi:10.1016/j.optlaseng.2008.07.009

Nicoletto, G., Marin, T., Anzelotti, G., & Roncella, R. (in press). Application of high magnification digital image correlation technique to micromechanical strain analysis. *Strain.*

Noori, M. (2003). *Traditional education or learning with computer.* Paper presented at the Virtual University Conference at Kashan Payam-e Noor College, Kashan.

Nor Azan, M. Z. (2007). Learning Activities Management System (SPAP). In *Proceedings of International Conference on Electrical Engineering and Informatics, Institut Teknologi Bandung, Indonesia.*

Núñez, R. E., Edwards, L. D., & Matos, J. F. (1999). Embodied cognition as grounding for situatedness and context in mathematics education. *Educational Studies in Mathematics, 39*(1-3), 45–65. doi:10.1023/A:1003759711966

Ogot, M., & Okudan, G. (2007). A student-centered approach to improving course quality using quality function deployment. *International Journal of Engineering Education, 23*(5), 916–928.

Olivier, B. A. (2003). Learning content interoperability standards . In Littlejohn, A. (Ed.), *Reusing Online Resources: A Sustainable Approach to eLearning. London: Kogan Page.* Liber, O.

Oprean, C. (2006). The Romanian contribution to the development of Engineering Education. *Global Journal of Engineering Education, 10,* 45–50.

Osguthorpe, R. T., & Graham, C. R. (2003). Blended learning environments: Definitions and directions. *The Quarterly Review of Distance Education, 4*(3), 227–233.

Overview of e3L. (2005). Retrieved from http://e3learning.edc.polyu.edu.hk/main.htm

Padilla-Melendez, A., Garrido-Moreno, A., & Aguila-Obra, A. R. (2008). Factors affecting e-collaboration technology use among management students. *Computers & Education, 51,* 609–623. doi:10.1016/j.compedu.2007.06.013

Page, E. B. (1968). The Use of the Computer in Analyzing Student Essays. *International Review of Education, 14,* 210–225. doi:10.1007/BF01419938

Page, E. B. (1994). New computer grading of student prose, using modern concepts and software. *Journal of Experimental Education, 62*(2), 127–142.

Page, E. B. (1996). *Grading essay by computer: Why the controversy?* Handout for NCME Invited Symposium.

Palmer, J., Williams, R., & Dreher, H. (2002). Automated essay grading system applied to a first year university subject- How can we do it better. In *Proceedings of the Informing Science and IT Education (InSITE) Conference, Cork, Ireland* (pp. 1221-1229).

Pang, M. F., & Marton, F. (2005). Learning theory as teaching resource: Enhancing students' understanding of economic concepts. *Instructional Science, 33*(2), 159–191. doi:10.1007/s11251-005-2811-0

Popescu, S., Brad, S., & Popescu, D. (2006). The engineering study program as a customized product: barriers and directions for intervention, In E. Westkaemper (Ed.), 1st CIRP International Seminar on Assembly Systems, (pp. 289-294). Stuttgart: IFF Univ. Stuttgart.

Proceedings of 2002 European Workshop on Mobile and Contextual Learning. University of Birmingham.

Pryor, C. R., & Bitter, G. G. (2008). Using multimedia to teach inservice teachers: Impacts on learning, application, and retention. *Computers in Human Behavior, 24,* 2668–2681. doi:10.1016/j.chb.2008.03.007

Pulichino, (2006). The trends in blended learning research report. Santa Rosa, CA: The eLearning Guild.

Purcell-Roberston, R. M., & Purcell, D. F. (2000). *Interactive distance learning. Distance Learning Technologies: Issues, Trends and Opportunities.* Hershey, PA: IGI Global.

Pyrz, R., Olhoff, N., Lund, E., & Thomsen, O. T. (2003). Mechanics for the future. Retrieved from http://me.aau.dk/GetAsset.action?contentId=1987778&assetId=3356989

Raharjo, H., Xie, M., Goh, T. N., & Brombacher, A. C. (2007). A methodology to improve higher education quality using the quality function deployment and analytic hierarchy process. *Total Quality Management & Business Excellence, 18*(10), 1097–1115. doi:10.1080/14783360701595078

Ramachandran, R. P., Schmazel, J., & Mandayam, S. (1999). Engineering Principles of an Electric Toothbrush. ASEE annual conference and Exhibition, Charlotte, North Carolina, Session 2253, June 20-23, 1999.

Ramasundaram, V., Grunwald, S., Mangeot, A., Comerford, N. B., & Bliss, C. M. (2005). Development of an environmental virtual field laboratory. *Computers & Education, 45,* 21–34. doi:10.1016/j.compedu.2004.03.002

Reiner, M. (2000). Thought experiments and embodied cognition . In Gilbert, J. K., & Boulter, C. J. (Eds.), *Developing Models in Science Education* (pp. 157–176). Netherlands: Kluwer Academic Publishers.

Reiner, M., Slotta, J. D., Chi, M. T. H., & Resnick, L. B. (2000). Naïve physics reasoning: A commitment to substance-based conceptions. *Cognition and Instruction, 18*(1), 1–34. doi:10.1207/S1532690XCI1801_01

Rittle-Johnson, B., & Star, J. (2007). Does comparing solution methods facilitate conceptual and procedural knowledge? An experimental study on learning to solve equations. *Journal of Educational Psychology, 99*(3), 561–574. doi:10.1037/0022-0663.99.3.561

Rittle-Johnson, B., & Star, J. (in press). Compared to what? The effects of different comparisons on conceptual knowledge and procedural flexibility for equation solving. *Journal of Educational Psychology.*

Rizzotti, S., & Burkhart, H. (2006). Web-based test and assessment system: Design principles and case study. In V. Uskov (Ed.), *5th IASTED International Conference on Web-based Education* (pp. 37-42). Puerto Vallarta: Acta Press.

Rogers, E. M. (2005). *Diffusion of innovations.* Glencoe: Free Press.

Rojko, A., Hercoq, D., & Jezemik, K. (2008). Educational aspects of mechatronic control course design for collaborative remote laboratory. In IEEE (Ed.), *IEEE 13th International Power Electronics and Motion Control Conference* (pp. 2349-2353). Poznan: Poznan Univ. Press.

Romero, M., & Wareham, J. (2009). Just-in-time mobile learning model based on context awareness information. *Learning Technology publication of IEEE Computer Society Technical Committee on Learning Technology (TCLT) . Learning Technology Newsletter, 11*(1-2), 4–6.

Rossett, A., Douglis, F., & Frazee, R. V. (2003). Strategies for building blended learning. *Learning Circuits.* Retrieved June 13, 2009 from http://www.astd.org/LC/2003/0703_rossett.htm

Roth, W. M., & Lawless, D. V. (2002). Scientific investigations, metaphorical gestures, and the emergence of abstract scientific concepts. *Learning and Instruction, 12*, 285–304. doi:10.1016/S0959-4752(01)00023-8

Rudner, L. M., & Liang, T. (2002). Automated essay scoring using Bayes' Theorem. *The Journal of Technology . Learning and Assessment, 1*(2), 3–21.

Rugarcia, A., Felder, R. M., Woods, D. R., & Stice, J. E. (2000). The future of engineering education I: a vision for a new century. *Chemical Engineering Education, 349*(1), 16–25.

Ryder, M. (2008). Instructional design models. Retrieved from http://carbon.cudenver.edu/~mryder/ itc_data/idmodels.html#modern

Saaty, T. L., & Vargas, L. G. (2001). *Models, methods, concepts & applications of the analytic hierarchy process.* Secaucus, NJ: Springer.

Salihbegovic, A., & Tanovic, O. (2008). Internet based laboratories in engineering education. In V. Luzar, V. Dobric, & Z. Bekic (Ed.), *30th International Conference on Information Technology Interfaces* (pp. 163-170). Cavtat: Zagreb Univ. Press.

Salmon, W. C. (1998). *Causality and Explanation.* Oxford University Press. doi:10.1093/0195108647.001.0001

Saygin, C., & Kahraman, F. (2004). A Web-based programmable logic controller laboratory for manufacturing engineering education. *International Journal of Advanced Manufacturing Technology, 24*(7-8), 590–598. doi:10.1007/s00170-003-1787-7

Schmalzel, J. L., Marchese, A. J., Mariappan, J., & Mandayam, S. (1998). The Engineering Clinic: A Four-Year Design Sequence. *2nd Annual Conference of National Collegiate Inventors and Innovators Alliance,* Washington, DC.

Schnotz, W., & Bannert, M. (2003). Construction and interference in learning from multiple representation. *Learning and Instruction, 13*(2), 141–156. doi:10.1016/S0959-4752(02)00017-8

Schwartz, D., & Bransford, J. (1998). A time for telling. *Cognition and Instruction, 16*(4), 475–522. doi:10.1207/s1532690xci1604_4

Schwartz, D., Biswas, G., Bransford, J., Bhuva, B., Balac, T., & Brophy (2000). Computer tools that link assessment and instruction: Investigating what makes electricity hard to learn. In S. Lajoie (Ed.), *Computers as Cognitive Tools, Vol. II* (pp. 273-307). Mahwah, NJ: Lawrence Erlbaum Associates.

Science (1941). The Engineers' Council for Professional Development. *Science, 94*(2446), 456.

Sessink, O., van der Schaaf, H., Beeftink, H., Hartog, R., & Tramper, J. (2007). Web-based education in bioprocess engineering. *Trends in Biotechnology, 25*(1), 16–23. doi:10.1016/j.tibtech.2006.11.001

Shee, D. Y., & Wang, Y. (2008). Multi-criteria evaluation of the web-based e-learning system: A methodology based on learner satisfaction and its applications. *Computers & Education, 50*, 894–905. doi:10.1016/j.compedu.2006.09.005

Shih, W. C., Tseng, S. S., & Yang, C. T. (2008). Wiki-based rapid prototyping for teaching-material design in e-Learning grids. *Computers & Education, 51*, 1037–1057. doi:10.1016/j.compedu.2007.10.007

Siau, K., Sheng, H., & Nah, F. F.-H. (2006). Use of a classroom response system to enhance classroom interactivity. *IEEE Transactions on Education, 49*(3), 398–403. doi:10.1109/TE.2006.879802

Siemens, G. (2002). *Instructional design in e-learning.* Retrieved from http://www.elearnspace.org/ Articles/InstructionalDesign.htm

Singh, H. (2003). Building effective blended learning programs. *Educational Technology, 43*(6), 51–54.

Sirinaruemitr, P. (2004). Trends and forces for eLearning in Thailand. *International Journal of The Computer, the Internet and Management, 12*, 132-137.

Siritongthaworn, S., & Krairit, D. (2004). Use of Interactions in E-learning: A Study of Undergraduate Courses in Thailand. *International Journal of The Computer . The Internet and Management, 12*, 162–170.

Smith, J. M. (2001) Blended Learning An old friend gets a new name. Retrieved June 13, 2009 from http://www.gwsae. org/Executiveupdate/2001/March/blended

Smith, R. J. (2009). *Engineering*. Retrieved from http://www.britannica.com/EBchecked/topic/187549/engineering

Sperka, M. (2004). *Web 3D and new forms of human computer interaction*. Paper presented at the 2ed International Symposium of Interactive Media Design, Istanbul, Turkey.

Suliman, S. M. A. (2006). Application of QFD in engineering education curriculum development and review. *International Journal of Continuing Engineering Education and Lifelong Learning, 16*(6), 482–492. doi:10.1504/IJCEELL.2006.011892

Sun, P., Cheng, H. K., Lin, T., & Wang, F. (2008). A design to promote group learning in e-learning: Experiences from the field. *Computers & Education, 50*, 661–677. doi:10.1016/j.compedu.2006.07.008

Tabakov, S. (2008). e-Learning development in medical physics and engineering. *Biomedical Imaging and Intervention Journal, 4*(1), e27. doi:10.2349/biij.4.1.e27

Takago, D., Matsuishi, M., Goto, H., & Sakamoto, M. (2007). Requirements for a web 2.0 course management system of engineering education. In IEEE Computer Society (Ed.), *9th IEEE International Symposium on Multimedia* (pp. 435-440). Taichung: Asia Univ. Press.

Tan, A. C. C., Tang, T., & Paterson, G. (2008). *Web-based Remote Vibration Experimental Laboratory.* Paper presented at the Engineering Education, Loughbrough, London.

Thompson, C. (2001). Can computers understand the meaning of words? Maybe, in the new of latent semantic analysis. *ROB Magazine*. Retrieved from http://www.vector7.com/client_sites/ROB_preview/html/thompson.html

Thornton, P., & Houser, C. (2005). Using mobile phones in English education in Japan. *Journal of Computer Assisted Learning, 21*, 217–228. doi:10.1111/j.1365-2729.2005.00129.x

Toral, S., Barrero, F., & Martinez-Torres, M. (2007). Analysis of utility and use of a web-based tool for digital signal processing teaching by means of a technological acceptance model. *Computers & Education, 49*(4), 957–975. doi:10.1016/j.compedu.2005.12.003

Toth, P. (2008). *New Method in Quality Assurance of Electronic-based Teaching Materials.* Paper presented at the Engineering Education, Loughbrough, London.

Troha, F. J. (2002) Bulletproof instructional design: A model for blended learning. *USDLA Journal, 16*(5).

Tsai, S., Yeh, T. L., Lee, C. K., & Wang, N. C. (2008). *Teaching Strategies for Design Realization in Engineering Design Education.* Paper presented at the Engineering Education, Loughbrough, England.

UI. (2004). *Engineering Outreach Distance Education.* Retrieved from http://www.uidaho.edu/evo

UNESCO. (2006a). *UIS statics in brief-India*. Retrieved from http://stats.uis.unesco.org/unesco/TableViewer/document.aspx?ReportId=121&IF_Language=eng&BR_Country=3560

Upitis, G., Maziais, J., & Rudnevs, J. (2008). *Problen-Oriented E-tools for Studies in Mechanical Engineering.* Paper presented at the Engineering Education, Loughborough, England.

Vadillo, M. A., & Matute, H. (2009). Learning in virtual environments: Some discrepancies between laboratory- and Internet-based research on associative learning. *Computers in Human Behavior, 25*, 402–406. doi:10.1016/j.chb.2008.08.009

Valenti, S., Cucchiarelli, A., & Panti, M. (2000). Web based assessment of student learning . In Aggarwal, A. (Ed.), *Web-based Learning & Teaching Technologies, Opportunities and Challenges* (pp. 175–197). Idea Group Publishing.

Valenti, S., Cucchiarelli, A., & Panti, M. (2002). Computer based assessment systems evaluation via the ISO9126 quality model. *Journal of Information Technology Education, 1*(3), 157–175.

Valiathan, P. (2003) Blended learning models. *Learning Circuits, ASDT Online Magazine.* Retrieved June 13, 2009 from http://www. learningcircuits.com/2002/aug2002/valiathan.html

van Raaij, E. M., & Schepers, J. J. L. (2008). The acceptance and use of a virtual learning environment in China. *Computers & Education, 50,* 838–852. doi:10.1016/j.compedu.2006.09.001

Vargas, H., Sanchez, J., Duro, N., Dormido, R., Farias, G., & Dormido, S. (2008). A systematic two-layer approach to develop web-based experimentation environments for control engineering education. *Intelligent Automation and Soft Computing, 14*(4), 505–524.

Vate-U-Lan, P. (2007). Readiness of eLearning connectivity in Thailand. *Special Issue of the International Journal of the Computer, the Internet and Management, 15,* 21-27.

Wang, X., Dannenhoffer, J., Davidson, B., & Spector, M. (2005). Design issues in a cross-institutional collaboration on a distance education course. *Distance Education, 26*(3), 405–423. doi:10.1080/01587910500291546

Wankat, P. C., Felder, R. m., Smith, K. S., & Oreovicz, F. S. (2002). *The scholarship of teaching and learning in engineering.* Washangton: AAHE/Carnegie Foundation for the Advancement of Teaching.

Wild, R. H., Griggs, K. A., & Downing, T. (2002). A framework for e-learning as a tool for knowledge management. *Industrial Management & Data Systems, 102*(7), 371–381. doi:10.1108/02635570210439463

Willis, B. (2008). *Instance education in glance.* Retrieved from http://www.uiweb.uidaho.edu/eo/dist3.html

Willis, B. D. (1993). *Distance Education- A Practical Guide.* Englewood Cliffs: Educational Technology Publications.

Wintermantel, K. (1999). Process and product engineering Ð achievements, present and future challenges. *Chemical Engineering Science, 54,* 1601–1620. doi:10.1016/S0009-2509(98)00412-6

Witsawakiti, N., Suchato, A., & Punyabukkana, P. (2006). Thai language E-training for the hard of hearing. *Special Issue of the International Journal of the Computer . The Internet and Management, 14,* 41–46.

Xuelian, H. (2008). Research and practice on web-based distance education of master of engineering. In L. Maoqing (Ed.), *3rd International Conference on Computer Science & Education* (pp. 1511-1514). Kaifeng: Xiamen Univ. Press.

Yeo, R. (2008). Brewing service quality in higher education: Characteristics of ingredients that make up the receipe. *Quality Assurance in Education, 16*(3), 266–286. doi:10.1108/09684880810886277

YilDirim. S. (2009). *Web-Based Training: Design and Implementation Issues.* Retrieved from http://ocw.metu.edu.tr/informatics-institute/web-based-training-design-and-implementation

Yu, C. Y., & Shaw, D. T. (2006). *Fostering Creativity and Innovation in Engineering Students.* Paper presented at the 2006 International Mechanical Engineering Education Conference, joined Conference by ASME and CMES, Beijing, China.

Zenger, J., & Uehlein, C. (2001) Why Blended Will Win. *T+D, 55*(8), 54-62.

Zhiting, Z. (2004). The development and applications of elearning technology standards in China. *International Journal of The Computer, the Internet and Management, 12,* 100-104.

Zio, E. (2009). Reliability engineering: Old problems and new challenges. *Reliability Engineering & System Safety, 94,* 125–141. doi:10.1016/j.ress.2008.06.002

Zohar, A., & Ginossar, S. (1998). Lifting the taboo regarding teleology and anthropomorphism in biology education: Heretical suggestions. *Science Education, 82,* 679–697. doi:10.1002/(SICI)1098-237X(199811)82:6<679::AID-SCE3>3.0.CO;2-E

Zwicky, F. (1966). *Entdecken* (p. 174). München, Zürich: Erfinden, Forschen im Morphologischen Weltbild, Droemer/Knaur.

About the Contributors

Donna Russell is an Assistant Professor at the University of Missouri-Kansas City and co-owner of Arete' Consulting, LLC. She has a bachelors and masters degree in Education specializing in instructional design. Her PhD is in Educational Psychology with an emphasis on cognition and technology. She is Co-PI on the National Science Foundations grant, Achieving Recruitment, Retention and Outreach With STEM, developing science, technology and engineering programs for urban high school students. She is chair of the Problem-Based Education Special Interest Group committee for the American Education Researchers Association. She has published several articles and book chapters on virtual learning including Online Professional Development for Educators; A Case Study Analysis using Cultural Historical Activity Theory, Implementing an Innovation Cluster in Educational Settings to Develop Constructivist-based Learning Environments, Transformation in an Urban School: Using Systemic Analysis to Understand an Innovative Urban Teacher's Implementation of an Online Problem-Based Unit, Group Collaboration in an Online Problem-based University Course.. In Creativity and Problem-Based Learning. and Understanding the Effectiveness of Collaborative Activity in Online Professional Development with Innovative Educators through Inter-Subjectivity in Information and Communication Technology for Enhanced Education and Learning: Advanced Applications and Development.

A.K.Haghi holds a BSc in urban and environmental engineering from University of North Carolina (USA), a MSc in mechanical engineering from North Carolina A&T State University (USA), a DEA in applied mechanics, acoustics, and materials from Université de Technologie de Compiègne (France), and a PhD in engineering sciences from Université de Franche-Comté (France). He is the author and editor of 35 books, as well as 550 papers in various journals and conference proceedings. Dr. Haghi has received several grants, consulted for a number of major corporations, and is a frequent speaker to national and international audiences. Since 1983, he served as professor in several universities. He is currently Editor-in-chief of International Journal of Chemoinformatics and Chemical Engineering and on the Editorial Boards of many International journals.

* * *

Attila Somfai works as associate professor at Szechenyi István University in Győr, Hungary. He graduated from Budapest University of Technology and Economics, Department of Public Building Design in 1996 and spent the next 13 years as university lecturer and architect with an interval during that time as a PhD researcher. In his borntown Győr he teaches courses and authors publications on computer aided architectural design, building constructions and regional and urban planning studies.

Somfai is a member of the Public Body of the Hungarian Academy of Sciences and also a member of the International Council for Research and Innovation in Building and Construction.

Dr. Noroozi has 5 years of experience in bench scale research and chemical analysis with various analytical instruments in textile and environmental chemistry. His research interests include: processes for textile wastewater treatment such as ozone, electrochemical and adsorption by different adsorbents in single and multicomponent systems, modeling of adsorption systems (equilibrium and dynamics), use of biopolymers (chitin and chitosan) in production process of textiles and production of activated carbon, activated carbon fiber and nanofiber for controlling the quallity of water pollutants. At the University of Guilan, Dr Noroozi teaches graduate courses in the field of advanced topics in textile chemistry and principles of textile environmental problems. He also teaches undergraduate courses on textile waste-water treatment systems, textile fibre chemistry, and physical chemistry of dye adsorption. Now he is director of industry relation department at the university of Guilan.

Prof. Sorial has 25 years of experience in bench scale and pilot scale research and chemical analysis with various analytical instruments. His research interests include: air biofiltration, electrochemical processes for destruction of organic contaminants, activated carbon adsorption (micropollutant removal from drinking water, adsorption of micropollutants by activated carbon and alternative adsorbents, characterization of adsorbent materials and natural organic matter, interactions between micropollutants and humic substances), modeling of adsorption systems (equilibrium and dynamics), remediation of contaminated soils, modeling of mercury speciation and chemical interactions in sediments and aquatic systems, development of analytical methods for analysis, and development of protocols for US EPA in effectiveness of surfactants on oil spills. At the University of Cincinnati, Professor Sorial teaches graduate courses in the field of Advanced Topics in Environmental Chemistry, Chemical Principles of Environmental Systems, and a graduate laboratory course on Environmental Instrumentation. He also teaches an undergraduate course on Environmental Material Balances and coordinates in the under-graduate Capstone Course "Integrated Design Sequence".

Siddhartha Ghosh works as an Associate Professor in the Dept. of Computer Science & Engineering (CSE) in G. Narayanamma Institute of Technology and Science (GNITS) for women, Shaikpet, Hyder-abad, Andhra Pradesh. Pursuing Ph.D. in the area of Natural Language Processing (NLP) and Artificial Intelligence (AI) from Dept. of CSE, Osmania University, Hyderabad. M.Tech. CSE, from Unversity of Hyderabad. Having about 10 years of Teaching Experience in B.Tech. & M.Tech. Computer Science courses and about two years of Industrial Experience in Computer Aided Design. Have many national and international publications. Attended and also conducted many workshops, seminars. Have worked as Dept. Placement Coordinator, ISO Internal Auditor, NBA Task force member for college. Given several guest lectures. Was responsible for development and maintenance of college websites. Member of IBM Academic Initiative Program. Have guided many B.Tech., M.Tech. projects and one M.Phil. student. Also attached with corporate training. Areas of interest NLP – Text Processing, Artificial Intelligence, Machine Learning, Web Technologies, Computing in Indian Languages, Linux., E-commerce.

Rochelle Jones is a Systems Engineer with Lockheed Martin Corporation. She obtained her Ph.D. in Industrial Engineering with a concentration in ergonomics and human factors from the University of Central Florida. Prior to her doctoral studies, she received a B.S. and M.S. in Systems Engineering from George Mason University. Her primary research interests are in the areas of distance education and mobile learning. Her research is centered on mobile learning and emerging distributed education technologies, specifically methods to enhance the user experience through the use of mobile devices and human engineering principles. Aside from academic and professional endeavors, she has mentored middle and high school aged girls regarding pursuing education and career goals.

Chandre Butler is a burgeoning Doctoral Candidate at the University of Central Florida. He has been awarded multi-year grant awards with the National Science Foundation (NSF) GK-12 science initiative. He has also published inquiry-based science lessons for middle and high school students in compliance with NSF program requirements. In addition, he is a recipient of the multi-year FEF-McKnight Doctoral Fellowship award. Some of his research interests include determining and defining the factors that impact human cognitive performance in game-based and video media environments. The method of modeling human performance he specializes in is Fuzzy Mathematical Theory, which has proven a viable approach within current human factors research. Aside from professional endeavors, Chandre's community service efforts include founding a pre-college chapter of the National Society of Black Engineers, servicing in the Orlando metro area, in addition to providing continued volunteer IT support and consultation for the Hi-Tech Tutoring Center partner.

Pamela Bush has been extremely involved in service throughout her career. She has served on department and university committees, as well as national and international organizations. In 2009, she served as a plenary speaker at the IEEE SOFA Conference, Romania. She has served as a member of the Women in Engineering Program Advocates Network (WEPAN). She has been involved in professional societies serving as an advisor, member of conference committees and serving on committees at the New University of Lisbon where she delivered a keynote address and workshops. Through her many service activities she has placed an emphasis on community service to the surrounding community, and national efforts to promote an interest in engineering education among youth. These activities have enhanced her technical reputation and have resulted in national recognition for her commitment to community service. She is nationally sought as a motivational speaker and advocate for engineering education.

Douglas Holton is an Assistant Professor at Utah State University. Dr. Holton has developed a number of educational software applications and tools. He is actively research engineering education, conceptual change, online education, and the design of interactive learning environments.

Amit Verma is an Assistant Professor at Texas A&M University-Kingsville. Besides conducting research in engineering education, Dr. Verma is also very actively involved in cutting-edge nanotechnology research. This involves modeling and characterization of nano-materials, including carbon nanotubes and silicon nanowires, and devices composed of those materials.

Kauser Jahan a Professor of Civil and Environmental Engineering in the College of Engineering at Rowan University. She completed her Ph.D. studies in the Department of Civil and Environmental Engineering at the University of Minnesota, Minneapolis in 1993. Dr. Jahan is a registered Professional Civil Engineer in Nevada and is actively involved in environmental engineering education and outreach for K-12 students/educators. Her research interests include pollution prevention, alternate energy and teaching pedagogy. She has been recognized as an outstanding educator by various professional organizations.

Jess W. Everett is a Professor of Civil and Environmental Engineering in the College of Engineering at Rowan University. Dr. Everett is a registered Professional Civil Engineer and is actively involved in environmental research and education. Dr. Everett received B.S.E., M.S., and Ph.D degrees in Civil and Environmental Engineering from Duke University in 1984, 1986, and 1991, respectively.

Stephanie Farrell is an Associate Professor of Chemical Engineering at Rowan University. Prior to joining Rowan in 1998, she was a faculty member in Chemical Engineering at Louisiana Tech University. Her educational efforts currently focus on the development of innovative laboratory and classroom materials for sustainable engineering and biomedical systems. Stephanie received her Ph.D. in Chemical Engineering from New Jersey Institute of Technology. She also holds B.S. and M.S. Degrees from the University of Pennsylvania and Stevens Institute of Technology.

Ying Tang received the B.S. and M.S. degrees from the Northeastern University, P. R. China, in 1996 and 1998, respectively, and Ph. D degree from New Jersey Institute of Technology, Newark, NJ, in 2001. She is currently an Associate Professor of Electrical and Computer Engineering at Rowan University, Glassboro, NJ. Her research interests include modeling and scheduling of computer-integrated systems, Petri nets and applications, Artificial Intelligence, Reconfigurable systems design, and System-on-Chip education.

Hong Zhang obtained his PhD from Univ. of Pennsylvania and is an associate professor of Mechanical Engineering at Rowan University. He has years of experience on robotics, control, embedded system, and electro-mechanical design. He had also taught Mechatronics, Machine Design and Mechanical Design for years.

Angela Wenger, Executive Vice President & COO at the New Jersey Academy of Aquatic Sciences. Ms. Wenger has extensive science teaching and research experience and holds B.S. and M.S. degrees in biology (specialty: physiology) and marine science (specialty: malacology and aquaculture). During her tenure with the Academy, she has been involved with all facets of the organization, particularly exhibit design; public programming; family and classroom workshops; staff and teacher training courses; and programs for underserved audiences. Ms. Wenger is a member of the Committee on Conference Planning of the Association of Science and Technology Centers (ASTC). Ms. Wenger is a co-author of *Family Learning in Museums: The PISEC Perspective* and *Staying Connected.*

Majid Noori graduated from the University of Maine in 1987 with a Ph.D. degree in Physics. Dr. Noori taught Physics and Astronomy at Towson University in Maryland for ten yours followed by a five year appointment at Seton Hall University in South Orange, New Jersey. He is currently an Assistant professor of Physics/Engineering at Cumberland County College. During his academic career, Dr. Noori has taught many graduate and under graduate courses in physics and astronomy and several undergraduate courses in Engineering technology. His current research activities at CCC include promotion of science and engineering among minority high school students with participation in a Cumberland County College grant for STEM careers.

Bruno Moreira Trigo is a Computing Engineering student at Escola Politécnica da Universidade de São Paulo, Brazil. He develops under-graduation research on Computer Simulations and Auto-trade software programs for the financial market. He spent the last two years working at the Pedagogical Orientation Group at Escola Politécnica.

Giuliano Salcas Olguin is a Education Coordinator at Escola Politécnica da Universidade de São Paulo – Brazil. He received Physics Teaching degree (2002) and a MSc in Science Education (2005) from Instituto de Física da Universidade de São Paulo. Since 2006, he has been active as a researcher in the field of engineering education. The central focus of his work is the development of new educational strategies in the classroom. He also works with assessment in education and curriculum comparison among universities around the world (Bologna Process and others similar projects). Giuliano Salcas Olguin is a member of ABENGE, ASEE and SEFI.

Patricia H. L. S. Matai is a Professor at Escola Politécnica da Universidade de São Paulo, Brazil. She is a Chemistry Major (1978), MSc and PhD degrees at Escola Politécnica (respectively 1992 and 1998). She is responsible for the coordination of the first and second year engineering students at Escola Politécnica. Since 1990, she has been researching educational techniques of adult education and the cooperative education models.

Stelian Brad is Full Professor at the Technical University of Cluj-Napoca, Romania, leading the research group on Competitive Engineering in Design and Development. He is also the Director of the Department of Research, Development and Innovation Management of the same university. He has several years of experience in leading two offshore outsourcing software development companies, with customers in Europe, Northern America, Middle East and Far East. His main research fields of interest are: competitive engineering applied in robot design, software development, engineering education and industrial product design, science of complexity, engineering and management of innovation, as well as intelligent systems in industrial robotics.

Yih-Ruey Juang, as known Jack Juang, is an assistant professor of the Department of Information Management in Jinwen University of Science and Technology in Taiwan. He was graduated from the Department of Elementary Teacher Education in Taipei Municipal Teacher's College (now has been changed to Taipei Municipal University of Education), Taipei City, Taiwan, in 1988. He received his Master's degree from the Department of Computer Science and Information Engineering in National Chiao-Tung University, Xin-Chu, Taiwan, in 1995, and the Ph. D. degree from the department with the same name in National Central University, Jhoung-Li, Taiwan, in 2007. Jack's research interest is focused on technology enhanced teaching and learning, information and computer education, and computer supported collaborative learning.

Mehregan Mahdavi is an Assistant Professor in the Department of Computer Engineering, University of Guilan, Iran. He received his PhD from the School of Computer Science and Engineering, University of New South Wales, Australia in 2005 in the area of caching dynamic data for Web applications. His research interests are in the area of Web and Database including Web Caching, Web Portals, Web Services, Web Data Integration, Electronic Learning and Data Versioning. He has published some conference and journal papers on these topics. His research area also includes Identity and Access Management, and Human Computer Interaction which he has also published some papers. He is a member of IEEE.

Mohammad H. Khoobkar received his B.Sc. degree in Computer Engineering from Islamic Azad University of Lahijan, Iran in 2006. He received his M.Sc. degree in Computer Engineering from Islamic Azad University of Qazvin, Iran in 2009. He is currently a research assistant at the Department of Computer Engineering, Islamic Azad University of Qazvin. His area of interest includes Parallel and Distributed Computing, Grid Computing, Resource Management, E-Learning and Peer-to-Peer systems.

Masoumeh Valizadeh was born on October 1978 in Iran. She has 7 years of experience in textile engineering training and research, in University of Guilan, Isfahan University of Technology and Islamic Azad University- branch of Rasht. She got graduated in Textile chemistry-Textile chemistry & Fibre Science from University of Guilan in 2001 and she obtained her MSc from Isfahan University of Technology in 2003. Masoumeh joined to the University of Guilan, Rasht in 2003 and spent 2 years working in Faculty of Engineering teaching undergraduate students. Starting her PhD studies in Isfahan University of Technology in 2005, she continued her cooperation with University of Guilan. After her PhD, she is the member of academy in the Faculty of Engineering, University of Guilan. Her research interests include: Mechanic of fabrics and yarns, Finite Element Modeling (FEM), Nanotechnology for fibers and multilayer, electrospinning of biopolymers and E-learning. At the University of Guilan, Masoumeh teaches graduate courses in the field of numerical Mathematic and Mechanic and Dynamic of Filament Yarns. She also teaches undergraduate courses on Physical chemistry, Polymer chemistry.

Giancarlo Anzelotti was born on February 1982 in Italy. He got graduated in Mechanical Engineering from University of Parma (Italy) in 2004 and he obtained his MSc from this university in 2006. He started his PhD studies at the Industrial Engineering Department of the University of Parma in 2007, working on the field of optical methods for strain measurements, especially in composite materials. Giancarlo has 5 years of experience in industrial design and composite material engineering training and research. He is interested in making software mainly in Mechanic and material engineering fields for both academic and industrial purposes. His research interests include: contact and non contact methods for strain measurements, finite element modeling and digital image correlation. Developing a software for digital image correlation is his recent activity. At the University of Parma Giancarlo teaches graduated courses in the field of computer aided design, mechanics of materials, finite element modeling.

Arezou Sedigheh Salehi was born on March 21, 1977 in Tehran, Iran. In June 1994 she obtained a diploma in Mathematics and got graduated as Materials Science Engineer from the Tehran University (Iran) in January 1999. After 4 years working in industry, mainly in the fields of electro-plating and welding, in Iran (February 1999 till September 2003), Arezou started a Master in Materials Engineering (MME) at the Department of Metallurgy and Materials Engineering of K.U.Leuven in October 2003. She obtained her Master degree in Materials Engineering, METCER option, in July 2004 and completed a master thesis on the characterization of DLC (diamond like carbon) coatings. Form August 2004 till August 2007, she was a research scientist at the Department of Metallurgy and Materials Engineering of K.U.Leuven, conducting research on electro-erodable ceramic composites within the framework of the European Moncerat project. She obtained her PhD in Dec. 2008. Arezou S. Salehi currently is working for Bekaert company as R&D technology manager.

Attila Koppány works as professor at Széchenyi István University in Győr, Hungary. He graduated from Budapest University of Technology in 1971 and spent 38 years in education as university lecturer, associate professor and from 1998 as a university professor. He is the head of Department of Architecture and Building Construction. In Győr he teaches courses on building construction. Prof. Attila Koppány is a member of the Public Body of the Hungarian Academy of Sciences (HASc) and president of Committee of Architectural Science of HASc., also a member of the International Council for Research and Innovation in Building and Construction (CIB). He works as associate member of W86 Working Committee Building Pathology (in CIB).

Index